Lecture Notes in Computer Science 6087

Commenced Publication in 1973
Founding and Former Series Editors:
Gerhard Goos, Juris Hartmanis, and Jan van Leeuwen

Editorial Board

David Hutchison
Lancaster University, UK

Takeo Kanade
Carnegie Mellon University, Pittsburgh, PA, USA

Josef Kittler
University of Surrey, Guildford, UK

Jon M. Kleinberg
Cornell University, Ithaca, NY, USA

Alfred Kobsa
University of California, Irvine, CA, USA

Friedemann Mattern
ETH Zurich, Switzerland

John C. Mitchell
Stanford University, CA, USA

Moni Naor
Weizmann Institute of Science, Rehovot, Israel

Oscar Nierstrasz
University of Bern, Switzerland

C. Pandu Rangan
Indian Institute of Technology, Madras, India

Bernhard Steffen
TU Dortmund University, Germany

Madhu Sudan
Microsoft Research, Cambridge, MA, USA

Demetri Terzopoulos
University of California, Los Angeles, CA, USA

Doug Tygar
University of California, Berkeley, CA, USA

Gerhard Weikum
Max-Planck Institute of Computer Science, Saarbruecken, Germany

M. Anwar Hasan   Tor Helleseth (Eds.)

Arithmetic of Finite Fields

Third International Workshop, WAIFI 2010
Istanbul, Turkey, June 27-30, 2010
Proceedings

 Springer

Volume Editors

M. Anwar Hasan
University of Waterloo, Department of Electrical and Computer Engineering
Waterloo, Ontario N2L 3G1, Canada
E-mail: ahasan@uwaterloo.ca

Tor Helleseth
University of Bergen, Department of Informatics, HIB
PB 7803, 5020, Bergen, Norway
E-mail: tor.helleseth@ii.uib.no

Library of Congress Control Number: 2010928209

CR Subject Classification (1998): I.1, G.2, E.3, K.6.5, D.4.6, F.2.1

LNCS Sublibrary: SL 1 – Theoretical Computer Science and General Issues

ISSN   0302-9743
ISBN-10   3-642-13796-2 Springer Berlin Heidelberg New York
ISBN-13   978-3-642-13796-9 Springer Berlin Heidelberg New York

springer.com

© Springer-Verlag Berlin Heidelberg 2010
Printed in Germany

Typesetting: Camera-ready by author, data conversion by Scientific Publishing Services, Chennai, India
Printed on acid-free paper   06/3180

Preface

These are the proceedings of WAIFI 2010, the Third International Workshop on the Arithmetic of Finite Fields, held in Istanbul, Turkey, during June 27-30, 2010. The first workshop, WAIFI 2007, was held in Madrid, Spain, and then WAIFI 2008 was held in Siena, Italy. In 2008, the workshop series was made biannual and it is now being held every even year, bringing together mathematicians, computer scientists, engineers and physicists who are doing research on various aspects of finite field arithmetic.

This year the workshop received 33 submissions, each of which was reviewed by at least three reviewers who were either members of the Program Committee of the workshop or external reviewers chosen by the members. Once the review phase was over, the Program Committee had online discussions over a period of several days. In the end, a total of 15 papers representing both theoretical and practical aspects of finite field arithmetic were accepted for presentation. These accepted papers are part of these proceedings. In addition to the presentations of these papers, we were fortunate to have three invited talks given by P. Vijay Kumar, Alfred Menezes and Henning Stichtenoth. The papers, which the invited talks were based on, are also part of the proceedings.

We are very grateful to the members of the Program Committee for their dedication, professionalism and careful work with the review and selection process. We also sincerely thank the external reviewers who contributed with their special expertise to review papers for this workshop.

We deeply thank General Co-chairs Çetin Kaya Koç and Ferruh Özbudak for their support of the Program Committee and their hard work in leading the overall organization of the workshop and holding it in the historic city of Istanbul– a joint European Capital of Culture for year 2010. We are also very grateful to José Luis Imaña for diligently maintaining the workshop website, and to Claude Carlet for making workshop announcements. Our very special thanks go to Murat Cenk, Gökay Saldamli and Zülfükar Saygi for dealing with various local arrangements with a lot of care.

We would also like to sincerely thank members of the Steering Committee of the workshop series for their constant support and encouragement in our efforts to create a stimulating scientific program, leading to the proceedings of WAIFI 2010. Special thanks go to Jean-Jacques Quisquater for making arrangements with Springer to publish the proceedings as a volume of *Lecture Notes in Computer Science*.

The process of paper submission, review and online discussion was carried out using the EasyChair conference management system, which we found to be very useful. The system was also used for dealing with final versions of the accepted and invited papers and towards the preparation of the proceedings. So, thank you EasyChair! We would also like to acknowledge Istanbul Şehir University for being a sponsor of the workshop.

Finally, but most importantly, we deeply thank the authors from all over the world who submitted their papers to the workshop. It was their hard work and endeavor to advance the field of knowledge that made the workshop a stimulating forum. We also thank the participants of the workshop for making it a very successful event.

June 2010   M. Anwar Hasan
                                                             Tor Helleseth

Organization

Steering Committee

Claude Carlet   University of Paris 8, France
Jean-Pierre Deschamps   University Rovira i Virgili, Spain
José Luis Imaña   Complutense University of Madrid, Spain
Çetin Kaya Koç   University of California Santa Barbara, USA,
     & Istanbul Şehir University, Turkey
Christof Paar   Ruhr University of Bochum, Germany
Jean-Jacques Quisquater   Université Catholique de Louvain, Belgium
Berk Sunar   Worcester Polytechnic Institute, USA
Gustavo Sutter   Autonomous University of Madrid, Spain

Executive Committee

General Co-chairs

Çetin Kaya Koç   University of California Santa Barbara, USA,
     & Istanbul Şehir University, Turkey
Ferruh Özbudak   Middle East Technical University, Turkey

Program Co-chairs

M. Anwar Hasan   University of Waterloo, Canada
Tor Helleseth   University of Bergen, Norway

Financial, Local Arrangements Chairs

Murat Cenk   Middle East Technical University, Turkey
Gökay Saldamli   Boğaziçi University, Turkey
Zülfükar Saygi   TOBB ETU, Turkey

Publicity Chair

Claude Carlet   University of Paris 8, France

Program Committee

Daniel Augot   INRIA, France
Roberto Avanzi   Ruhr-University Bochum, Germany
Jean-Claude Bajard   LIP6 CNRS/Université Pierre et Marie Curie, France
Luca Breveglieri   Politecnico di Milano, Italy
Stephen Cohen   University of Glasgow, Scotland, UK

Cunsheng Ding	Hong Kong University of Science and Technology, China
Serdar Erdem	Gebze Institute of Technology, Turkey
Haining Fan	Tsinghua University, China
Olav Geil	Aalborg University, Denmark
Guang Gong	University of Waterloo, Canada
Jorge Guajardo	Philips Research, The Netherlands
Darrel Hankerson	Auburn University, USA
M. Anwar Hasan	University of Waterloo, Canada
Tor Heleseth	University of Bergen, Norway
José Luis Imaña	Complutense University of Madrid, Spain
Alexander Kholosha	University of Bergen, Norway
P. Vijay Kumar	Indian Institute of Science, Bangalore, India
Tanja Lange	Technical University of Eindhoven, The Netherlands
Julio López	UNICAMP, Brazil
Gary McGuire	University College Dublin, Ireland
Eiji Okamoto	University of Tsukuba, Japan
Alexander Pott	University of Magdeburg, Germany
Francisco Rodríguez-Henríquez	Cinvestav, Mexico
Erkay Savas	Sabanci University, Turkey
Igor Semaev	University of Bergen, Norway
Patrick Solé	Télécom ParisTech, France

External Reviewers

Selcuk Baktir	Conrado Gouvêa	Gerardo Pelosi
Daniel J. Bernstein	Honggang Hu	Massimiliano Sala
Jean-Luc Beuchat	Yiyuan Luo	Nazar Abbas Saqib
Lilya Budaghyan	Subhamoy Maitra	Sumanta Sarkar
Stanislav Bulygin	Ryutaroh Matsumoto	Peter Schwabe
Ricardo Dahab	Wilfried Meidl	Arnaud Tisserand
Jean-Pierre Deschamps	Hiroyoshi Morita	Arne Winterhof
Xinxin Fan	Carlos Munuera	Bo-Yin Yang
Fung-Wei Fu	Harald Niederreiter	

Sponsoring Institution

Istanbul Şehir University, Turkey

Table of Contents

Functions, Equations and Modular Multiplication

Finite Field Arithmetic for Pairing Based Cryptography

Invited Talk 3

Finite Fields, Cryptography and Coding

Recursive Towers of Function Fields over Finite Fields

Henning Stichtenoth

Sabancı University, MDBF
34956 Tuzla, İstanbul, Turkey
henning@sabanciuniv.edu

Abstract. The theory of recursive towers of function fields over finite fields was developed by A. Garcia and the author since 1995. We give a survey about the main ideas and results, and we propose some problems for future work.

Keywords: Function fields, towers of function fields, curves with many points, AG codes.

1 Introduction

Families of algebraic curves which have many rational points over a finite field $\mathbb{F}_q$, became popular around 1980 after the seminal papers by V.D.Goppa [8], who introduced algebraic geometry (AG) codes, and by M.A.Tsfasman, S.G.Vladut and T.Zink [13]. They used modular curves for the construction of long AG codes over $\mathbb{F}_q$ whose limit parameters are better than the Gilbert-Varshamov bound (for q square, $q \geq 49$). The key point of their work is the existence of a sequence of algebraic curves $(\mathcal{C}_i)_{i \geq 0}$, defined over $\mathbb{F}_q$, such that the sequence $N(\mathcal{C}_i)/g(\mathcal{C}_i)$ has a strictly positive limit. Here $N(\mathcal{C})$ (resp. $g(\mathcal{C})$) denotes the number of $\mathbb{F}_q$-rational points (resp. the genus) of the curve $\mathcal{C}$. Certain modular curves yield such families.

Recursive towers of function fields over $\mathbb{F}_q$ provide a more *elementary* (yet non-trivial) and *explicit* approach to families of curves with many rational points, and hence to the Tsfasman-Vladut-Zink theorem.

2 Notations and Definitions

We denote by $\mathbb{F}_q$ the finite field of cardinality q. Rather than dealing with curves, we use the notion of function fields, which is essentially equivalent to the concept of algebraic curves [10]. By definition, an algebraic function field F over $\mathbb{F}_q$ is a finite extension of the rational function field $\mathbb{F}_q(x)$ such that no element $z \in F \setminus \mathbb{F}_q$ is algebraic over $\mathbb{F}_q$. The theory of function fields is very similar to the theory of algebraic number fields (i.e.; finite extensions of the field $\mathbb{Q}$ of rational numbers). The analogon to a P-adic valuation of a number field K (which corresponds to a prime ideal P in the ring of integers of K) is, in the

M.A. Hasan and T. Helleseth (Eds.): WAIFI 2010, LNCS 6087, pp. 1–6, 2010.

functior field case, a *valuation* $v : F \to \mathbb{Z} \cup \{\infty\}$, satisfying $v(yz) = v(y) + v(z)$ and $v(y+z) \geq \min\{v(y), v(z)\}$ for all $y, z \in F$. The corresponding *valuation ring* $\mathcal{O} = \{z \in F \mid v(z) \geq 0\}$ has a unique maximal ideal $P = \{z \in F \mid v(z) > 0\}$, which is called a *place* of F. As $\mathbb{F}_q \subseteq \mathcal{O}$ and $P \cap \mathbb{F}_q = \{0\}$, the residue class field $\mathcal{O}/P$ can be considered as an extension field of $\mathbb{F}_q$. In fact, this is a finite field extension. P is called a *rational place* (or place of degree one) if $\mathcal{O}/P = \mathbb{F}_q$. The number of rational places of $F/\mathbb{F}_q$ is finite and it is denoted by $N(F)$.

An important numerical invariant of a function field F is its *genus* $g(F)$. This is a non-negative integer which measures - in some sense - how complicated the function field is. The rational function field $\mathbb{F}_q(x)$ has genus 0, and a non-rational function field always has genus $g(F) \geq 1$.

For every sequence $(F_i)_{i \geq 0}$ of function fields over $\mathbb{F}_q$ with $g(F_i) \to \infty$, one has that

$$\limsup_{i \to \infty} N(F_i)/g(F_i) \leq \sqrt{q} - 1 ;$$

this is the so-called *Drinfeld-Vladut bound*. The sequence $(F_i)_{i \geq 0}$ is called *asymptotically good* if $\limsup_{i \to \infty} N(F_i)/g(F_i) > 0$, and *asymptotically optimal* if this upper limit attains the Drinfeld-Vladut bound.

The function fields corresponding to certain families of modular curves are asymptotically optimal (over fields of square cardinality $q = \ell^2$). This property of modular curves is the main ingredient in the proof of the Tsfasman-Vladut-Zink theorem. However, it is a non-trivial task to produce asymptotically good families in an *explicit* manner. If you try to do so, you will see that, most likely, either the genera $g(F_i)$ of your sequence of function fields grow too fast, or the numbers of rational places $N(F_i)$ do not grow fast enough, so that you will obtain $\lim_{i \to \infty} N(F_i)/g(F_i) = 0$.

3   Towers of Function Fields

A sequence $\mathcal{F} = (F_i)_{i \geq 0}$ of function fields over $\mathbb{F}_q$ is called a *tower*, if $F_0 \subseteq F_1 \subseteq F_2 \subseteq \ldots$, and all extensions F_{i+1}/F_i are separable of degree $[F_{i+1} : F_i] > 1$. Moreover we assume that $g(F_i) \to \infty$ for $i \to \infty$. One shows easily that the following limits exist in $\mathbb{R}_+ \cup \{\infty\}$:

$$\gamma(\mathcal{F}/F_0) := \lim_{i \to \infty} g(F_i)/[F_i : F_0] , \quad \text{the } \textit{genus} \text{ of } \mathcal{F}/F_0,$$

$$\nu(\mathcal{F}/F_0) := \lim_{i \to \infty} N(F_i)/[F_i : F_0] , \quad \text{the } \textit{splitting rate} \text{ of } \mathcal{F}/F_0,$$

$$\lambda(\mathcal{F}) := \lim_{i \to \infty} N(F_i)/g(F_i) = \nu(\mathcal{F}/F_0)/\gamma(\mathcal{F}/F_0) , \quad \text{the } \textit{limit} \text{ of } \mathcal{F}.$$

By the Drinfeld-Vladut bound we know that $0 \leq \lambda(F) \leq \sqrt{q} - 1$. The tower is asymptotically good if $\lambda(F) > 0$, and it is asymptotically optimal if $\lambda(F) = \sqrt{q} - 1$. It is also clear that $\mathcal{F}$ is asymptotically good if and only if $\gamma(\mathcal{F}/F_0) < \infty$ and $\nu(\mathcal{F}/F_0) > 0$. In order to determine the limit $\lambda(\mathcal{F})$, one usually studies the genus $\gamma(\mathcal{F}/F_0)$ and the splitting rate $\nu(\mathcal{F}/F_0)$ separately.

Before doing this, we recall some facts about extensions of function fields. Let E/F be a separable extension of function fields of degree $[E : F] = n$. A place Q of E is called an *extension* of the place P of F, if $P \subseteq Q$. We write then $Q|P$. Every place of F has at least one, but at most n extensions in E. If a rational place P of F has exactly n distinct extensions in E/F, then these places are rational places of E, and we say that P *splits completely* in E/F.

Assume that P is a place of F and Q is a place of E with $Q|P$. For the corresponding valuations v_P and v_Q there exists an integer $e \geq 1$ such that $v_Q(z) = e \cdot v_P(z)$ for all $z \in F$. We call $e =: e(Q|P)$ the *ramification index* of $Q|P$. The place P is said to be *unramified* in E/F if $e(Q|P) = 1$ for all $Q|P$, otherwise P is *ramified* in E/F. If $e(Q|P)$ is relatively prime to q (the cardinality of $\mathbb{F}_q$), for all $Q|P$, then P is called *tame* in E. Otherwise, P is *wild* in E. The number of ramified places in a separable extension E/F is always finite.

The genera $g(E)$ and $g(F)$ are related by the *Hurwitz genus formula*

$$2g(E) - 2 = [E : F](2g(F) - 2) + d(E/F) \; ,$$

where $d(E/F) \geq 0$ is the degree of the different of the extension E/F. Roughly speaking, $d(E/F)$ is small if only few places of F are ramified in E, and if their ramification is tame or not 'too wild'.

Now we can discuss the genus $\gamma(\mathcal{F}/F_0)$ and the splitting rate $\nu(\mathcal{F}/F_0)$ of a tower $\mathcal{F} = (F_i)_{i \geq 0}$ over $\mathbb{F}_q$.

A. We define the *ramification locus* $V(\mathcal{F}/F_0) := \{P \mid P \text{ is a place of } F_0 \text{ which is ramified in some extension } F_k/F_0, k \geq 1\}$. If $V(\mathcal{F}/F_0)$ is finite and all ramification in the tower is tame, it follows from the Hurwitz genus formula that the genus $\gamma(\mathcal{F}/F_0)$ is finite (which is a necessary condition for $\mathcal{F}$ to be asymptotically good). The same conclusion holds if some places ramify wildly, but not too wildly (one can make this statement more precise). In the case of wild ramification, it is often a difficult task to understand if ramification is not too wild (see the examples in Section 4 below).

B. We say that a rational place P of F_0 *splits completely* in the tower $\mathcal{F}$, if P splits completely in all extensions $F_k/F_0, k \geq 1$. The places Q of F_k with $Q|P$ are then also rational. Define the *splitting locus* of $\mathcal{F}/F_0$ as $Z(\mathcal{F}/F_0) := \{P \mid P \text{ is a rational place of } F_0 \text{ which splits completely in } \mathcal{F}\}$. It is clear that $\nu(\mathcal{F}/F_0) \geq |Z(\mathcal{F}/F_0)|$. Therefore a non-empty splitting locus implies that the splitting rate satisfies $\nu(\mathcal{F}/F_0) > 0$ (which is a necessary condition for an asymptotically good tower).

The art of finding an asymptotically good tower $\mathcal{F} = (F_i)_{i \geq 0}$ is therefore to ensure that it has a non-empty splitting locus $Z(\mathcal{F}/F_0)$, and that at the same time, the ramification locus $V(\mathcal{F}/F_0)$ is finite, having only tame or not too wild ramification.

4   Recursive Towers

A tower $\mathcal{F} = (F_i)_{i \geq 0}$ is said to be *recursive* if there exist a non-zero polynomial $f(x, y) \in \mathbb{F}_q[x, y]$ and elements $x_0, x_1, x_2, \ldots$ such that

$$F_0 = \mathbb{F}_q(x_0) \ , F_{i+1} = F_i(x_{i+1}) \text{ and } f(x_i, x_{i+1}) = 0 \ ,$$

for all $i \geq 0$. We call then the equation $f(x, y) = 0$ a *defining equation* for $\mathcal{F}$. Often it is convenient to write a defining equation in the form $\varphi(x, y) = \psi(x, y)$ with rational functions $\varphi(x, y), \psi(x, y)$. It is clear that such an equation can be rewritten in the form $f(x, y) = 0$ with some polynomial $f(x, y)$.

We know many examples of recursive towers, some of which are asymptotically good, some are asymptotically optimal. Here are some of these examples (we give only the defining equations for the towers, and the finite field, over which the towers are considered).

Example 1. (see [7]) The *Fermat tower*, defined by the equation

$$y^m + (x + 1)^m = 1$$

over the field $\mathbb{F}_q$ with $q = \ell^e$, $e \geq 2$ and $m = (q - 1)/(\ell - 1)$. The Fermat tower is asymptotically good; it is asymptotically optimal over the field $\mathbb{F}_4$, for $\ell = e = 2$,

Example 2. (see [5]) This tower is recursively defined by the equation

$$y^2 = (x^2 + 1)/2x$$

over a finite field $\mathbb{F}_q$ of odd characteristic $p > 2$. The tower is asymptotically optimal for $q = p^2$.

Example 3. (see [4]) This is a 'wild' tower (i.e., there are wildly ramified places) over $\mathbb{F}_q$ with $q = \ell^2$. The defining equation is

$$y^\ell + y = x^\ell/(x^{\ell-1} + 1) \ .$$

The tower is asymptotically optimal over $\mathbb{F}_{\ell^2}$ for any prime power ℓ.

Example 4. (see [1]) An asymptotically good tower over any field $\mathbb{F}_q$ with $q = \ell^3$ is defined by the equation

$$(y^\ell - y)^{\ell-1} + 1 = -x^{\ell(\ell-1)}/(x^{\ell-1} - 1)^{\ell-1} \ .$$

Each tower above has finite ramification locus and a non-empty splitting locus. However, the problems in proving that the towers are asymptotically good (resp. optimal) are of quite different type:

In Example 1, to prove that the ramification locus is finite and the splitting locus is non-empty, is rather easy. But the tower is far from being asymptotically optimal for $q \neq 4$.

In Example 2 one can easily determine the ramification locus, but it is difficult to show that the splitting locus is non-empty.

The towers in Example 3 and 4 are both wild. In both cases it is not very difficult to show that the ramification locus is finite and the splitting locus is non-empty. But it is hard to control wild ramification and to show that it is not 'too wild'.

5   Problems

Here we propose some open problems on recursive towers of function fields.

Problem 1. One knows that there are towers with strictly positive splitting rate, whose splitting locus is empty [3]. But the following is not known: Is there a *recursive* tower $\mathcal{F} = (F_i)_{i \geq 0}$ over $\mathbb{F}_q$ with splitting rate $\nu(\mathcal{F}/F_0) > 0$ such that its splitting locus $Z(\mathcal{F}/F_k)$ is empty for all $k \geq 0$?

Problem 2. One knows that there are towers with finite genus $\gamma(\mathcal{F}/F_0)$, whose ramification locus is infinite [3]. But one does not know if there exists a *recursive* tower $\mathcal{F} = (F_i)_{i \geq 0}$ over $\mathbb{F}_q$ such that $\gamma(\mathcal{F}/F_0) < \infty$ and its ramification locus $V(\mathcal{F}/F_0)$ is infinite.

Problem 3. For $q = \ell^k$ with $k = 2s + 1 \geq 5$ odd, find *recursive* towers over $\mathbb{F}_q$ having a limit $\lambda(\mathcal{F}) \geq c \cdot q^s$ (with some constant $c > 0$). For $q = \ell^3$, such towers are provided in Example 4 above.

Problem 4. Find an asymptotically good *recursive* tower over a prime field $\mathbb{F}_p$ (p a prime number). So far, one only knows that there *exist* asymptotically good towers over $\mathbb{F}_p$. This was proved in [11], using class field theory.

Problem 5. Construct explicit bases of Riemann-Roch spaces, in some asymptotically good recursive tower over a finite field. This is an important task if one wants to construct the corresponding AG codes explicitly. See [9].

Problem 6. Recent applications of towers in cryptography [2] ask for towers $(F_i)_{i \geq 0}$ such that the function fields F_i have a small p-rank ($p = \text{char}(\mathbb{F}_q)$). Hence one should study more systematically the behaviour of the ℓ-part of the divisor class group in towers (ℓ any prime number).

For an exhaustive list of references we refer to the survey article [6].

References

1. Bassa, A., Garcia, A., Stichtenoth, H.: A New Tower over Cubic Finite Fields. Moscow Math. Journal 8(3), 401–418 (2008)
2. Cramer, R.: Private Communication (2010)
3. Duursma, I., Poonen, B., Zieve, M.: Everywhere ramified towers of global function fields. In: Mullen, G.L., Poli, A., Stichtenoth, H., et al. (eds.) Fq7 2003. LNCS, vol. 2948, pp. 148–153. Springer, Heidelberg (2004)
4. Garcia, A., Stichtenoth, H.: On the Asymptotic Behaviour of Some Towers of Function Fields over Finite Fields. J. Number Theory 61, 248–273 (1996)

5. Garcia, A., Stichtenoth, H.: On Tame Towers over Finite Fields. J. Reine Angew. Math. 557, 53–80 (2003)
6. Garcia, A., Stichtenoth, H.: Explicit Towers of Function Fields over Finite Fields. In: Topics in Geometry, Coding Theory and Cryptography, pp. 1–58. Springer, Heidelberg (2006)
7. Garcia, A., Stichtenoth, H., Thomas, M.: On Towers and Composita of Towers of Function Fields over Finite Fields. Finite Fields and Appl. 3, 257–274 (1997)
8. Goppa, V.D.: Codes on Algebraic Curves. Soviet Math. Dokl. 24(1), 170–172 (1981)
9. Hu, X., Maharaj, H.: On the qth Power Algorithm. Finite Fields and Appl. 14, 1068–1082 (2008)
10. Niederreiter, H., Xing, C.P.: Algebraic Geometry in Coding Theory and Cryptography. Princeton University Press, Princeton (2009)
11. Serre, J.-P.: Sur le Nombre des Points Rationnels d'une Courbe Algébrique sur un Corps Finis. C. R. Acad. Sci. Paris 296, 397–402 (1983)
12. Stichtenoth, H.: Algebraic Function Fields and Codes, 2nd edn. Graduate Texts in Mathematics, vol. 254. Springer, Heidelberg (2009)
13. Tsfasman, M.A., Vladut, S.G., Zink, T.: Modular Curves, Shimura Curves, and Goppa Codes, Better than the Varshamov-Gilbert Bound. Math. Nachr. 109, 21–28 (1982)

High-Performance Modular Multiplication on the Cell Processor

Joppe W. Bos

Laboratory for Cryptologic Algorithms
EPFL, Station 14, CH-1015 Lausanne, Switzerland

Abstract. This paper presents software implementation speed records for modular multiplication arithmetic on the synergistic processing elements of the Cell broadband engine (Cell) architecture. The focus is on moduli which are of special interest in elliptic curve cryptography, that is, moduli of bit-lengths ranging from 192- to 521-bit. Finite field arithmetic using primes which allow particularly fast reduction is compared to Montgomery multiplication. The special primes considered are the five recommended NIST primes, as specified in the FIPS 186-3 standard, and the prime used in the elliptic curve *curve25519*. While presented and benchmarked on the Cell architecture, the proposed techniques to efficiently implement the modular multiplication algorithms are suited to run on any architecture which is able to compute multiple computations concurrently; e.g. graphics processing units.

Keywords: Cell Broadband Engine, Curve25519, Elliptic Curve Cryptography (ECC), Montgomery Multiplication, NIST primes.

1 Introduction

Elliptic curve cryptography (ECC) [20,24] is an approach to public-key cryptography which enjoys increasing popularity since its invention in the mid 1980s. The attractiveness of small key-sizes [22] has placed this public-key cryptosystem as the preferred alternative to the widely used RSA public-key cryptosystem [30]. This is emphasized by the current migration away from 80-bit to 112-bit security where, for instance, the United States' National Security Agency restricts the use of public key cryptography in "Suite B" [27] to ECC.

In this paper we present performance results for one of the key operations in ECC: modular multiplication. The performance results are obtained when running on the heterogeneous, multi-core, single instruction, multiple data (SIMD) Cell broadband engine (Cell) architecture. As far as we know, our performance results set new speed records for generic moduli, using interleaved Montgomery multiplication [25], and *special* modular multiplication for moduli ranging from 192 to 521 bits. This range covers the current standardized parameters for ECC cryptosystems as specified by National Institute of Standards (NIST) [34].

The special primes considered in this work are the recommended primes of special form by NIST [34] and the prime used in *curve25519* as proposed by

M.A. Hasan and T. Helleseth (Eds.): WAIFI 2010, LNCS 6087, pp. 7–24, 2010.
© Springer-Verlag Berlin Heidelberg 2010

Bernstein [2]. These special primes are used to enhance the performance of ECC-based schemes in practice by exploiting the special form of the primes to construct a fast reduction step. Typically, the multiplication and special reduction are performed sequentially. For the separated multiplication step we consider schoolbook and Karatsuba multiplication [18] techniques. We use the straight-forward methods to implement the fast reduction for the NIST recommended primes (see [32]). For the special prime in *curve25519* we use a different approach in order to compare with the proposed fast reduction from [2].

The performance results are obtained by using the features of SIMD architectures. The implementations are optimized for the Cell and take both the advantages (e.g., the rich instruction set and large register file) and disadvantages (e.g., the "small" $16 \times 16 \rightarrow 32$-bit multiplier) of this architecture into account. Furthermore, multiple streams of computations are interleaved to increase throughput. Multi-stream modular multiplication computations are useful in both a cryptanalytic and cryptographic setting. For instance, one could use multi-stream modular multiplication routines, either the generic or special variant, to speedup batch decryption for ECC-based schemes. Additionally, this work shows the practical benefit of using the special over generic prime moduli on the Cell.

The paper is organized as follows. Section 2 introduces the Cell broadband engine architecture. Section 3 recalls some basic facts about elliptic curves, Montgomery multiplication and discusses the special primes used in this work. Section 4 describes the cryptographic and cryptanalytic applications where multi-stream modular multiplications can be used. Section 5 describes how the different modular multiplication methods can be combined into a multi-stream high-performance implementation on the Cell. Section 6 presents and discusses our performance results and compares them to implementations by others on the Cell. Section 7 concludes the paper.

2   The Cell Broadband Engine

The Cell architecture [15], jointly developed by Sony, Toshiba, and IBM, is equipped with one dual-threaded, 64-bit in-order "Power Processing Element" (PPE), which can offload work to the eight "Synergistic Processing Elements" (SPEs) [33]. The SPEs are the workhorses of the Cell processor which can be found in the PlayStation 3 (PS3) game console. Each SPE, running at 3.2 GHz in the PS3, consists of a Synergistic Processing Unit (SPU), 256 kilobyte of private memory called Local Store (LS) and a Memory Flow Controller.

Most SPU instructions are 128-bit wide single instruction, multiple data (SIMD) operations performing sixteen 8-bit, eight 16-bit, four 32-bit, or two 64-bit computations in parallel. Each SPU is equipped with a large register file containing 128 registers of 128 bits each, providing space for unrolling and software pipelining of loops, hiding the relatively long latencies of its instructions. Unlike the processor in the PPE, the SPUs are asymmetric processors, having two pipelines (denoted by the odd and the even pipeline) which are designed to execute two disjoint sets of instructions (denoted by odd and even instructions). In the ideal case, two instructions (one odd and one even) can be dispatched per cycle. The SPUs are in-order processors and have no hardware branch-prediction. Instead, the programmer (or compiler) can tell the instruction fetch unit in advance where a (single) branch instruction will jump to.

Each SPE has access to a rich instruction set which operates simultaneously on 8-, 16- or 32-bit words. Instructions of particular interest are `shuffle` (odd instruction) and `select` (even instruction). The $d = \texttt{shuffle}(a, b, c)$ instruction uses the pattern given in c to shuffle 16 of the 32 bytes of a and b to the output d, in such a way that the jth byte of c determines the jth byte of d, either as a copy of a byte of a or b or as one of the constants $\{\texttt{0x00}, \texttt{0xFF}, \texttt{0x80}\}$, and where duplicate copies are allowed. The $d = \texttt{select}(a, b, p)$ instruction acts as a 2-way multiplexer; depending on the input pattern p the corresponding bit from either a or b is selected as the output bit in d. The SPEs are equipped with a 4-way SIMD multiplier (even instruction) which can compute four 16-bit integer multiplications simultaneously per clock cycle. In addition, an even 4-way SIMD multiply-and-add instruction, which performs a $16 \times 16 \rightarrow 32$-bit unsigned multiplication and an addition of a 32-bit unsigned operand to the 32-bit product, is available and has the same latency as a multiplication without the addition. Note that carries are not generated for this instruction.

3   Preliminaries

In this section the required background about elliptic curves, the various (modular) multiplication techniques and the special primes are recalled. We want to compute the product $C \equiv A \cdot B \bmod M$, by either first applying schoolbook or Karatsuba multiplication and next a fast reduction, or $C \equiv A \cdot B \cdot r^{-n} \bmod M$ using Montgomery multiplication, with $A, B, C, r, n, M \in \mathbb{Z}$. Here, M is an n-word, odd modulus such that $r^{n-1} \leq M < r^n$. In practice $r = 2^w$ with w the bit-length of a word, for the algorithms implemented for the SPE we either use $w = 32$ or $w = 16$ (cf. Section 5).

Elliptic Curves. Let $p > 3$ be a prime, then any $a, b \in \mathbb{F}_p$ such that $4a^3 + 27b^2 \neq 0$ define an elliptic curve $E_{a,b}$ over $\mathbb{F}_p$. The zero point, the so-called point at infinity, together with the set of points $(x, y) \in \mathbb{F}_p \times \mathbb{F}_p$ which satisfy the shortened affine Weierstrass equation $y^2 = x^3 + ax + b$, form an Abelian group $E_{a,b}(\mathbb{F}_p)$ [31] (usually written additively). Repeated point addition is called scalar multiplication and a single instance of point addition can be computed using multiple operations in $\mathbb{F}_p$. Besides the affine Weierstrass representation one can use a whole range of different representations. An overview of the costs, expressed in arithmetic operations in the underlying field, is given by Bernstein and Lange in [5].

Montgomery Multiplication. The Montgomery modular multiplication method is introduced in [25] and can be used to replace the conventional modular multiplication. In order to be used, the operands need to be converted: given an

Algorithm 1. Schoolbook (left), Karatsuba (middle) and interleaved Montgomery (right) multiplication algorithms.

Input: $\begin{cases} A = \sum_{i=0}^{n-1} a_i r^i, \\ B = \sum_{i=0}^{n-1} b_i r^i \end{cases}$ **Output:** $\begin{cases} C = A \cdot B \\ \quad = \sum_{i=0}^{2n-1} c_i r^i \end{cases}$ 1. $C = A \cdot b_0$ 2. **for** $i = 1$ to $n-1$ **do** 3. $\quad C = C + r^i (A \cdot b_i)$ 4. **return** C	**Input:** $\begin{cases} A = \sum_{i=0}^{n-1} a_i r^i, \\ B = \sum_{i=0}^{n-1} b_i r^i, \\ T : \text{ some threshold for} \\ \quad \text{switching to schoolbook} \\ \quad \text{multiplication.} \\ \text{Let } \widetilde{r} = r^{\lceil n/2 \rceil}. \end{cases}$ **Output:** $\begin{cases} C = A \cdot B \\ \quad = \sum_{i=0}^{2n-1} c_i r^i \end{cases}$ 1. **if** $n < T$ **then** 2. $\quad$ **return** $C = \text{schoolbook}(A, B)$ 3. $A = A_0 + A_1 \widetilde{r}, \quad 0 \leq A_0, A_1 < \widetilde{r}$ 4. $B = B_0 + B_1 \widetilde{r}, \quad 0 \leq B_0, B_1 < \widetilde{r}$ 5. $T_0 = \text{Karatsuba}(A_0, B_0)$ 6. $T_1 = \text{Karatsuba}(A_1, B_1)$ 7. $T_2 = \text{Karatsuba}(A_0 + A_1, B_0 + B_1) - T_0 - T_1$ 8. **return** $C = (T_0 + T_2 \cdot \widetilde{r} + T_1 \cdot \widetilde{r}^2)$	**Input:** $\begin{cases} A = \sum_{i=0}^{n-1} a_i r^i, B, \\ M, \mu \text{ such that} \\ 0 \leq A, B < r^n, \\ r^{n-1} \leq M < r^n, \\ 2 \nmid M, \quad \gcd(r, M) = 1 \\ \mu = -M^{-1} \bmod r, \end{cases}$ **Output:** $\begin{cases} C \equiv A \cdot B \cdot r^{-n} \bmod M \\ \text{such that } 0 \leq C < r^n \end{cases}$ 1. $C = 0$ 2. **for** $i = 0$ to $n - 1$ **do** 3. $\quad C = C + a_i \cdot B$ 4. $\quad q = \mu \cdot C \bmod r$ 5. $\quad C = (C + q \cdot M)/r$ 6. **if** $C \geq r^n$ **then** 7. $\quad C = C - M$ 8. **return** C

integer X, the Montgomery residue of this integer is defined as $\widetilde{X} = X \cdot r^n \bmod M$ with $r^{r-1} \leq M < r^n$. The constant r^n is the Montgomery radix such that $\gcd(r^n, M) = 1$. The Montgomery product is defined as $\widetilde{X} \cdot \widetilde{Y} \cdot r^{-n} \bmod M$, addition and subtraction remain unchanged. Since converting to and from Montgomery form requires computational effort, the Montgomery multiplication is mostly used in settings where the computation of a sequence of modular operations is required. See Algorithm 1 for a high-level description of the interleaved Montgomery multiplication method.

Fast Reduction. One way to speed up elliptic curve arithmetic is to enhance the performance of the finite field arithmetic by using a prime of a special form. The structure of such a prime is exploited by constructing a fast reduction method, applicable to this prime only. Typically, the multiplication and reduction are in two sequential phases. For the multiplication phase we consider the so-called schoolbook, or textbook, multiplication and the asymptotically faster Karatsuba multiplication techniques (see Algorithm 1 for a high-level description).

NIST PRIMES. In the FIPS 186-3 standard [34] NIST recommends the use of five prime fields when using the elliptic curve digital signature algorithm. These primes allow fast reduction, see Appendix A for the algorithms optimized for a machine word (limb) size of 32 bits, based on the work by Solinas [32]. The five recommended primes are

$$p_{192} = 2^{192} - 2^{64} - 1, \qquad\qquad p_{224} = 2^{224} - 2^{96} + 1,$$
$$p_{256} = 2^{256} - 2^{224} + 2^{192} + 2^{96} - 1, \quad p_{384} = 2^{384} - 2^{128} - 2^{96} + 2^{32} - 1,$$
$$p_{521} = 2^{521} - 1.$$

An extensive study of a software implementation of the NIST-recommended elliptic curves over prime fields on the x86 architecture is given by Brown et al. [8].

CURVE25519. The elliptic curve *curve25519* is proposed by Bernstein in [2]. Besides offering high-speed arithmetic, a list of other advantages can be found in the original article [2]. This curve is over $\mathbb{F}_{p_{255}}$ with $p_{255} = 2^{255} - 19$, an element $x \in \mathbb{F}_{p_{255}}$ can be represented as $x = \sum_{i=0}^{9} x_i 2^{\lceil 25.5i \rceil}$. Bernstein proposes to implement the arithmetic using floating point instructions and therefore representation inside a CPU is achieved by using floating-point registers. The original article gives performance data obtained on a Pentium M.

4   Applications

To increase throughput the 4-way SIMD instructions of the SPE are used to implement a modular multiplication routine which operates on 4 streams, or a small multiple of 4 by interleaving these streams, in parallel. When a sequence of multiplications has to be computed, for instance in elliptic curve scalar multiplication, the algorithm performs the same operations in SIMD-mode on all inputs. When the scalar multipliers are different, a square-and-multiply algorithm needs to perform a different sequence of point additions and doublings, since this depends on the binary expansion of the scalar multiplier. Performing the same computations on multiple streams concurrently, when multiplying with different scalars, in a SIMD fashion might be suboptimal since all streams which are being processed in parallel need to perform the same computations. In this section we present some applications in cryptography and cryptanalysis where SIMD modular multiplication algorithms can be beneficial; i.e., where the same multiplier is used in multiple independent instances.

Cryptography. Cryptographic schemes often need to perform exponentiations with a randomly selected exponent, or scalar multiplications when using the additive group law as in the elliptic curve setting. If this exponent is used several times, in independent calculations, these operations can be performed in parallel in a SIMD fashion. For instance, in elliptic curve public-key schemes the ability to process multiple streams of modular multiplication computations can be used to speedup batch decryption. Examples of such schemes are the elliptic curve integrated encryption scheme (ECIES), proposed by Bellare and Rogaway [1] and standardized in [9], and the provably secure encryption curve scheme (PSEC), based on the work by Fujisaki and Okamoto [13] and standardized in [17]. The decryption of a message consist of multiplying an elliptic curve point, as specified by the ciphertext, by the private key d in PSEC or by $h \cdot d$ in the case of ECIES, where h is the cofactor of the group order and is constant for a given private key. When many messages need to be decrypted, using the same private key, SIMD algorithms as described in this article can be used to speedup computations.

In other settings, where the bitsize of the modulus is usually larger compared to the ECC setting, multi-stream modular multiplication computations can be useful as well. ElGamal encryption schemes [12] require two exponentiations with the same random exponent. Other related methods perform more exponentiations with the same exponent. The double base variant of ElGamal by Damgård, often referred to as Damgård ElGamal [11], performs three exponentiations. The "double" hybrid Damgård ElGamal, as proposed by Kiltz et al. [19], requires four exponentiations with the same exponent in every encryption.

Cryptanalysis. In cryptanalysis, multi-stream modular multiplication computations, for moduli sizes as considered in this article, can be used to enhance the performance of the Pollard rho discrete logarithm algorithm [29], a method to solve the elliptic curve discrete logarithm problem (ECDLP) which is essential to the security of ECC. In practice, modular inversions in the Pollard rho algorithm are traded for modular multiplications, to increase speed, by using the Montgomery simultaneous inversion technique [26]. This technique allows one to trade, when running N computations in parallel, N inversions for roughly $3N$ modular multiplications and one inversion. For example, this technique is used in [7] to solve a 112-bit ECDLP on the SPE architecture by working concurrently on 400 computations. Here, 70 percent of the total run-time is spent on the computation of modular multiplications [7].

Another cryptanalytic application is factoring integers. The integer factorization problem is essential to cryptographic algorithms as RSA. The fastest known method to factor integers is the number field sieve [28,21]. This method can use the elliptic curve factorization method (ECM) [23] in a co-factorization phase. Performing elliptic curve arithmetic on multiple points allows the use of multi-stream modular multiplication methods. Related work by Bernstein et al. [4] gives performance details of a high-performance multi-stream implementation of modular arithmetic in the ECM on graphics cards.

5   Multiplication on the SPE Architecture

The (modular) multiplication operations in this work are designed to operate on relatively small (≤ 521 bits) integers. On the widely available x86 and x86-64 architectures the threshold for switching from schoolbook multiplication to methods with a lower asymptotic run-time complexity (e.g. Karatsuba multiplication) is > 800 bits [14]. On these architectures the size of the operands on which the multiplication and addition instructions work is typically the same (either 32 or 64 bits).

On the Cell "only" a $16 \times 16 \to 32$ bits multiplication instruction is available, performing four multiplications in parallel, while the size of the 4-way SIMD operands to the addition instruction is 32 bits. Unlike the x86 architecture an integer multiply-and-add instruction is available. This allows the addition of two extra 16-bit values to a result of a 16-bit multiplication without generating a carry, since if $0 \leq a, b, c, d < 2^{16}$, then $a \cdot b + c + d < 2^{32}$. We consider both the schoolbook and Karatsuba multiplication for the special modular multiplication routines.

Integer Representation on the Cell. For a high-performance implementation of arithmetic algorithms on the Cell, vectorization techniques are applied and data are represented using the 4-way SIMD organization of the SPEs. Using m 128-bit registers $x[0], x[1], \ldots, x[m-1]$ a four-tuple (x_1, x_2, x_3, x_4) of integers is represented. Each x_i is a wm-bit integer, where w is either 16 or 32 depending on the setting; typically we use $w = 16$ for multiplication and $w = 32$ for

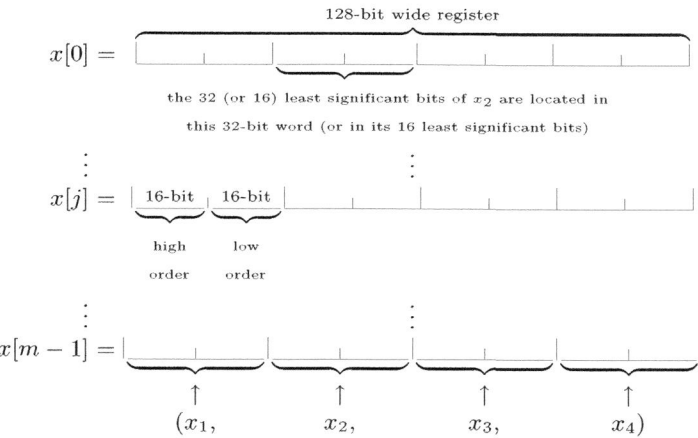

Fig. 1. A four-tuple (x_1, x_2, x_3, x_4) of $32m$-bit (or $16m$-bit) integers arranged in m 128-bit registers

addition and subtraction to match the bit-lengths of the corresponding 4-way SIMD instructions. Every element of the four-tuple is represented in a radix-2^w system:

$$x_i = \sum_{j=0}^{m-1} x[j]_i 2^{wj},$$

for $i = 1, 2, 3, 4$. The four 32-bit words of the 128-bit register $x[j]$ are denoted by $x[j]_i$. The representation of such a four-tuple (x_1, x_2, x_3, x_4) is depicted in Figure 1.

Multiplication. Algorithm 2 depicts schoolbook multiplication designed to run on SIMD architectures and is optimized for architectures with a native multiply-and-add instruction. After trivially unrolling the for-loops the algorithm is branch-free. Algorithm 2 splits the operands in 16-bit words, to take advantage of the 16-bit multiplier on the Cell, but this can be modified to work with any other word size on different architectures. Hence, on the SPE, Algorithm 2 operates on four-tuples of inputs simultaneously using the data representation from Fig. 1.

After the multiply-and-add, and a possible extra addition of one 16-bit word, the 32-bit result z is split into the 16 most and 16 least significant bits, x and y respectively. This is denoted by $\mathtt{split}(z) = (\lfloor \frac{z}{2^{16}} \rfloor, z \bmod 2^{16})$. On the SPE this splitting can be implemented in different ways, i.e. by using two odd $\mathtt{shuffle}$ instructions, or one even $\mathtt{and}$ and one odd $\mathtt{shuffle}$ instruction, or two even $\mathtt{and}$ instructions. The appropriate $\mathtt{splitting}$ implementation is chosen to balance the number of odd and even instructions, reducing the total number of required cycles. Note that when $i = 1$ the extra addition of d_{i+1} can be omitted. Hence, Algorithm 2 requires $n^2 \times \mathtt{split}$, $n^2 \times \mathtt{muladd}$ and $n(n-2) \times \mathtt{add}$ (when multiplying two $16n$-bit integers); this can be computed in $2n(n - \frac{3}{4})$ cycles,

---

Algorithm 2. Radix-2^{16} schoolbook multiplication algorithm.

Input: $\begin{cases} \text{Integer } a = (a_{n-1}, \ldots, a_1, a_0), & \text{each } a_i \text{ is a 16-bit word.} \\ \text{Integer } b = (b_{n-1}, \ldots, b_1, b_0), & \text{each } b_i \text{ is a 16-bit word.} \end{cases}$

Output: Integer $c = (c_{2n-1}, \ldots, c_1, c_0) = a \cdot b$, each c_i is a 16-bit word.

1. $d_i = 0, \quad i \in [1, n]$
2. **for** $j = 0$ to $n - 1$ **do**
3. $\quad (e_0, D_j) = \mathtt{split}(a_0 \cdot b_j + d_1)$
4. $\quad$ **for** $i = 1$ to $n - 1$ **do**
5. $\quad\quad (e_i, d_i) = \mathtt{split}(a_i \cdot b_j + e_{i-1} + d_{i+1})$
6. $\quad d_r = e_{n-1}$
7. **return** $(c = (d_n, d_{n-1}, \ldots, d_1, D_{n-1}, D_{n-2}, \ldots, D_0))$

---

optimistically assuming all odd and even pairs can be dispatched simultaneously. Furthermore, this approximation ignores the function-call overhead and loading and storing the in- and output from the local store. This leads to an optimistic approximation for the computation of a single $16n \times 16n \rightarrow 32n$-bit schoolbook multiplication in $\frac{n}{2}\left(n - \frac{3}{4}\right)$ cycles (when processing 4 streams in parallel).

A branch-free (when unrolled) Karatsuba multiplication algorithm optimized for vector architectures is given in Algorithm 3. This algorithm works on 32-bit words, which is the word size of the even 4-way SIMD addition and subtraction instructions on the SPE. Just as with the schoolbook multiplication this word size can trivially be modified. Algorithm 3 assumes that the bitsize of the input values is a multiple of 64 to split the operands evenly in two 32-bit multiples. These parts are multiplied using another multiplication routine mul, which is either a schoolbook or Karatsuba multiplication, which operates on inputs of half the size.

The $2m$-bit multiplication is split into two $m \times m$-bit and one $(m+1) \times (m+1)$-bit multiplications (see Alg. 1). In order to avoid the use of a probably more expensive multiplication by an extra limb, three $m \times m$-bit multiplications are used. The correct result, for the $(m+1) \times (m+1)$-bit multiplication, is computed by creating select-masks from the most significant bit of each of the two operands. These are used to select the appropriate value (one of the inputs) or zero, which is added to the result of the $m \times m$-bit multiplication. Note that the initial borrow values, in line 21, are (counterintuitively) set to one. An extra subtraction of one is performed when the borrow is zero and no subtraction is performed when the borrow is one on the SPE.

Special Reduction. The special reduction algorithms, see Appendix A, do not fully reduce the input to the range $[0, p\rangle$ but to $[0, t \cdot p\rangle$, where p is the prime modulus used and t a small positive integer. In order to reduce a four-tuple of integers simultaneously using SIMD instructions, different approaches can be applied. Obviously the reduction algorithm can be applied again. A most likely faster approach, when t is sufficiently small, is to subtract p repeatedly until the result is in the desired range. The repeated subtracting is done by masking the value appropriately before subtracting, which needs to be performed up to $t - 1$ times since multiple integer values are processed in parallel.

Algorithm 3. Radix-2^{32} Karatsuba multiplication algorithm for architectures which support vector instructions, n is even.

Input: $\begin{cases} \text{Integer } X = (x_{n-1}, \ldots, x_0), & \text{each } x_i \text{ is a 32-bit word.} \\ \text{Integer } Y = (y_{n-1}, \ldots, y_0), & \text{each } y_i \text{ is a 32-bit word.} \end{cases}$

Output: Integer $Z = (z_{2n-1}, \ldots, z_0) = X \cdot Y$, each z_i is a 32-bit word.

1. $(B_{n-1}, \ldots, B_0) = \text{mul}((x_{n-1}, \ldots, x_{n/2}), (y_{n-1}, y_{n/2}))$
2. $(C_{n-1}, \ldots, C_0) = \text{mul}((x_{n/2-1}, \ldots, x_0), (y_{n/2-1}, \ldots, y_0))$
3. $zero = \text{carry}_1 = \text{carry}_2 = \{0\}$
4. **for** $i = 0$ to $n/2 - 1$ **do**
5. $\quad X_i = \text{add\_extended}(x_{n/2+i}, x_i, \text{carry}_1)$
6. $\quad Y_i = \text{add\_extended}(y_{n/2+i}, y_i, \text{carry}_2)$
7. $\quad \text{carry}_1 = \text{gen\_carry\_extended}(x_{n/2+i}, x_i, \text{carry}_1)$
8. $\quad \text{carry}_2 = \text{gen\_carry\_extended}(y_{n/2+i}, y_i, \text{carry}_2)$
9. $\text{mask}_1 = \text{cmpgt}(\text{carry}_1, 0), \text{ mask}_2 = \text{cmpgt}(\text{carry}_2, 0)$
10. **for** $i = 0$ to $n/2 - 1$ **do**
11. $\quad s_i = \text{select}(zero, Y_i, \text{mask}_1), \; t_i = \text{select}(zero, X_i, \text{mask}_2)$
12. $c_1 = \text{select}(zero, \text{carry}_1, \text{mask}_2)$
13. $(z_{n-1}, \ldots, z_{n/2}, A_{n/2-1}, \ldots, A_0) = \text{mul}((X_{n/2-1}, \ldots, X_0), (Y_{n/2-1}, \ldots, Y_0))$
14. $\text{carry}_1 = \text{carry}_2\{0\}$
15. **for** $i = n/2$ to $n - 1$ **do**
16. $\quad T = \text{add\_extended}(z_i, s_{i-n/2}, \text{carry}_1)$
17. $\quad A_i = \text{add\_extended}(T, t_{i-n/2}, \text{carry}_2)$
18. $\quad \text{carry}_1 = \text{gen\_carry\_extended}(z_i, s_{i-n/2}, \text{carry}_1)$
19. $\quad \text{carry}_2 = \text{gen\_carry\_extended}(T, t_{i-n/2}, \text{carry}_2)$
20. $A_n = \text{add\_extended}(\text{carry}_1, \text{carry}_2, c_1)$
21. $\text{borrow}_1 = \text{borrow}_2 = \{1\}$
22. **for** $i = 0$ to $n - 1$ **do**
23. $\quad T = \text{sub\_extended}(A_i, B_i, \text{borrow}_1)$
24. $\quad E_i = \text{sub\_extended}(T, C_i, \text{borrow}_2)$
25. $\quad \text{borrow}_1 = \text{gen\_borrow\_extended}(A_i, B_i, \text{borrow}_1)$
26. $\quad \text{borrow}_2 = \text{gen\_borrow\_extended}(T, C_i, \text{borrow}_2)$
27. $E_n = \text{sub}(A_n, zero, \text{borrow}_1), \; E_n = \text{sub}(A_n, zero, \text{borrow}_2)$
28. $\text{carry}_1 = 0$
29. **for** $i = n/2$ to $n - 1$ **do**
30. $\quad Z_i = \text{add\_extended}(C_i, E_{i-n/2}, \text{carry}_1)$
31. $\quad \text{carry}_1 = \text{gen\_carry\_extended}(C_i, E_{i-n/2}, \text{carry}_1)$
32. **for** $i = n$ to $n + n/2 - 1$ **do**
33. $\quad Z_i = \text{add\_extended}(B_{i-n}, E_{i-n/2}, \text{carry}_1)$
34. $\quad \text{carry}_1 = \text{gen\_carry\_extended}(B_{i-n}, E_{i-n/2}, \text{carry}_1)$
35. $Z_{n+n/2} = \text{add\_extended}(B_{n/2}, E_n, \text{carry}_1)$
36. $\text{carry}_1 = \text{gen\_carry\_extended}(B_{n/2}, E_n, \text{carry}_1)$
37. **for** $i = n + n/2 + 1$ to $2n - 1$ **do**
38. $\quad Z_i = \text{add}(B_{i-n}, \text{carry}_1)$
39. $\quad \text{carry}_1 = \text{gen\_carry}(B_{i-n}, \text{carry}_1)$
40. **return** $Z = (Z_{2n-1}, \ldots, Z_{n/2}, C_{n/2-1}, \ldots, C_0)$

Table 1. The values of the 32-bit unsigned limbs of $t \cdot p_{224}$, c_7 and c_0 are the most and least significant limb respectively. In order to avoid using a look-up table the value $t \cdot p_{224}$ can be computed efficiently. Given t, $c_0 = t$, $c_1 = c_2 = 0$, $c_3 = 0 - t$, the values for c_4, c_5, c_6, c_7 can be constructed (using the `select` instruction) depending on t.

t	$t \cdot p_{224} = \{c_7, \ldots, c_0\}$							
	c_7	c_6	c_5	c_4	c_3	c_2	c_1	c_0
0	0	0	0	0	0	0	0	0
1	0	$2^{32} - 1$	$2^{32} - 1$	$2^{32} - 1$	$2^{32} - 1$	0	0	1
2	1	$2^{32} - 1$	$2^{32} - 1$	$2^{32} - 1$	$2^{32} - 2$	0	0	2
3	2	$2^{32} - 1$	$2^{32} - 1$	$2^{32} - 1$	$2^{32} - 3$	0	0	3
4	3	$2^{32} - 1$	$2^{32} - 1$	$2^{32} - 1$	$2^{32} - 4$	0	0	4

An additional performance gain is possible when the modulus is constant. Select the desired multiple of p, which needs to be subtracted, from a look-up table and perform a single subtraction. This can be achieved, when operating on multiple integer values in parallel, using the `select` instruction. If reduction to $[0, 2^m)$, for an m-bit modulus p, is allowed, the most significant word, containing the possible carry, has to be inspected only to determine the multiple of p to subtract. Note that an extra single subtraction might be needed in the unlikely situation that the result after the subtraction is $> 2^m$. This rare case is implemented by a branch which is hinted to be false to reduce the branch-overhead. The partially reduced numbers can be used as input to the same modular multiplication routines and if reduction to $[0, p)$ is required this can be achieved at the cost of a single multi-limb comparison and subtraction.

For the moduli of special form more instructions can be saved. For example consider the modulus $p_{224} = 2^{224} - 2^{96} + 1$. As described in Algorithm 6, in Appendix A, the algorithm returns with $(s_1 + s_2 + s_3 - s_4 - s_5)$, where all the s_i are 224-bit integers. At the implementation level we work with unsigned integers and prefer not to work with negative numbers. This is achieved by subtracting $s_4 + s_5$ from $2p_{224}$. We can bound the return value d by $d = s_1 + s_2 + s_3 + (2p_{224} - s_4 - s_5) < 5p_{224}$. To reduce d to $[0, 2^{224})$ the value $t \cdot p_{224}$, for some $t \in [0, 5)$, must be subtracted for four possibly different values of t in parallel after inspection of the most significant word. As can be seen from the representation in Table 1, when using a 2^{32} radix system, selecting the correct value for the different limbs is computationally easy. This allows the computation of $t \cdot p_{224}$ on-the-fly without the need to use and load from a look-up table. The reductions for the other special NIST primes can be done in a similar fashion.

We propose a different approach for calculating the reduction step for the special prime $p_{255} = 2^{255} - 19$ compared to the floating point approach from [2] (see Section 3). This approach is similar to the special reduction technique applied to the 112-bit prime modulus in [6, Appendix A]. A redundant representation modulo $\widetilde{P}_{255} = 2 \cdot p_{255} = 2^{256} - 38$ is used. Let $R = r_h \cdot 2^{256} + r_l$ be the 512-bit result after multiplication. Next, the first reduction step is performed by computing $S = r_l - 38 \cdot r_h \equiv R \bmod \widetilde{P}_{255}$; note that $S < 2^{262}$. Next, the same computation

Algorithm 4. Radix-2^{16} Montgomery Multiplication Algorithm.

Input:
$\begin{cases}
\text{Integer } a = (a_{n-1}, \ldots, a_1, a_0), \text{ each } a_i \text{ is a 16-bit word.} \\
\text{Integer } b = (b_{n-1}, \ldots, b_1, b_0), \text{ each } b_i \text{ is a 16-bit word.} \\
\text{Integer } M = (M_{n-1}, \ldots, M_1, M_0), \text{ each } M_i \text{ is a 16-bit word and } M \text{ is odd.} \\
\text{An 16-bit integer } \widetilde{m} = -M^{-1} \bmod 2^{16}.
\end{cases}$

Output: Integer $c = (c_{n-1}, \ldots, c_1, c_0) \equiv a \cdot b \cdot 2^{-16n} \bmod M$.

1. $d_i = 0, \quad i \in [0, n]$
2. **for** $i = 0$ to $n - 1$ **do**
3. $(e_0, d_0) = \mathtt{split}(a_0 \cdot b_i + d_0)$
4.   **for** $j = 1$ to $n - 1$ **do**
5. $(e_j, d_j) = \mathtt{split}(a_j \cdot b_i + d_j + e_{j-1})$
6. $d_n = d_n + e_{n-1}$
7. $(*, q) = \mathtt{split}(d_0 \cdot \widetilde{m})$
8. $(e_0, d_0) = \mathtt{split}(M_0 \cdot q + d_0)$
9.   **for** $j = 1$ to $n - 1$ **do**
10. $(e_j, d_{j-1}) = \mathtt{split}(M_j \cdot q + d_j + e_{j-1})$
11. $(d_n, d_{n-1}) = \mathtt{split}(d_n + e_{n-1})$
12. **if** $d_n > 0$ **then**
13. $(d_{n-1}, \ldots, d_1, d_0) = (d_n, d_{n-1}, \ldots, d_1, d_0) - (M_{n-1}, \ldots, M_1, M_0)$
14. **return** $(c = (d_{n-1}, \ldots, d_1, d_0))$

is repeated on $S = s_h \cdot 2^{256} + s_l$: $T = s_l + 38 \cdot s_h \equiv S \equiv R \bmod \widetilde{P}_{255}$. This is computationally faster since $s_h < 2^6$, note that the resulting $T < 2^{257}$. Similar techniques as described for the NIST primes are used to reduce the result to $[0, 2^{256})$.

Montgomery Multiplication. The interleaved Montgomery multiplication, optimized for the use on vector architectures, is given in Algorithm 4. As presented, it uses 16-bit limbs and on the Cell four-tuples of inputs are processed concurrently (but Alg. 4 can trivially be modified to operate on any radix size). A conditional subtraction step is needed at the end of the algorithm to ensure that the result is $< 2^{16n}$, for 16n-bit inputs. This conditional subtraction is replaced by a comparison which creates a select mask, using this mask the value zero or the value of the modulus is selected and subtracted. This eliminates a branch which is to be avoided when processing multiple integer values in a SIMD fashion. For efficiency, the integer representation is switched to a 2^{32} radix system when doing the final masking and subtraction.

The same notation for the split function is used as in Section 5. Hence, Algorithm 4 requires $2n(n + 1) \times \mathtt{split}$, $2n(n + 1) \times \mathtt{muladd}$ (when counting the multiplication in line 8 as an multiply-and-add) and $2n(n - 1) \times \mathtt{add}$ since the addition of d_j in line 5 when $j = 1$ can be omitted. For the conditional subtraction we first convert the integer representation to a 2^{32} radix system using $\lceil \frac{n}{2} \rceil$ $\mathtt{shuffle}$ instructions. Next we compare the carry (one $\mathtt{cmpgt}$ instruction) and mask the value which we are going to subtract using $\lceil \frac{n}{2} \rceil$ $\mathtt{and}$ instructions. The subtraction requires $\lceil \frac{n}{2} \rceil$ (extended) subtraction instructions and $\lceil \frac{n}{2} \rceil - 1$ (extended) generate borrow instructions.

Counting the number of instructions required in Algorithm 4 gives $4n^2 + 3\lceil \frac{n}{2} \rceil$ even and $\lceil \frac{n}{2} \rceil$ odd instructions plus $2n(n + 1)$ times the split function. Hence, an optimistic estimate of the number of cycles, ignoring overhead and assuming perfect scheduling, for a single computation of Montgomery multiplication on $16n$-bit inputs, when computing four computations in parallel, using Algorithm 4 on a single SPE is $n^2 + \frac{9n}{16}$ cycles.

6   Results

We implemented the proposed generic and special modular multiplication algorithms using the C-programming language for the SPEs on the Cell architecture. Four, or a small multiple of four, computations are processed in parallel. The performance benchmarks are performed on a single SPE in the PlayStation 3 game console. We summarize these results, together with other (single and multi-stream computation) modular multiplication results,7 obtained from the literature, in Table 2. The metric of our performance results is the number of cycles for a single modular multiplication computation. Our performance results are obtained by averaging over long sequences, hundreds of millions, of different modular multiplications and include the timing benchmark overhead, the function call overhead, loading and storing the in- and output from the local store and possibly converting the in- and output from the different integer representations (from radix-2^{32} to radix-2^{16} and vice-versa).

Performance Comparison. Performance results obtained with the Multi-Precision Math (MPM) Library [16], provided by IBM in the example API for the Cell, are given in Table 2 for different bit-sizes. The MPM library implements a single-stream Montgomery multiplication computation. In order to obtain a faster implementation for specific bit-lengths (to make a fair comparison) we unrolled the various loops inside the MPM library. These unrolled versions are significantly faster compared to the standard MPM implementation; e.g., the unrolled 256-bit Montgomery multiplication is 1.4 times faster compared to the unmodified MPM implementation. Our multi-stream implementations have a higher latency compared to the unrolled MPM library but process multiple streams resulting in fewer cycles per single multiplication. For instance, in the setting of 256-bit moduli the unrolled MPM requires 877 cycles for a single multiplication while our implementation requires 1188 cycles to compute four multiplications in parallel. This is a speedup of almost a factor of three per single multiplication.

In [10] Costigan and Schwabe implement elliptic curve arithmetic aimed at *curve25519* on the SPE architecture. The representation used differs slightly, but is based on, the one proposed in [2]; an element $x \in \mathbb{F}_{p_{255}}$ is represented as $x = \sum_{i=0}^{19} x_i 2^{\lceil 12.75i \rceil}$. A multi-stream version working on four streams in parallel is implemented and hand-optimized in assembly and "perfectly" scheduled with the surrounding code in a larger function implementing elliptic curve arithmetic. This multi-stream implementation is estimated to compute a single modular multiplication in around 168 cycles [10], this does not include any overhead for

Table 2. Performance results of Montgomery multiplication or modular multiplication modulo the special prime p_i. The latter uses a separate multiplication (schoolbook (S) or Karatsuba (K)) and a fast reduction phase. The benchmarks are performed on a single SPE on a Cell in the PlayStation 3 game console. The stated number of cycles are for a single modular multiplication (when processing the reported number of streams in parallel) and the optimistic estimates are from the formulas from Section 5 and do not include the special reduction cost.

From	Bitsize of the modulus	Method	Streams	Performance (cycles)	Estimate (cycles)
This article	192	p_{192} (K)	8	105	
This article	192	p_{192} (S)	8	126	68
This article	192	Montgomery	8	176	151
Bernstein et al. [3]	195	Montgomery	6	189	
This article	224	p_{224} (K)	8	139	
This article	224	p_{224} (S)	8	143	93
This article	224	Montgomery	4	234	204
Costigan and Schwabe [10]	255	p_{255} (S)	4	168^1	
This article	255	p_{255} (K)	8	175	
This article	255	p_{255} (S)	8	182	122
This article	256	p_{256} (S)	8	192	122
This article	256	p_{256} (K)	4	193	
This article	256	Montgomery	4	297	265
MPM unrolled [16]	256	Montgomery	1	877	
MPM [16]	256	Montgomery	1	1188	
This article	384	p_{384} (K)	4	389	
This article	384	p_{384} (S)	4	391	279
This article	384	Montgomery	4	665	590
MPM unrolled [16]	384	Montgomery	1	1610	
MPM [16]	384	Montgomery	1	2092	
This article	521	p_{521} (S)	4	622	500
This article	521	p_{521} (K)	4	723	
This article	512	Montgomery	4	1393	1042
MPM unrolled [16]	512	Montgomery	1	2700	
MPM [16]	512	Montgomery	1	3275	

saving and storing the in- and output registers to and from the local store, function call overhead and overhead due to benchmarking. In comparison, our implementation requires 175 cycles for a single modular multiplication using a

---

[1] This is the required number of cycles for an in-register implementation, no loading of input and storing of output from the local store is performed, and excludes benchmark and function call overhead.

different approach for the special reduction (see Section 5). This includes loading and storing the in- and output, function call and benchmarking overhead and additional latencies because not all code can be scheduled perfectly (especially at the beginning and end of the function where stalls occur). Comparing the performance of the two different approaches for the reduction step is difficult since the reported performance results of two versions are in different settings; ours is a stand-alone multiplication function while the implementation from [10] is an inline version working on registers only. In [10] it is estimated that the time to load and store the in- and output requires 56 cycles in the setting of a single modular multiplication. When considering this cost our approach using the redundant representation looks preferable (since $175 < 168 + 56$), especially since we did not use any fine-tuned assembly code to achieve these results.

Improved multi-stream modular multiplication computations results, compared to [4], are given by Bernstein et al. in [3]. Here, not only results for GPUs are reported but also for the Cell architecture as used in the PlayStation 3. In this setting Montgomery multiplication is implemented and optimized for one bit size: a 195-bit generic modulus. A radix-2^{13} system is used to represent 195-bit integers using 15 limbs, this has the advantage of accumulating multiple carries before an overflow occurs (on the SPE architecture) compared to a radix-2^{16} system but requires more limbs to represent the integers. When quadratically scaling our 192-bit performance result, in a similar fashion as done in [3], this leads to an estimate of $176 \cdot \left(\frac{195}{192}\right)^2 = 182$ cycles; this is slightly faster compared to the 189 required cycles reported in [3].

Discussion. The performance data from Table 2 show that the modular multiplication using the special primes are in almost all cases, with the exception of p_{256} and p_{521}, roughly 1.7 times faster compared to the Montgomery multiplication implementations targeting the same bit-lengths. Our results show that p_{256} is 1.55 times faster than 256-bit Montgomery multiplication while p_{521} is 2.2 times faster compared to 512-bit Montgomery multiplication. This can be partially explained by the relatively complicated and easy structure of p_{192} and p_{521} respectively.

For p_{192} the version using Karatsuba multiplication is significantly (20 percent) faster compared to the version using schoolbook multiplication. For p_{224}, p_{255}, p_{256} and p_{384} the performance is roughly the same while for p_{521} schoolbook multiplication is 16 percent faster. These differences can be explained due to extra load and store operations from and to the local store. For the smaller bitsizes almost all operations can be performed, after the initial loading from the inputs, on registers. For the larger values the available 128 registers are not sufficient and extra load and store instructions, leading to more instructions and possibly extra stalls, are required. This also explains why processing four streams instead of eight gives a higher performance for p_{384} and p_{521}.

The number of cycles required for the Montgomery multiplication is 12 to 17 percent higher compared to the estimations for all special primes except p_{521}. This overhead is mainly caused by extra load and stores and due to the fact that the estimates are too optimistic (not every cycle a pair of instructions can be dispatched due to instruction dependencies). For the special prime p_{521} more than 33 percent of the estimated number of cycles is needed. After compiling our code to assembly inspection shows that the significant overhead is as expected due to the extra loads and stores. Note that loading the two input values in registers (after conversion to radix-2^{16}) requires 66 registers which is already more than half of the available register space.

7   Conclusions

In this paper we presented techniques to efficiently implement modular multiplication algorithms to SIMD architectures (such as the Cell or GPUs). We considered Montgomery multiplication and various special reduction routines which are of interest for elliptic curve cryptography. The modular multiplication implementations, which use these faster reduction schemes, are at least 1.5 times faster compared to general purpose Montgomery multiplication for the same bitsize. The performance results of our multi-stream modular multiplication implementations for the synergistic processing elements of the Cell broadband engine architecture set new performance records for moduli of bit-length in the range [192, 521] on this platform. These high-performing modular multiplication, generic or special, implementations can be used to speed up public-key cryptography; e.g. in batch elliptic curve decryption.

References

1. Bellare, M., Rogaway, P.: Minimizing the use of random oracles in authenticated encryption schemes. In: Han, Y., Quing, S. (eds.) ICICS 1997. LNCS, vol. 1334, pp. 1–16. Springer, Heidelberg (1997)
2. Bernstein, D.J.: Curve25519: New Diffie-Hellman speed records. In: Yung, M., Dodis, Y., Kiayias, A., Malkin, T.G. (eds.) PKC 2006. LNCS, vol. 3958, pp. 207–228. Springer, Heidelberg (2006)
3. Bernstein, D.-J., Chen, H.-C., Chen, M.-S., Cheng, C.-M., Hsiao, C.-H., Lange, T., Lin, Z.-C., Yang, B.-Y.: The Billion-Mulmod-Per-Second PC. In: SHARCS 2009, pp. 131–144 (2009)
4. Bernstein, D.J., Chen, T.-R., Cheng, C.-M., Lange, T., Yang, B.-Y.: ECM on graphics cards. In: Joux, A. (ed.) EUROCRYPT 2009. LNCS, vol. 5479, pp. 483–501. Springer, Heidelberg (2010)
5. Bernstein, D.J., Lange, T.: Analysis and optimization of elliptic-curve single-scalar multiplication. In: Finite Fields and Applications. Contemporary Mathematics Series, vol. 461, pp. 1–19 (2008)
6. Bos, J.W., Kaihara, M.E., Kleinjung, T., Lenstra, A.K., Montgomery, P.L.: On the Security of 1024-bit RSA and 160-bit Elliptic Curve Cryptography. Cryptology ePrint Archive, Report 2009/389 (2009), http://eprint.iacr.org/
7. Bos, J.W., Kaihara, M.E., Montgomery, P.L.: Pollard rho on the PlayStation. In: SHARCS 2009, vol. 3, pp. 35–50 (2009)

8. Brown, M., Hankerson, D., López, J., Menezes, A.: Software implementation of the NIST elliptic curves over prime fields. In: Naccache, D. (ed.) CT-RSA 2001. LNCS, vol. 2020, pp. 250–265. Springer, Heidelberg (2001)

9. Certicom Research: Standards for Efficient Cryptography 1: Elliptic Curve Cryptography. Standard SEC1, Certicom (2000)

10. Costigan, N., Schwabe, P.: Fast elliptic-curve cryptography on the Cell broadband engine. In: Preneel, B. (ed.) AFRICACRYPT 2009. LNCS, vol. 5580, pp. 368–385. Springer, Heidelberg (2009)

11. Damgård, I.: Towards practical public key systems secure against chosen ciphertext attacks. In: Feigenbaum, J. (ed.) CRYPTO 1991. LNCS, vol. 576, pp. 445–456. Springer, Heidelberg (1992)

12. El Gamal, T.: A public key cryptosystem and a signature scheme based on discrete logarithms. In: Blakely, G.R., Chaum, D. (eds.) CRYPTO 1984. LNCS, vol. 196, pp. 10–18. Springer, Heidelberg (1985)

13. Fujisaki, E., Okamoto, T.: Secure integration of asymmetric and symmetric encryption schemes. In: Wiener, M. (ed.) CRYPTO 1999. LNCS, vol. 1666, pp. 537–554. Springer, Heidelberg (1999)

14. Granlund, T.: GMP small operands optimization. In: SPEED 2007 (2007)

15. Hofstee, H.P.: Power efficient processor architecture and the Cell processor. In: HPCA 2005, pp. 258–262 (2005)

16. IBM: Multi-precision math library, Example Library API Reference, https://www.ibm.com/developerworks/power/cell/documents.html

17. ISO/IEC 18033-2: Information technology – Security techniques – Encryption algorithms – Part 2: Asymmetric ciphers (2006)

18. Karatsuba, A., Ofman, Y.: Multiplication of many-digital numbers by automatic computers. In: Proceedings of the USSR Academy of Science, vol. 145, pp. 293–294 (1962)

19. Kiltz, E., Pietrzak, K., Stam, M., Yung, M.: A new randomness extraction paradigm for hybrid encryption. In: Joux, A. (ed.) EUROCRYPT 2009. LNCS, vol. 5479, pp. 590–609. Springer, Heidelberg (2010)

20. Koblitz, N.: Elliptic curve cryptosystems. Mathematics of Computation 48, 203–209 (1987)

21. Lenstra, A.K., Lenstra Jr., H.W.: The Development of the Number Field Sieve. Lecture Notes in Mathematics, vol. 1554. Springer, Heidelberg (1993)

22. Lenstra, A.K., Verheul, E.R.: Selecting cryptographic key sizes. Journal of Cryptology 14(4), 255–293 (2001)

23. Lenstra Jr., H.W.: Factoring integers with elliptic curves. Annals of Mathematics 126, 649–673 (1987)

24. Miller, V.S.: Use of elliptic curves in cryptography. In: Williams, H.C. (ed.) CRYPTO 1985. LNCS, vol. 218, pp. 417–426. Springer, Heidelberg (1986)

25. Montgomery, P.L.: Modular multiplication without trial division. Mathematics of Computation 44(170), 519–521 (1985)

26. Montgomery, P.L.: Speeding the Pollard and elliptic curve methods of factorization. Mathematics of Computation 48, 243–264 (1987)

27. National Security Agency: Fact sheet NSA Suite B Cryptography (2009), http://www.nsa.gov/ia/programs/suiteb_cryptography/index.shtml

28. Pollard, J.M.: Factoring with cubic integers. In: [21], pp. 4–10

29. Pollard, J.M.: Monte Carlo methods for index computation (mod p). Mathematics of Computation 32, 918–924 (1978)

30. Rivest, R.L., Shamir, A., Adleman, L.: A method for obtaining digital signatures and public-key cryptosystems. Communications of the ACM 21, 120–126 (1978)

31. Silverman, J.H.: The Arithmetic of Elliptic Curves. In: Gradute Texts in Mathematics. Springer, Heidelberg (1986)
32. Solinas, J.A.: Generalized Mersenne numbers. Technical Report CORR 99-39, Centre for Applied Cryptographic Research, University of Waterloo (1999)
33. Takahashi, O., Cook, R., Cottier, S., Dhong, S.H., Flachs, B., Hirairi, K., Kawasumi, A., Murakami, H., Noro, H., Oh, H., Onish, S., Pille, J., Silberman, J.: The circuit design of the synergistic processor element of a Cell processor. In: ICCAD 2005, pp. 111–117. IEEE Computer Society, Los Alamitos (2005)
34. U.S. Department of Commerce and National Institute of Standards and Technology: Digital Signature Standard (DSS) (2009),
    http://csrc.nist.gov/publications/fips/fips186-3/fips_186-3.pdf

A   NIST Reduction

Algorithm 5. Fast reduction modulo $p_{192} = 2^{192} - 2^{64} - 1$.

Input: Integer $c = (c_{11}, \ldots, c_1, c_0)$, each c_i is a 32-bit word, and $0 \leq c < p_{192}^2$.
Output: Integer $d \equiv c \bmod p_{192}$.
  Define 192-bit integers:
  $s_1 = (c_5, c_4, c_3, c_2, c_1, c_0),$ $s_2 = (0, 0, c_7, c_6, c_7, c_6),$
  $s_3 = (c_9, c_8, c_9, c_8, 0, 0),$ $s_4 = (c_{11}, c_{10}, c_{11}, c_{10}, c_{11}, c_{10});$
  return $(d = s_1 + s_2 + s_3 + s_4);$

Algorithm 6. Fast reduction modulo $p_{224} = 2^{224} - 2^{96} + 1$.

Input: Integer $c = (c_{13}, \ldots, c_1, c_0)$, each c_i is a 32-bit word, and $0 \leq c < p_{224}^2$.
Output: Integer $d \equiv c \bmod p_{224}$.
  Define 224-bit integers:
  $s_1 = (\ c_6,\ c_5,\ c_4,\ c_3,\ c_2,\ c_1,\ c_0),$ $s_2 = (\ c_{10},\ c_9,\ c_8,\ c_7,\ 0,\ 0,\ 0),$
  $s_3 = (\ 0, c_{13}, c_{12}, c_{11},\ 0,\ 0,\ 0),$ $s_4 = (\ c_{13}, c_{12}, c_{11}, c_{10}, c_9, c_8, c_7)$
  $s_5 = (\ 0,\ 0,\ 0,\ 0, c_{13}, c_{12}, c_{11});$
  return $(d = s_1 + s_2 + s_3 - s_4 - s_5);$

Algorithm 7. Fast reduction modulo $p_{256} = 2^{256} - 2^{224} + 2^{192} + 2^{96} - 1$.

Input: Integer $c = (c_{15}, \ldots, c_1, c_0)$, each c_i is a 32-bit word, and $0 \le c < p_{256}^2$.
Output: Integer $d \equiv c \bmod p_{256}$.

Define 256-bit integers:

$$s_1 = (\ c_7,\ c_6,\ c_5,\ c_4,\ c_3,\ c_2,\ c_1,\ c_0), s_2 = (\ c_{15}\ c_{14}, c_{13}, c_{12}, c_{11},\ \ 0,\ \ 0,\ \ 0),$$
$$s_3 = (\ \ 0, c_{15}, c_{14}, c_{13}, c_{12},\ \ 0,\ \ 0,\ \ 0), s_4 = (\ c_{15}, c_{14},\ \ 0,\ \ 0,\ \ 0, c_{10},\ c_9,\ c_8),$$
$$s_5 = (\ c_8, c_{13}, c_{15}, c_{14}, c_{13}, c_{11}, c_{10},\ c_9), s_6 = (\ c_{10},\ c_8,\ \ 0,\ \ 0,\ \ 0, c_{13}, c_{12}, c_{11}),$$
$$s_7 = (\ c_{11},\ c_9,\ \ 0,\ \ 0, c_{15}, c_{14}, c_{13}, c_{12}), s_8 = (\ c_{12},\ \ 0, c_{10},\ c_9,\ c_8, c_{15}, c_{14}, c_{13}),$$
$$s_9 = (\ c_{13},\ \ 0, c_{11}, c_{10},\ c_9,\ \ 0, c_{15}, c_{14});$$

$\quad$ **return** $(d = s_1 + 2s_2 + 2s_3 + s_4 + s_5 - s_6 - s_7 - s_8 - s_9)$;

Algorithm 8. Fast reduction modulo $p_{384} = 2^{384} - 2^{128} - 2^{96} + 2^{32} - 1$.

Input: Integer $c = (c_{23}, \ldots, c_1, c_0)$, each c_i is a 32-bit word, and $0 \le c < p_{384}^2$.
Output: Integer $d \equiv c \bmod p_{384}$.

Define 384-bit integers:

$$s_1 = (\ c_{11}, c_{10},\ c_9,\ c_8,\ c_7,\ c_6,\ c_5,\ c_4,\ c_3,\ c_2,\ c_1,\ c_0),$$
$$s_2 = (\ \ 0,\ \ 0,\ \ 0,\ \ 0,\ \ 0, c_{23}, c_{22}, c_{21},\ \ 0,\ \ 0,\ \ 0,\ \ 0),$$
$$s_3 = (\ c_{23}, c_{22}, c_{21}, c_{20}, c_{19}, c_{18}, c_{17}, c_{16}, c_{15}, c_{14}, c_{13}, c_{12}),$$
$$s_4 = (\ c_{20}, c_{19}, c_{18}, c_{17}, c_{16}, c_{15}, c_{14}, c_{13}, c_{12}, c_{23}, c_{22}, c_{21}),$$
$$s_5 = (\ c_{19}, c_{18}, c_{17}, c_{16}, c_{15}, c_{14}, c_{13}, c_{12}, c_{20},\ \ 0, c_{23},\ \ 0),$$
$$s_6 = (\ \ 0,\ \ 0,\ \ 0,\ \ 0, c_{23}, c_{22}, c_{21}, c_{20},\ \ 0,\ \ 0,\ \ 0,\ \ 0),$$
$$s_7 = (\ \ 0,\ \ 0,\ \ 0,\ \ 0,\ \ 0,\ \ 0, c_{23}, c_{22}, c_{21},\ \ 0,\ \ 0, c_{20}),$$
$$s_8 = (\ c_{22}, c_{21}, c_{20}, c_{19}, c_{18}, c_{17}, c_{16}, c_{15}, c_{14}, c_{13}, c_{12}, c_{23}),$$
$$s_9 = (\ \ 0,\ \ 0,\ \ 0,\ \ 0,\ \ 0,\ \ 0,\ \ 0, c_{23}, c_{22}, c_{21}, c_{20},\ \ 0),$$
$$s_{10} = (\ \ 0,\ \ 0,\ \ 0,\ \ 0,\ \ 0,\ \ 0,\ \ 0, c_{23}, c_{23},\ \ 0,\ \ 0,\ \ 0);$$
$\quad$ **return** $(d = s_1 + 2s_2 + s_3 + s_4 + s_5 + s_6 + s_7 - s_8 - s_9 - s_{10})$;

Algorithm 9. Fast reduction modulo $p_{521} = 2^{521} - 1$.

Input: Integer $c = (c_{33}, \ldots, c_1, c_0)$, each c_i is a 32-bit word, and $0 \le c < p_{521}^2$.
Output: Integer $d \equiv c \bmod p_{521}$.

Define 521-bit integers:

$$s_1 = (c_{16}, \ldots, c_1, c_0), \quad s_2 = (c_{32}, \ldots, c_{17}, c_{16});$$
$\quad$ **return** $(d = s_1 \bmod 2^{521} + \frac{s_2}{2^{23}})$;

A Modified Low Complexity Digit-Level Gaussian Normal Basis Multiplier

Reza Azarderakhsh and Arash Reyhani-Masoleh

Department of Electrical and Computer Engineering
The University of Western Ontario
London, ON, Canada, N6A 5B9
{razarder,areyhani}@uwo.ca

Abstract. Gaussian normal bases have been included in a number of standards, such as IEEE [1] and NIST [2] for elliptic curve digital signature algorithm (ECDSA). Among different finite field operations used in this algorithm, multiplication is the main operation. In this paper, we consider type T Gaussian normal basis (GNB) multipliers over $GF(2^m)$, where m is odd. Such fields include five binary fields recommended by NIST for ECDSA. A modified digit-level GNB multiplier over $GF(2^m)$ is proposed in this paper. For $T > 2$, a complexity reduction algorithm is proposed to reduce the number of XOR gates without increasing the gate delay of the digit-level multiplier. The original and modified digit-level GNB multipliers are implemented on the Xilinx® Virtex5™ FPGA family for different digit sizes. It is shown that the modified digit-level GNB multiplier requires lower space complexity with almost the same delay as compared to the original type T, $T > 2$, GNB multiplier. Moreover, the bit-parallel GNB multiplier obtained from the proposed modified digit-level multiplier has the least space and time complexities among the existing fast bit-parallel type T GNB multipliers for $T > 2$.

Keywords: Finite field, Gaussian normal basis, digit-serial multiplier, complexity reduction.

1 Introduction

Elliptic curve cryptosystem, which is proposed independently by Miller [3] and Koblitz [4], requires extensive finite field operations for the point multiplication. Multiplication is the main operation and its structure depends strongly on the representation of the field element. There are a number of ways to represent field elements. Among them, the most common bases are the polynomial basis and the normal basis representations [1]. In normal basis representation, squaring of a field element is free in hardware. Recently, implementation of the point multiplication for elliptic curve cryptography (ECC) using normal basis has received attention in the literature, see for example [5], [6] and [7].

The first normal basis multiplier over $GF(2^m)$ was invented by Massey and Omura [8]. This bit-serial multiplier, which has a parallel-in serial-out structure,

M.A. Hasan and T. Helleseth (Eds.): WAIFI 2010, LNCS 6087, pp. 25–40, 2010.

generates a bit of the result in each clock cycle. Therefore, the coordinates of the multiplication are generated after m clock cycles with the least complexity. There are also other bit-serial multipliers with parallel outputs, see for example [9], [10]. To make a fast hardware implementation, a bit-parallel multiplier is proposed in [11] by having m copies of identical bit-serial structure of [8] with shifted inputs. In such a multiplier, once $2m$ bits of two inputs are received, m bits of the product are obtained after propagation delay through gates. Various efficient bit-parallel architecture for normal basis multiplication over $GF(2^m)$ have been developed in the literature, see for example [11], [12] and [13] for arbitrary normal basis as well as [14], [15], [16], and [17] for special classes of normal basis.

Bit-parallel multipliers require a lot of silicon area and it is impractical for resource constrained environments such as smart cards. To obtain an optimum multiplier for such applications, a digit-level multiplier can be utilized, where the digit size can be chosen depending on the available resources. Using a digit-level multiplier allows the designers to trade-off between speed and area. Among different digit-level normal basis multipliers available in the literature, the ones with the parallel outputs run at much higher frequency than the other ones. Such a multiplier is proposed in [15] and [7] for GNB over $GF(2^m)$, where m is odd.

A special classes of normal basis called GNBs, have been included in the recent standards, such as IEEE and NIST for ECDSA. In this paper, a complexity reduction algorithm is proposed to reduce the number of XOR gates for the original parallel-output type T digit-level GNB multiplier proposed in [15] and [7] for $T > 2$. This algorithm uses sub-expression sharing without increasing the gate delay of the multiplier. It is noted that no such common terms is obtained for $T = 2$. Thus, the algorithm is coded using MATLAB for the $GF(2^{163})$ ($T = 4$) and $GF(2^{283})$ ($T = 6$) finite fields in terms of different digit sizes. Then, based on the results obtained from the algorithm, the original digit-level multiplier structure is modified. The modified GNB multiplier requires fewer number of XOR gates without impacting the gate delay. Both the original $GF(2^{163})$ and $GF(2^{283})$ multipliers and the modified ones are compared in terms of number of XORs for different digit sizes. To obtain the actual implementation results, the original and modified structures are coded in VHDL and they are implemented on a Xilinx® Virtex5™ field-programmable gate array (FPGA) device for different digit sizes. The comparison results show that the modified structure outperforms the original one in terms of area without significantly affecting the multiplication time. It is also shown that for the highest digit size, the bit-parallel multiplier obtained from the modified digit-level multiplier requires the least number of XOR gates with the same gate delay compared to the existing fast GNB multipliers.

The organization of the remaining parts of this paper is as follows. In Section 2, we state the preliminaries required in this paper. Also, the original digit-level Gaussian normal basis multiplier with parallel output is presented in this section. A modified version of this multiplier is proposed in Section 3 using a complexity reduction algorithm. Moreover, in this section a bit-parallel GNB multiplier obtained from the proposed digit-level GNB multiplier is presented and compared with its counterparts in terms of time and area complexities. Results of the hardware implementations of the proposed multiplier on the Xilinx® Virtex5™ FPGA are presented in Section 4. We finally conclude the paper in Section 5.

2   Preliminaries

It is well known that there is always a normal basis $N = \{\beta, \beta^2, \beta^{2^2}, \cdots, \beta^{2^{m-1}}\}$, for a finite field $GF(2^m)$ over $GF(2)$ for any positive integer m, where β is called normal element [18]. The elements of N are linearly independent and each element, say $A = (a_0,\ a_1, \cdots, a_{m-1})$, can be represented as a linear combination of the elements in N, as $A = \sum_{i=0}^{m-1} a_i \beta^{2^i}$, where coefficients $a_i \in GF(2)$, $0 \leq i \leq m-1$, denote the coordinates of A. The merit of the normal basis is that, squaring of an element A in the normal basis representation is a right cyclic shift of its coordinates, i.e., $A^2 = (a_{m-1},\ a_0,\ a_1, \cdots, a_{m-2})$, and it is free in hardware.

Definition 1. *Let* $p = mT + 1$ *be a prime number and* $\gcd(mT/k,\ m) = 1$, *where* k *is the multiplication order of 2 module* p. *Then, the normal basis* $N = \{\beta, \beta^2, \cdots, \beta^{2^{m-1}}\}$ *over* $GF(2^m)$ *is called the Gaussian normal basis of type* T, $T > 0$.

It is noted that the GNBs exist over $GF(2^m)$ whenever m is not divisible by 8 [1]. In this paper, we only consider the GNBs with odd values of m. This implies that T is an even number. It is noted that such GNBs are important since they include the five binary fields, i.e., $m \in \{163, 233, 283, 409, 571\}$, recommended by NIST for ECDSA [2]. The corresponding types for these fields are $T = 4, 2, 6, 4$, and 10, respectively.

2.1   Normal Basis Multiplication

Let $A = (a_0,\ a_1, \cdots, a_{m-1}) = \sum_{i=0}^{m-1} a_i \beta^{2^i}$ and $B = (b_0,\ b_1, \cdots, b_{m-1}) = \sum_{j=0}^{m-1} b_j \beta^{2^j}$ be two field elements over $GF(2^m)$. Let $C \in GF(2^m)$ be their product, i.e., $C = (c_0, c_1, \cdots, c_{m-1}) = AB = \sum_{i=0}^{m-1} \sum_{j=0}^{m-1} a_i b_j \beta^{2^i+2^j}$. Let us represent the field element $\beta^{2^i+2^j} \in GF(2^m)$, $0 \leq i, j \leq m-1$, with respect to N as $\beta^{2^i+2^j} = \sum_{l=0}^{m-1} \mu_{i,j}^{(l)} \beta^{2^l}$. Then, one can find C as

$$C = \sum_{i=0}^{m-1} \sum_{j=0}^{m-1} a_i b_j \sum_{l=0}^{m-1} \mu_{i,j}^{(l)} \beta^{2^l} = \sum_{l=0}^{m-1} \sum_{i=0}^{m-1} \sum_{j=0}^{m-1} a_i b_j \mu_{i,j}^{(l)} \beta^{2^l}. \tag{1}$$

By representing C with respect to N, i.e., $C = \sum_{l=0}^{m-1} c_l \beta^{2^l}$, and equating it with (1), the l-th coordinate of C can be written as $c_l = \sum_{i=0}^{m-1} \sum_{j=0}^{m-1} a_i b_j \mu_{i,j}^{(l)}$. Then, it can be written in a matrix form as

$$c_l = \underline{a} \mathbf{M}^{(l)} \underline{b}^{tr},\ 0 \leq l \leq m-1, \tag{2}$$

where $\mathbf{M}^{(l)} = [\mu_{i,j}^{(l)}]_{i,j=0}^{m-1}$, $\mu_{i,j}^{(l)} \in GF(2)$, $0 \leq i, j \leq m-1$, $\underline{a} = [a_0, a_1, \cdots, a_{m-1}]$ and $\underline{b}^{tr}$ denotes the matrix transpose of row vector $\underline{b} = [b_0, b_1, \cdots, b_{m-1}]$. In (2), $\mathbf{M}^{(l)}$ is obtained from the l-fold right and down circular shifts of the *multiplication matrix* $\mathbf{M} = \mathbf{M}^{(0)}$. The computation of entries of $\mathbf{M}$ can be found from [1]. Massey and Omura in [8] proposed the bit-serial multiplier by implementing (2) for one coordinate, say $c_0 = \underline{a}\mathbf{M}\underline{b}^{tr} = F(A, B)$. Then, the lth coordinate of C can be obtained by cyclic shifts of the coordinates of A and B, i.e., $c_l = F(A \ll l, B \ll l)$ [8].

The number of non-zero entries in the multiplication matrix $\mathbf{M} = \mathbf{M}^{(0)}$ in (2) is called the complexity of the normal basis and is denoted by C_N [19]. This can be used to estimate the area complexity of hardware implementation of the multiplier. Gao et al. in [20] proved that $C_N \geq 2m - 1$. The normal basis is said to be optimal if $C_N = 2m - 1$. The optimal normal bases are extended to another class of low complexity normal basis called Gaussian normal basis by Ash et. al [21]. For type T GNB, $T \geq 2$, the complexity of multiplication matrix $\mathbf{M}$ satisfies $C_N \leq mT - 1$ [21]. A slightly tighter upper bound for C_N is found in [10] as $C_N = mT - T + 1$. Therefore, if there is no optimal normal basis for a given m, the GNB with the least value of T is an alternative for choosing normal bases.

2.2   Digit-Level Gaussian Normal Basis Multiplier with Parallel Output

Let $A = (a_0, a_1, \cdots, a_{m-1})$ and $B = (b_0, b_1, \cdots, b_{m-1})$ be the GNB elements over $GF(2^m)$, and let d, $1 \leq d \leq m$, be the digit size. Reyhani-Masoleh in [15] and Kim et al. in [7] proposed a digit-level Gaussian normal basis multiplier with parallel output (DLGMp). It requires q, $1 \leq q \leq m$, clock cycles to generate all m coordinates of $C = AB$ simultaneously at the end of the final clock cycle. The original multiplier structure of DLGMp is shown in Figure 1. Let $X = (x_0, x_1, \cdots, x_{m-1})$ and $Y = (y_0, y_1, \cdots, y_{m-1})$ be the input registers of this multiplier. Then, it implements [15]

$$J(X, Y) = \sum_{k=0}^{m-1} x_{m-k} s_0'(k, Y)\beta^{2^i}, \tag{3}$$

where

$$s_0'(k, Y) = \sum_{i \in R_k} y_{i-k}, \tag{4}$$

and R_k is a set containing the locations of non-zero entries of row $2k$, $0 \leq 2k \leq m - 1$, of the multiplication matrix $\mathbf{M} = \mathbf{M}^{(0)}$ defined in (2). Based on the properties of $\mathbf{M}$ for GNB, one can find $s_0'(0, Y) = y_1$ and $s_0'(k, Y) = s_0'(m-k, Y)$, $1 \leq k \leq \frac{m-1}{2}$ [15]. Also, it is shown in [10] and [15] that the number of elements in R_k is even and less than or equal to T, i.e., $|R_k| \leq T$. The J block in Figure 1 performs (3) using m AND gates. For the multiplication operation, the registers X and Y of this figure are initially loaded by the coordinates of A

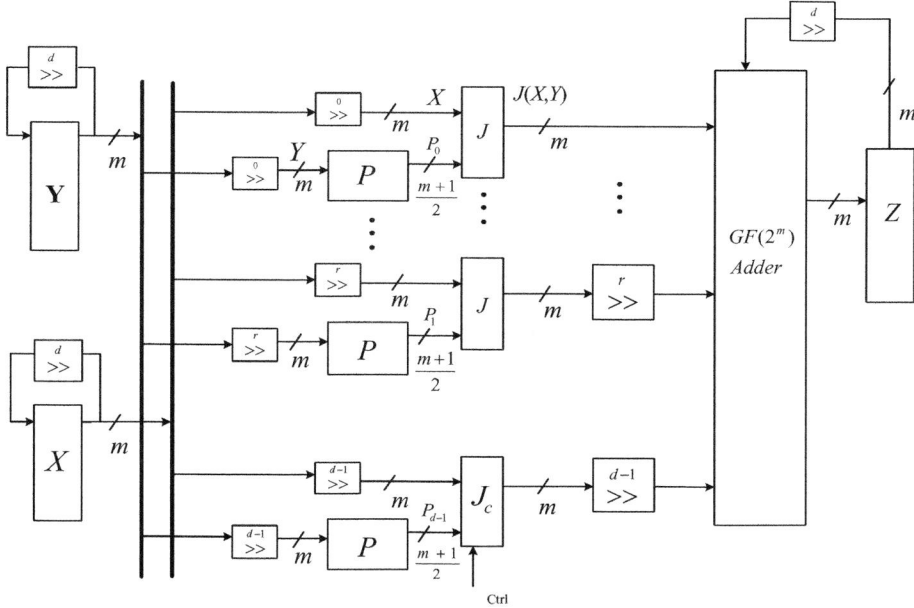

Fig. 1. Digit-serial Gaussian normal basis multiplier proposed in [15], [7], where the i-fold right cyclic shift is denoted by $\underset{\gg}{i}$ and r is a number $0 \leq r \leq d-1$ such that $m = qd - r$

and B, respectively. Also, the output register Z should be cleared before starting the multiplication operation. Then, after q clock cycles, the output register Z contains the coordinates of $C = AB$. In the following section, we modify this multiplier to reduce the number of XOR gates.

3   Modified Digit-Level GNB Multiplier

The number of XOR gates of the DLGMp multiplier presented in the previous section can be reduced by reusing the common terms appeared at the outputs of the P blocks. The complexity reduction scheme presented in [15] cannot be applied for a practical field, such as, $GF(2^{163})$ and $GF(2^{283})$. For the $GF(2^7)$ example used in [15], the P block is optimized first and then the same block is copied for all P blocks used in the multiplier. It is interesting to note that for type 4 GNB over $GF(2^{163})$, no common pair can be found in the P block of Figure 1 if one applies the method presented in [15]. For this purpose, we modify this multiplier by replacing all P blocks that generate $P_1, \cdots, P_d$ in Figure 1 with one block. As seen in Figure 1, the number of outputs of an unoptimized P block in this figure is $\frac{m+1}{2}$. These are based on the following signals [15]

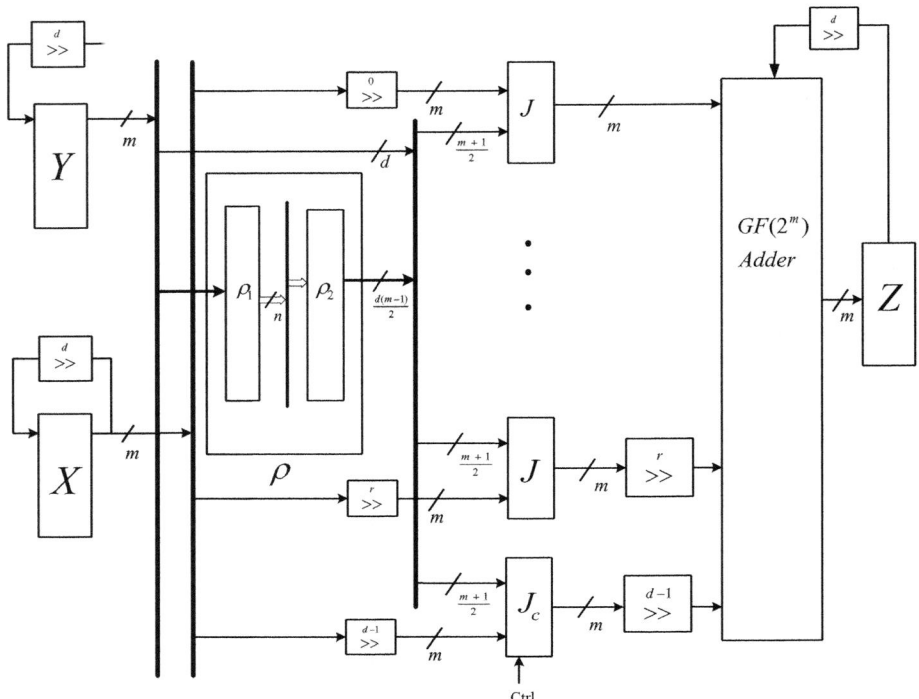

Fig. 2. Modified Digit-level Gaussian Normal Basis Multiplier (MDLGMp)

$$P_k(Y) = (y_{1-k}, \; s_0'(1, Y \ll k), \; s_0'(2, Y \ll k), \cdots$$
$$, \cdots, s_0'(\tfrac{m-1}{2}, \; Y \ll k)), \; 0 \le k \le d-1, \tag{5}$$

for the F block that generates $P_k(Y)$. The modified digit-level multiplier is shown in Figure 2. The combination of all P blocks in Figure 1 is shown by ρ in Figure 2. All signals in (5) are used to build the block ρ in Figure 2. As shown in this figure, y_{1-k}s are removed from the block ρ. To reduce the complexity of the ρ block in Figure 2, we divide the ρ block in two blocks ρ_1 and ρ_2, where ρ_1 includes all common pairs used to generate all signals in (5). In the following section, a complexity reduction algorithm is presented for a given GNB to obtain the optimized blocks of ρ_1 and ρ_2 so that the time delay (in terms of gate delays) of the original block ρ is the same as the one in the modified multiplier, i.e., the addition of gate delays of the two blocks ρ_1 and ρ_2.

3.1   A Complexity Reduction Algorithm

In this section, an approach for reducing the area complexity of the modified digit-level GNB multiplier is proposed. It is noted that unlike the complexity reduction schemes available in the literature, see for example [22], the proposed algorithm does not increase the gate delay of the modified structure as compared to the original one. The complexity reduction algorithm to reduce the number of XOR gates in the block ρ of Figure 2 is summarized as follows.

Input: The multiplication matrix $\mathbf{M}$ and digit size d for type T GNB over $GF(2^m)$.

Output: A pairset which contains all the pairs that should be implemented in the block ρ_1. This set will be used to obtain the formulations for the implementation of the modified multiplier.

1. Corresponding to the output signals of the P block in Figure 1, an $\frac{m-1}{2} \times T$ matrix denoted by $\mu = [\mu_k]_{k=1}^{\frac{m-1}{2}}$ is constructed, where μ_k is the row k, $1 \leq k \leq \frac{m-1}{2}$ of the matrix μ. The entries of μ_k are at most T integers in the range of $[0, m-1]$ and can be found from (4) which can be written as $s_0'(k, Y) = \sum_{j \in \mu_k} y_j, 1 \leq k \leq \frac{m-1}{2}$.

2. Based on the matrix μ and the given digit-size d, a matrix denoted by ρ is obtained by appending the $d-1$ matrices of $\mu - [i] \mod m$ to μ as follows:

$$\rho = \begin{bmatrix} \mu \\ \mu - & [1] \mod m \\ \mu - & [2] \mod m \\ \vdots \vdots & \vdots \\ \mu - & [d-1] \mod m \end{bmatrix}_{(d \times \frac{m-1}{2}) \times T,} \tag{6}$$

where $[i]$, $1 \leq i \leq d$, denotes an $\frac{m-1}{2} \times T$ matrix whose all entries are i.

3. Let ρ_i be a set which contains the entries in row i of the matrix ρ. Then, all signals

$$s_j = \sum_{j \in \rho_i} y_j, \ 1 \leq j \leq d\frac{(m-1)}{2} \tag{7}$$

should be implemented by the block ρ shown in Figure 2.

4. We want to find the common addition pairs to realize (7) with the least number of XOR gates without changing the delay of the modified multiplier as compared with the original one. Therefore, a pairset is generated to form all pairs that should be implemented in the block ρ_1. This set initially contains all pairs with only two entries in the rows of the matrix ρ. We update the ρ matrix by removing such pairs from the matrix. Then, go to Step 3.

5. The scheme will be terminated if no common terms will be obtained.

6. Finally, based on common pairs stored in the pairset, the ρ_1 inside the ρ is generated. By reusing the output of the block ρ_1, we can generate all signals from the block ρ_2 in Figure 2.

In the following section, we present an illustrative example for the proposed complexity reduction algorithm.

3.2   An Example over $GF(2^7)$

To better understand the complexity reduction algorithm, we illustrate an example for the proposed algorithm for type 4 digit-level multiplier over $GF(2^7)$ when the digit-size is $d = m = 7$. The matrix $\mathbf{M}$ for type 4 GNB over $GF(2^7)$ is

$$
\mathbf{M} = \begin{pmatrix}
0 & 1 & 0 & 0 & 0 & 0 & 0 \\
1 & 0 & 1 & 0 & 0 & 1 & 1 \\
0 & 1 & 0 & 1 & 1 & 1 & 0 \\
0 & 0 & 1 & 0 & 0 & 1 & 0 \\
0 & 0 & 1 & 0 & 0 & 0 & 1 \\
0 & 1 & 1 & 1 & 0 & 0 & 1 \\
0 & 1 & 0 & 0 & 1 & 1 & 1
\end{pmatrix}_{7 \times 7}.
$$

The matrix μ can be generated according to the output of the P blocks in Figure 1 as $s_0'(1, Y) = y_{1-1} + y_{3-1} + y_{4-1} + y_{5-1} = y_0 + y_2 + y_3 + y_4$, $s_0'(2, Y) = y_{2-2} + y_{5-2} = y_0 + y_4$, and $s_0'(3, Y) = y_{1-3} + y_{4-3} + y_{5-3} + y_{6-3} = y_5 + y_1 + y_2 + y_3$. Then μ can be written as

$$
\mu = \begin{pmatrix}
0 & 2 & 3 & 4 \\
0 & 4 & - & - \\
5 & 1 & 2 & 3
\end{pmatrix}_{3 \times 4}.
$$

$$
\rho = \begin{pmatrix}
0 & 2 & 3 & 4 \\
0 & 4 & - & - \\
5 & 1 & 2 & 3 \\
6 & 1 & 2 & 3 \\
6 & 3 & - & - \\
4 & 0 & 1 & 2 \\
5 & 0 & 1 & 2 \\
5 & 2 & - & - \\
3 & 6 & 0 & 1 \\
4 & 6 & 0 & 1 \\
4 & 1 & - & - \\
2 & 5 & 6 & 0 \\
3 & 5 & 6 & 0 \\
3 & 0 & - & - \\
1 & 4 & 5 & 6 \\
2 & 4 & 5 & 6 \\
2 & 3 & - & - \\
0 & 3 & 4 & 5 \\
1 & 3 & 4 & 5 \\
1 & 5 & - & - \\
6 & 2 & 3 & 4
\end{pmatrix}_{21 \times 4}
\quad
\text{Pairset1} = \left\{
\begin{array}{c}
y_{04} \\
y_{63} \\
y_{52} \\
y_{41} \\
y_{30} \\
y_{26} \\
y_{15}
\end{array}
\right.
\quad
\rho^{(1)} = \begin{pmatrix}
0 & 2 & 3 & 4 \\
5 & 1 & 2 & 3 \\
6 & 1 & 2 & 3 \\
4 & 0 & 1 & 2 \\
5 & 0 & 1 & 2 \\
3 & 6 & 0 & 1 \\
4 & 6 & 0 & 1 \\
2 & 5 & 6 & 0 \\
3 & 5 & 6 & 0 \\
1 & 4 & 5 & 6 \\
2 & 4 & 5 & 6 \\
0 & 3 & 4 & 5 \\
1 & 3 & 4 & 5 \\
6 & 2 & 3 & 4
\end{pmatrix}
\quad
\rho^{(2)} = \begin{pmatrix}
2 & 3 \\
1 & 3 \\
1 & 2 \\
1 & 2 \\
0 & 1 \\
0 & 1 \\
6 & 0 \\
6 & 0 \\
5 & 0 \\
5 & 6 \\
4 & 6 \\
4 & 5 \\
3 & 5 \\
2 & 4
\end{pmatrix}
\quad
\text{Pairset2} = \left\{
\begin{array}{c}
y_{23} \\
y_{13} \\
y_{12} \\
y_{01} \\
y_{60} \\
y_{50} \\
y_{56} \\
y_{46} \\
y_{45} \\
y_{35} \\
y_{24}
\end{array}
\right.
$$

Based on the digit-size $d = 7$ and the matrix $\mu_{(3 \times 4)}$, the matrix $\rho_{(21 \times 4)}$ can be generated corresponding the complexity reduction algorithm. One can obtain from the matrix $\rho_{(21 \times 4)}$ in which 7 rows of the matrix have just two entries. Therefore, the pairs corresponding to these rows should be implemented as collected in the pairset1. The matrix ρ is updated to $\rho^{(1)}$ by deleting all the two entries mentioned in the pairset1. Then the elements of the pairset1 should be searched in $\rho^{(1)}$ and all common pairs are removed and $\rho^{(1)}$ is updated to $\rho^{(2)}$. This iteration is repeated until there is no rows with more than two entries. As a result, all the remaining pairs as mentioned in the pairset2 should be implemented and repeated pairs (which are underlined in the updated $\rho^{(2)}$ matrix) are removed. The union of pairset1 and pairset2 includes the total of 18 pairs that should be implemented for the block ρ_1 as follows:

$$\text{pairset}=\{y_{04}, y_{63}, y_{52}, y_{41}, y_{30}, y_{26}, y_{15}, y_{23}, y_{13}, y_{12}, y_{01}, y_{60},$$
$$y_{50}, y_{56}, y_{46}, y_{45}, y_{35}, y_{24}\},$$

where $y_{ij} = y_i + y_j$. In addition to the implementation of the ρ block which requires 18 XOR gates, one need $d\frac{m-1}{2} - d = 14$ (as, $d = m$) extra XOR gates for the block ρ_2 to construct its outputs. Therefore, the total number of XOR gates required to implement the ρ block will be $18 + 14 = 32$, whereas the unoptimized P blocks need 49 XOR gates and the scheme proposed in [15] requires 35 XOR gates.

It is noted that the other complexity reduction algorithms available in the literature may result in fewer number of gates at the expense of more delay as the one proposed in this paper. To compare our complexity reduction algorithm with the one proposed in [22], we have applied the complexity reduction algorithm proposed in [22] for the block ρ of this example. It decreases the number of XORs to 23 with the increase of critical path delay to $8T_X$ (eight level of XOR gates). Note that our scheme for this block results in the complexity of 32 XOR gates with the same critical path delay as the original one, i.e., $2T_X$.

3.3   Simulation Results for the Digit-Level GNB Multipliers over $GF(2^{163})$ and $GF(2^{283})$

To evaluate the efficiency of our complexity reduction algorithm, a MATLAB code is written to generate common pairs and signals used in the blocks ρ_1 and ρ_2 of Figure 2. It is noted that for type 2 GNB which is a type 2 optimal normal basis over $GF(2^m)$, there is no common terms to be reused in the block ρ. Therefore, the algorithm presented in this paper cannot reduce the number of XOR gates for $T = 2$. The simulation results of the algorithm for the modified digit-level GNB multipliers (MDLGMp) over $GF(2^{163})$ and $GF(2^{283})$ are obtained and plotted in Figures 3a and 3b. In these figures, we plot the number of required XOR gates versus the digit size for the fields $GF(2^{163})$ ($T = 4$) and $GF(2^{283})$ ($T = 6$) recommended by NIST for ECDSA [2] as compared to ones of the original digit-level multiplier with parallel-output (DLGMp). For a given number of clock cycle, q, $1 \le q \le m$, the least value of digit sizes in the form of $d = \left\lceil \frac{m}{q} \right\rceil$, $1 \le d \le m$, is implemented so that the area complexity is optimized for both multipliers.

From Figures 3a and 3b, one can see that as the digit size increases, more common pairs will be found. As an example, in Figure 3a for the digit size $d = m = 163$, the total number of XOR gates required in the original DLGMp is 66178 gates whereas, the modified one, requires 50400 XOR gates for $GF(2^{163})$. It means that the complexity of the proposed MDLGMp is about 24% less than the original multiplier. More reduction can be found in Figure 3b

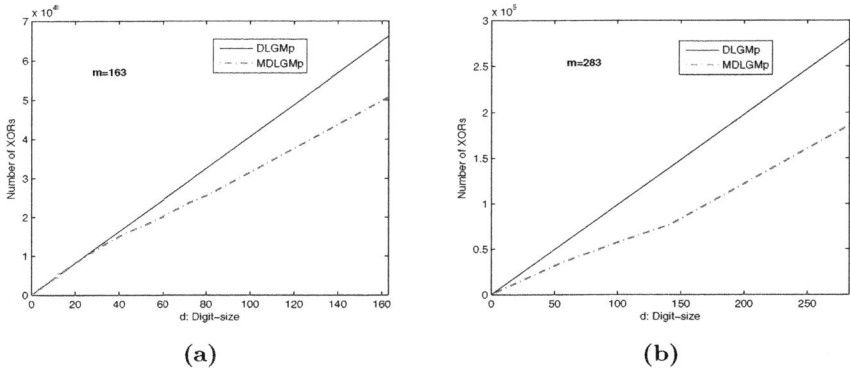

Fig. 3. Comparison between the number of XOR gates required in the DLGMp and the MDLGMp, for (a): $m = 163$ ($T = 4$), (b): $m = 283$ ($T = 6$)

for the $GF(2^{283})$ with $d = m = 283$. The number of XOR gates needed by the original DLGMp is 279,604, whereas the proposed MDLGMp requires 185,375 XOR gates which is about 34% less than that of the original multiplier.

The formulations for the output signals of the blocks ρ_1 and ρ_2 are coded in VHDL to obtain the actual FPGA implementation results. The implementation results are presented in Section 4.

3.4   An Extension to Bit-Parallel GNB Multiplier

To obtain the bit-parallel multiplier, one can implement (2) in hardware for all c_l, $0 \leq l \leq m - 1$. Thus, the hardware architecture of a bit-parallel multiplier is obtained by implementing m copies of identical structures used for c_0 with cyclic shifts of their inputs [11]. In this section, we extend the modified digit-level multiplier for $d = m$ to obtain a new bit-parallel GNB multiplier over $GF(2^m)$. Then, its complexities are obtained and compared with the ones of its counterparts.

Let n denote the total number of common pairs. Thus the block ρ_1 contains at most n XOR gates with the delay of an XOR gate. In the worst case, all combinations of two coordinates of A, i.e., $\binom{m}{2} = \frac{m(m-1)}{2}$ combinations, are required in the block ρ_1 for the bit-parallel multiplier and so, $n \leq \frac{m(m-1)}{2}$.

The block ρ_2 consists of XOR gates for the GNB, with $T > 2$. This is because there is no row in $\mathbf{M}$ with the number of 1s greater than 2 for type 2 GNB. Thus, for $T = 2$, $n = \frac{m(m-1)}{2}$ and the block ρ_2 connects its input bus to the next bus without using any XOR gates.

The exact complexities of ρ_1 and ρ_2 depend on the GNB. However, one can find the upper bound for the number of XOR gates and time delay of this structure as follows.

Proposition 1. *For Type T GNB over $GF(2^m)$, the proposed bit-parallel Gaussian normal basis multiplier architecture requires $m^2 AND$ gates and at most $(T+4)(\frac{m(m-1)}{4})$ XOR gates with the critical path delay of*

$$T_C = T_A + (\lceil \log_2 T \rceil + \lceil \log_2 m \rceil)T_X, \qquad (8)$$

where T_A and T_X are the time delay of a two-input AND gate and an XOR gate, respectively.

Proof. Let n be the the number of XOR gates which is the number of the pairs required to construct the block ρ_1. As mentioned earlier, one can see that the upper bound of n can be found from $n \leq \binom{m}{2} = \frac{m(m-1)}{2}$. Thus, the block ρ_1 contains at most $\frac{m(m-1)}{2}$ XOR gates. It is noted that each output of the block ρ_2 is modulo 2 addition of at most T coordinates of A which can be obtained by adding at most $\frac{T}{2}$ signals from the output of ρ_1. Therefore, the number of XOR gates required to construct the block ρ_2 of the bit-parallel multiplier is $(\frac{T}{2} - 1)(\frac{m-1}{2}) \times m = \frac{m(m-1)(T-2)}{4}$. The rest of Figure 2 requires m^2 AND gates and $m(m-1)$ XOR gates to implement all J blocks and the $GF(2^m)$ adder. By adding the number of XOR gates in the ρ_1, ρ_2 and other blocks, one can obtain the upper bound for the total number of XOR gates as $\frac{m(m-1)}{2} + \frac{m(m-1)(T-2)}{4} + m(m-1) = (T+4)(\frac{m(m-1)}{4})$.

The critical-path delay of the proposed architecture can be obtained by adding the delays of the three blocks of ρ_1, ρ_2, J, and the $GF(2^m)$ adder which are T_X, $\lceil \log_2 \frac{T}{2} \rceil T_X$, T_A, and $\lceil \log_2 m \rceil T_X$, respectively. This results in the total delay of $T_X + \lceil \log_2 \frac{T}{2} \rceil T_X + T_A + \lceil \log_2 m \rceil T_X = T_A + (\lceil \log_2 T \rceil + \lceil \log_2 m \rceil)T_X$, which completes the proof.

3.5   Comparison

The time and area complexities of the proposed bit-parallel GNB multiplier and the previous schemes are compared in Table 1 for general and special values of T. As shown in the table, the critical path delay of the proposed multiplier matches the fastest results available in the literature. For type $T = 2$ GNB, the number of XOR gates also matches the fastest result available in the open literature, i.e., $1.5m(m-1)$. However, it is much greater than the sub-quadratic results proposed in [17] and [16] which require much higher delay as compared to the one proposed here. It is interesting to note that for $T > 2$, the proposed multiplier outperforms its counterparts with the same delay in terms of number of XOR gates as shown in this table. It is noted the number of XOR gates required for the new bit-parallel GNB multiplier is still greater than the one required for the polynomial basis.

It should be noted that, to obtain the exact number of XOR gates for a given GNB, the exact value of n should be obtained by simulations. Using the complexity reduction algorithm proposed in Section 3.1, a comparison between the number of XOR gates of bit-parallel GNB multipliers is illustrated in Table 2 for $GF(2^{163})$ and $GF(2^{283})$ fields recommended by NIST for ECDSA.

Table 1. Area and time complexity comparison of bit-parallel GNB multipliers over $GF(2^m)$. Note that for Type T GNB: $C_N \leq Tm - T + 1$.

Multiplier	$T \geq 2$		
	#AND	#XOR	Critical path
Massey & Omura [8]	m^2	$m(C_N - 1)$	$T_A + \lceil \log_2 C_N \rceil T_X$
Gao & Sobelman[12]	m^2	$m(C_N - 1)$	$T_A + (\lceil \log_2 T \rceil + \lceil \log_2 m \rceil)T_X$
Reyhani-Masoleh & Hasan [13]	m^2	$\leq \frac{m}{2}(C_N + m - 2)$	$T_A + (\lceil \log_2(C_N + 1) \rceil)T_X$
DLGMp [15], [7] $(d = m)$	m^2	$\leq \frac{m}{2}(C_N + m)$	$T_A + (\lceil \log_2 T \rceil + \lceil \log_2(m) \rceil)T_X$
DLGMs [15] $(d = m)$	m^2	$\leq \frac{m(m-1)}{2}(T + 1)$	$T_A + (\lceil \log_2 T \rceil + \lceil \log_2(m) \rceil)T_X$
This work	m^2	$\leq (\frac{m(m-1)}{4})(T + 4)$	$T_A + (\lceil \log_2 T \rceil + \lceil \log_2(m) \rceil)T_X$
	T=2		
[8,12]	m^2	$2m(m - 1)$	$T_A + \lceil \log_2(2m - 1) \rceil T_X$
Koc & Sunar [14]	m^2	$1.5m(m - 1)$	$T_A + (1 + \lceil \log_2 m \rceil)T_X$
Fan & Hasan [16]	$2m^{1.6}$	$11m^{1.6} - 12m + 1$	$T_A + (2\log_2 m + 1)T_X$
Cathen et. al [17]	$2m^{1.6}$	$7.6m^{1.6} + \mathcal{O}(m \log m)$	$T_A + (2\log_2 m + 1)T_X$
[13,15,7], This work	m^2	$1.5m(m - 1)$	$T_A + (1 + \lceil \log_2 m \rceil)T_X$
	T=4		
[8], [12]	m^2	$4m^2 - 4m$	$T_A + (2 + \lceil \log_2(m) \rceil)T_X$
Reyhani-Masoleh & Hasan [13]	m^2	$2.5m^2 - 4.5m$	$T_A + \lceil 1 + \log_2(2m - 1) \rceil T_X$
DLGMp [15], [7] $(d = m)$	m^2	$2.5m^2 - 1.5m$	$T_A + (2 + \lceil \log_2(m) \rceil)T_X$
DLGMs [15] $(d = m)$	m^2	$2.5m^2 - 2.5m$	$T_A + (2 + \lceil \log_2(m) \rceil)T_X$
This work	m^2	$\leq 2m^2 - 2m$	$T_A + (2 + \lceil \log_2(m) \rceil)T_X$
	T=6		
[8], [12]	m^2	$6m^2 - 6m$	$T_A + (3 + \lceil \log_2(m) \rceil)T_X$
Reyhani-Masoleh & Hasan [13]	m^2	$3.5m^2 - 3.5m$	$T_A + (\lceil \log_2(6m - 4) \rceil)T_X$
DLGMp [15], [7] $(d = m)$	m^2	$3.5m^2 - 2.5m$	$T_A + (3 + \lceil \log_2(m) \rceil)T_X$
DLGMs [15] $(d = m)$	m^2	$3.5m^2 - 3.5m$	$T_A + (3 + \lceil \log_2(m) \rceil)T_X$
This work	m^2	$\leq 2.5m^2 - 2.5m$	$T_A + (3 + \lceil \log_2(m) \rceil)T_X$

Table 2. Comparison between bit-parallel GNB multipliers for $GF(2^{163})$ and $GF(2^{283})$

m	T	n	Number of XORs in DLGM for $d = m$ [15]	Number of XOR gates used in this work
163	4	10791	66178	50400
283	6	25763	279604	185375

4   FPGA Implementations

The architectures described in Sections 2.2 and 3 are written in VHDL. We have implemented the original (DLGMp) and the modified digit-level multipliers (MDLGMp) on the Xilinx® Virtex5™ FPGA family with target device xc5vlx330-2ff1760 for $GF(2^{163})$ and $GF(2^{283})$ fields. Correctness of the implementations is verified by performing functional simulations using the Quartus® II software. We have synthesized both multipliers for several different digit sizes

Table 3. FPGA implementation results for propagation delay (in terms of nano second) and area (in terms of number of slices) for different digit sizes with $m = 163$ and $T = 4$. Target device is Xilinx xc5vlx330-2ff1760.

Digit size (d)	# of cycles (q)	Delay [ns]		Area [# of Slice LUTs]	
		DLGMp	MDLGMp	DLGMp	MDLGMp
1	163	2.8	2.8	1221	1221
2	82	3.1	3.1	1282	1280
3	55	3.1	3.1	1347	1346
4	41	3.4	3.4	1406	1406
5	33	3.5	3.6	1564	1565
6	28	4.1	4.2	1751	1750
7	24	3.8	3.8	1960	1960
8	21	3.7	3.8	2104	2104
9	19	4.3	4.4	2157	2157
10	17	4.2	4.2	2309	2309
11	15	4.2	4.2	2385	2385
12	14	4.5	4.5	2567	2567
13	13	4.6	4.6	2785	2780
14	12	4.5	4.6	2852	2850
15	11	4.8	4.7	2923	2923
17	10	4.9	4.9	3164	3164
19	9	4.9	4.9	4048	4045
21	8	5.4	5.5	4146	4140
24	7	5.6	5.7	4593	4385
28	6	5.7	5.7	4730	4652
33	5	5.8	5.8	5288	5023
41	4	6.1	6.1	6129	5633
55	3	6.4	6.5	8115	6091
82	2	7.3	7.5	11187	7321
163	1	11.5	11.9	22917	14238

d, $1 \leq d \leq m$ using Xilinx synthesis technology (XST), and the timing analysis results via Xilinx ISE-9.1.03i after place and route (PAR) are illustrated in Tables 3 and 4 for $GF(2^{163})$ and $GF(2^{283})$, respectively. As seen in the tables, large digit sizes require more area in terms of number of slice look up tables (LUTs). Therefore, we chose the digit size, d, in such a way to decrease the critical path delay while increasing the area. Note that other values for d only increases area without decreasing latency. Our presented multiplier requires less area than the original one for the different digit size d. The total multiplication time can be calculated as the product of the minimum clock period and the number of clock cycles q presented in both tables. Obviously, as shown in the Tables 3 and 4, the time complexities of these structures are almost the same. It means that, our proposed multiplier reduces the required area having the same multiplication time.

Table 4. FPGA implementation results for propagation delay (in terms of nano second) and area (in terms of number of slice LUTs) for different digit sizes with $m = 283$ and $T = 6$. The target device is xc5vlx330-2ff1760.

Digit size (d)	# of cycles (q)	Delay [ns]		Area [# of Slice LUTs]	
		DLGMp	MDLGMp	DLGMp	MDLGMp
1	283	3.4	3.4	2118	1985
2	142	3.7	3.8	2252	2088
3	95	3.9	3.9	2388	2156
4	71	4.1	4.1	2603	2437
5	57	4.4	4.5	2829	2603
6	48	4.5	4.6	3216	2714
7	41	4.4	4.5	3358	2937
15	19	4.7	4.8	5803	3986
16	18	5.1	5.2	6086	4105
19	15	5.7	5.7	6309	4387
22	13	5.8	5.8	6429	4427
24	12	8.7	8.9	7039	4938
26	11	8.6	8.5	7325	5183
29	10	8.3	8.4	7723	5563
35	9	8.3	8.5	9398	6769
71	4	11.4	11.4	17224	10345
142	2	12.4	12.3	31395	22512

5   Conclusions

We have proposed a modified architecture for digit-level Gaussian normal basis multiplier over $GF(2^m)$. It is shown that this multiplier outperforms the original one in terms of number of XOR gates. Using a complexity reduction algorithm, the area complexity of the modified digit-level multiplier, is optimized. We have also presented a fast low complexity bit-parallel GNB multiplier over $GF(2^m)$. Its complexities have been derived and it is shown that it has fewer XOR gates for $T > 2$ and the same for $T = 2$ as compared to the fast GNB multipliers available in the literature. For practical applications, we have implemented the original and the modified digit-level GNB multipliers on the Xilinx® Virtex5™ FPGA family for different digit sizes. Our comparison results show that the modified digit-level GNB multiplier requires fewer area (slice LUTs) with almost the same delay as compared to the original one.

Acknowledgment

The authors of the paper would like to thank the reviewers for their comments. This work has been supported in part by an NSERC Discovery grant awarded to A. Reyhani-Masoleh.

References

1. IEEE Std 1363-2000: IEEE Standard Specifications for Public-Key Cryptography (January 2000)
2. U.S. Department of Commerce/NIST: Digital Signature Standards (DSS). Federal Information Processing Standards Publications (2000)
3. Miller, V.S.: Use of Elliptic Curves in Cryptography. In: Williams, H.C. (ed.) CRYPTO 1985. LNCS, vol. 218, pp. 417–426. Springer, Heidelberg (1986)
4. Koblitz, N.: Elliptic curve cryptosystems. Mathematics of Computation 48, 203–209 (1987)
5. Dimitrov, V.S., Järvinen, K.U., Jacobson Jr., M.J., Chan, W.F., Huang, Z.: Provably Sublinear Point Multiplication on Koblitz Curves and its Hardware Implementation. IEEE Transaction on Computers 57(11), 1469–1481 (2008)
6. Järvinen, K., Skyttä, J.: On Parallelization of High-Speed Processors for Elliptic Curve Cryptography. IEEE Transactions on Very Large Scale Integration (VLSI) Systems 16(9), 1162–1175 (2008)
7. Kim, C.H., Kwon, S., Hong, C.P.: FPGA Implementation of High Performance Elliptic Curve Cryptographic Processor over $GF(2^{163})$. Journal of System Architcture 54(10), 893–900 (2008)
8. Massey, J., Omura, J.: Computational Method and Apparatus for Finite Arithmetic. US Patent (4587627) (1986)
9. Agnew, G.B., Mullin, R.C., Onyszchuk, I.M., Vanstone, S.A.: An Implementation for a Fast Public-Key Cryptosystem. Journal of Cryptology 3(2), 63–79 (1991)
10. Kwon, S., Gaj, K., Kim, C.H., Hong, C.P.: Efficient Linear Array for Multiplication in $GF(2^m)$ using a Normal Basis for Elliptic Curve Cryptography. In: Joye, M., Quisquater, J.-J. (eds.) CHES 2004. LNCS, vol. 3156, pp. 76–91. Springer, Heidelberg (2004)
11. Wang, C.C., Truong, T.K., Shao, H.M., Deutsch, L.J., Omura, J.K., Reed, I.S.: VLSI Architectures for Computing Multiplications and Inverses in $GF(2^m)$. IEEE Transaction on Computers 34(8), 709–717 (1985)
12. Gao, L., Sobelman, G.E.: Improved VLSI Designs for Multiplication and Inversion in $GF(2^M)$ over normal bases. In: Proceedings of 13th Annual IEEE International ASIC/SOC Conference, pp. 97–101 (2000)
13. Reyhani-Masoleh, A., Hasan, M.A.: A New Construction of Massey-Omura Parallel Multiplier over $GF(2^m)$. IEEE Transactions on Computers 51(5), 511–520 (2002)
14. Koç, Ç.K., Sunar, B.: An Efficient Optimal Normal Basis Type II Multiplier over $GF(2^m)$. IEEE Transaction on Computers 50(1), 83–87 (2001)
15. Reyhani-Masoleh, A.: Efficient Algorithms and Architectures for Field Multiplication Using Gaussian Normal Bases. IEEE Transaction On Computers, 34–47 (2006)
16. Fan, H., Hasan, M.: Subquadratic computational complexity schemes for extended binary field multiplication using optimal normal bases. IEEE Transactions on Computers 56(10), 1435 (2007)
17. Gathen, J., Shokrollahi, A., Shokrollahi, J.: Efficient multiplication using type 2 optimal normal bases. In: Carlet, C., Sunar, B. (eds.) WAIFI 2007. LNCS, vol. 4547, pp. 55–68. Springer, Heidelberg (2007)
18. Lidl, R., Niederreiter, H.: Introduction to Finite Fields and Their Applications. Cambridge University Press, Cambridge (1994)

19. Mulin, R.C., Onyszchuk, I.M., Vanstone, S.A., Wilson, R.M.: Optimal Normal Bases in $GF(p^n)$. Discrete Appl. Math. 22(2), 149–161 (1989)
20. Gao S., Lenstra, H.W.: Optimal Normal Bases. Designs, Codes and Cryptography 2, 315–323 (1992)
21. Ash D.W., Blake, I.F., Vanstone, S.A.: Low Complexity Normal Bases. Discrete Applied Mathematics 25(3), 191–210 (1989)
22. Gustafsson, O., Olofsson, M.: Complexity reduction of constant matrix computations over the binary field. In: Carlet, C., Sunar, B. (eds.) WAIFI 2007. LNCS, vol. 4547, pp. 103–115. Springer, Heidelberg (2007)

Type-II Optimal Polynomial Bases*

Daniel J. Bernstein[1] and Tanja Lange[2]

[1] Department of Computer Science (MC 152)
University of Illinois at Chicago, Chicago, IL 60607–7053, USA
djb@cr.yp.to
[2] Department of Mathematics and Computer Science
Technische Universiteit Eindhoven, P.O. Box 513,
5600 MB Eindhoven, Netherlands
tanja@hyperelliptic.org

Abstract. In the 1990s and early 2000s several papers investigated the relative merits of polynomial-basis and normal-basis computations for $\mathbf{F}_{2^n}$. Even for particularly squaring-friendly applications, such as implementations of Koblitz curves, normal bases fell behind in performance unless a type-I normal basis existed for $\mathbf{F}_{2^n}$.

In 2007 Shokrollahi proposed a new method of multiplying in a type-II normal basis. Shokrollahi's method efficiently transforms the normal-basis multiplication into a single multiplication of two size-$(n+1)$ polynomials.

This paper speeds up Shokrollahi's method in several ways. It first presents a simpler algorithm that uses only size-n polynomials. It then explains how to reduce the transformation cost by dynamically switching to a 'type-II optimal polynomial basis' and by using a new reduction strategy for multiplications that produce output in type-II polynomial basis.

As an illustration of its improvements, this paper explains in detail how the multiplication overhead in Shokrollahi's original method has been reduced by a factor of 1.4 in a major cryptanalytic computation, the ongoing attack on the ECC2K-130 Certicom challenge. The resulting overhead is also considerably smaller than the overhead in a traditional low-weight-polynomial-basis approach. This is the first state-of-the-art binary-elliptic-curve computation in which type-II bases have been shown to outperform traditional low-weight polynomial bases.

Keywords: Optimal normal basis, ONB, polynomial basis, transformation, elliptic-curve cryptography.

1 Introduction

If $n+1$ is prime and 2 has order n modulo $n+1$ then the field $\mathbf{F}_{2^n} = \mathbf{F}_2[\zeta]/(\zeta^n + \cdots + \zeta + 1)$ has a "type-I optimal normal basis" $\zeta, \zeta^2, \zeta^4, \ldots$. It has been known

---

* Permanent ID of this document: 90995f3542ee40458366015df5f2b9de. Date of this document: 2010.04.12. This work has been supported in part by the European Commission through the ICT Programme under Contract ICT–2007–216676 ECRYPT-II and in part by the National Science Foundation under grant ITR–0716498.

M.A. Hasan and T. Helleseth (Eds.): WAIFI 2010, LNCS 6087, pp. 41–61, 2010.

for many years that this basis allows not only fast repeated squarings but also surprisingly fast multiplications, costing only $M(n) + 2n - 2$ bit operations where $M(n)$ is the minimum cost of multiplying n-coefficient polynomials. The idea is to permute the basis into $\zeta, \zeta^2, \zeta^3, \ldots, \zeta^n$, and to decompose multiplication into the following operations:

- $M(n)$ bit operations: multiply the polynomials $f_1\zeta + \cdots + f_n\zeta^n$ and $g_1\zeta + \cdots + g_n\zeta^n$ in $\mathbf{F}_2[\zeta]$.
- $n - 2$ bit operations: eliminate the coefficients of $\zeta^{n+2}, \ldots, \zeta^{2n}$ using the identities $\zeta^{n+2} = \zeta, \ldots, \zeta^{2n} = \zeta^{n-1}$; this requires additions to the existing coefficients of $\zeta^2, \ldots, \zeta^{n-1}$.
- n bit operations: eliminate the coefficient of ζ^{n+1} using the identity $\zeta^{n+1} = \zeta + \zeta^2 + \cdots + \zeta^n$.

An alternative introduced in [IT89] is to use a redundant representation, specifically coefficients of $1, \zeta, \ldots, \zeta^n$, with arithmetic modulo $\zeta^{n+1} + 1$. Multiplication then costs $M(n + 1) + n$ bit operations; this is worse than $M(n) + 2n - 2$ for small n, but it becomes better for large n, since $M(n)$ is subquadratic.

However, most integers n do not have type-I optimal normal bases. In particular, an odd prime n cannot have a type-I optimal normal basis. This poses severe problems for cryptographic applications that, for security reasons, prohibit composite values of n.

The conventional wisdom for many years was that type-I normal bases were a unique exception. For all other normal bases the best multiplication methods in the literature were quite slow. In particular, multiplication in a "type-II optimal normal basis" of $\mathbf{F}_{2^n}$ was asymptotically at least twice as expensive as multiplication in traditional low-weight polynomial bases (trinomial bases and pentanomial bases):

- Traditional normal-basis multipliers use $\Theta(n^2)$ bit operations.
- The type-II multiplier in [FH07] uses approximately $13 \cdot 3^{\lceil \log_2 n \rceil}$ bit operations.
- The type-II multiplier in [BG01, Section 4.1] uses approximately $3M(n)$ bit operations.
- The type-II multiplier in [GvzGP95] (see also [GvzGPS00]) uses approximately $2M(n)$ bit operations.

Normal bases were competitive only in extreme situations: applications where n was very small; applications having many repeated squarings and very few multiplications; and applications that imposed extremely small hardware-area requirements, effectively punishing polynomial bases by prohibiting fast-multiplication techniques.

The picture changed a few years ago when Shokrollahi introduced a new type-II multiplier using only $M(n) + O(n \log_2 n)$ operations. See Shokrollahi's thesis [Sho07, Chapter 4] and the subsequent WAIFI 2007 publication [vzGSS07] by von zur Gathen, Shokrollahi, and Shokrollahi. This new multiplier makes type-II normal bases competitive with traditional low-weight polynomial bases for a much wider variety of applications. The overhead term $O(n \log_2 n)$ is not quite as small as the $O(n)$ cost of low-weight polynomial reduction, but this difference is quite often outweighed by the benefit of fast repeated squarings.

In this paper we reduce the overhead in Shokrollahi's method in several ways. The overall reduction depends on the pattern of desired squarings and multiplications but can be as much as a factor of 2. We give a real-world example in which the overhead is reduced by more than a factor of 1.4.

1.1 Model of Computation

All of the algorithms in this paper are straight-line (branchless) sequences of bit operations. The "bit operations" allowed are two-input XORs (addition in $\mathbf{F}_2$) and two-input ANDs (multiplication in $\mathbf{F}_2$). We measure algorithm cost by counting the number of bit operations, as in [Sho07], [Ber09a], etc.

Optimizing bit operations is not the same as optimizing hardware area: a very small circuit can carry out many bit operations if the operations to be performed are sufficiently regular. Optimizing bit operations is also not the same as optimizing hardware latency. Furthermore, the word-level machine operations used by *software* are much more complicated than bit operations. For these reasons, optimizing bit operations is often believed to be of little relevance to optimizing hardware, and of even less relevance to optimizing software.

However, optimizing bit operations is very close to optimizing the *throughput* of unrolled, pipelined hardware; note that full unrolling eliminates all control logic, and pipelining hides latency. Furthermore, optimizing bit operations is very close to optimizing the performance of software built from *vectorized* bit operations. [Ber09a] recently set new software speed records for public-key cryptography by exploiting a synergy between "bitsliced" data structures, bit-operation-optimized polynomial-multiplication techniques, and the 128-bit vector operations available on common CPUs.

This work began as part of a larger project described in [BBB+09] to solve a cryptanalytic challenge, the Certicom "ECC2K-130" challenge [Cer97]. One of the surprises in [BBB+09] is that type-II bases save time for the ECC2K-130 computation on various *software* platforms, solidly outperforming low-weight (in this case pentanomial) polynomial bases. At the time of this writing, the latest version of [BBB+09] reports the speed of software that extends the synergy of [Ber09a] to include type-II bases, using Shokrollahi's approach together with the improvements described in Section 3 of this paper. The latest GPU software incorporates additional improvements described in subsequent sections of this paper.

A very recent pipelined FPGA implementation of the ECC2K-130 computation also uses this approach and is much faster than previous FPGA implementations; see [FBB+10]. We predict that the speedups discussed in this paper will turn out to be useful in other applications that need to maximize the number of $\mathbf{F}_{2^n}$ operations that can be carried out per second for a given chip area. We are now investigating the constructive use of the same techniques for fast Koblitz-curve cryptography — of course, at much larger sizes than ECC2K-130!

1.2   Outline of the Paper

Section 2 reviews Shokrollahi's algorithm for type-II normal-basis multiplication. Section 3 presents a streamlined algorithm for type-II normal-basis multiplication. The streamlined algorithm uses approximately $M(n) + 2n \log_2(n/2)$ bit operations. More precisely, if $n = 2^{n_0} + 2^{n_1} + \cdots$ with $n_0 > n_1 > \cdots$, then the streamlined algorithm uses $M(n) + \sum_i (2^{n_i}(2n_i - 2 + 4i) + 3)$ bit operations.

Section 4 presents an algorithm to multiply in a "type-II polynomial basis" using approximately $M(n) + n \log_2 n$ bit operations. The overhead $n \log_2 n$ saves almost half of the previous overhead $2n \log_2(n/2)$; about $0.5n \log_2(n/4)$ is saved by a new reduction method, and about $0.5n \log_2(n/4)$ is saved by a simple exercise in caching. Repeated squaring in a "type-II polynomial basis" is not as fast as repeated squaring in a normal basis but is still much faster than repeated squaring in a traditional polynomial basis.

We further reduce the total overhead of multiplications and repeated squarings by dynamically mixing a type-II polynomial basis P with a type-II normal basis N. In applications that contain only occasional multiplications, this mixture is tantamount to working purely in N, as in Section 3. In applications that contain only occasional repeated squarings, this mixture is tantamount to working purely in P. However, in many applications the mixture is better than any pure method. Section 5 uses ECC2K-130 as an illustrative example of this paper's overall improvements.

2   Review of the Original Shokrollahi Approach

In this section we review the normal-basis multiplier from [Sho07, Section 4] for the special case of binary fields $\mathbf{F}_{2^n}$.

In a nutshell the main achievement of [Sho07] is a map S between a normal-basis representation and a special polynomial-basis representation, taking $\Theta(n \log_2 n)$ bit operations instead of the usual $\Theta(n^2)$. The multiplication in normal basis can be performed as $S^{-1}(S(a_1) \cdot S(a_2))$: first use S on a_1 and a_2 to map to polynomial-basis representation, where efficient algorithms for polynomial multiplication can be used, and finally map the result back to normal basis representation.

The irreducible polynomial for this special polynomial basis is usually rather dense. To avoid reduction modulo this polynomial, Shokrollahi defines a double-length map that takes an unreduced polynomial product back to normal basis. The reduction is done on the normal-basis side, where it is simple addition of length-n vectors.

Before giving the details we review the construction of a type-II normal basis for $\mathbf{F}_{2^n}$ and establish notations P and N used throughout the paper. For more information on type-II normal bases, see Mullin et al. [MOVW89].

2.1   Type-II Normal Bases

For the rest of the paper we assume that n is a positive integer, that $2n + 1$ is prime, and that either (1) the order of 2 in the multiplicative group $\mathbf{F}_{2n+1}^*$ is $2n$

or (2) n is odd and the order of 2 in $\mathbf{F}_{2n+1}^*$ is n. In the latter case 2 generates the subgroup of squares in $\mathbf{F}_{2n+1}^*$. Since n is odd we have $2n + 1 \equiv 3 \pmod 4$ so -1 is not a square in $\mathbf{F}_{2n+1}^*$; i.e., -1 is not a power of 2.

The conditions on n imply that $2^{2n} = 1$ in $\mathbf{F}_{2n+1}$ and so there exists an element $\zeta \in \mathbf{F}_{2^{2n}}$ that is a primitive $(2n+1)$st root of unity. Now $2^n = \pm 1$ in $\mathbf{F}_{2n+1}$, since 2 has order $2n$ or n, so $\zeta^{2^n} \in \{\zeta, \zeta^{-1}\}$. Define $c = \zeta + \zeta^{-1}$. Then $c^{2^n} = \zeta^{2^n} + \zeta^{-2^n} = \zeta + \zeta^{-1} = c$, so $c \in \mathbf{F}_{2^n}$.

The elements $c, c^2, c^{2^2}, c^{2^3}, \ldots, c^{2^{n-1}}$ are distinct. Indeed, any repetition would (after some square roots) imply an equation of the form $c^{2^i} = c$ with $i \in \{1, 2, \ldots, n-1\}$, i.e., $(\zeta + \zeta^{-1})^{2^i} = \zeta + \zeta^{-1}$, i.e., $\zeta^{2^i} + \zeta^{-2^i} + \zeta + \zeta^{-1} = 0$. This equation factors as $(\zeta^{2^i} + \zeta)(1 + \zeta^{-2^i-1}) = 0$, so $\zeta^{2^i} = \zeta$ or $\zeta^{2^i+1} = 1$, so $2^i = \pm 1$ in $\mathbf{F}_{2n+1}$. This implies $2^{2i} = 1$ in $\mathbf{F}_{2n+1}$ which contradicts that the order of 2 is $2n$. If n is odd and the order of 2 is n then -1 is not a power of 2, contradicting $2^i = -1$, while $2^i = 1$ contradicts that the order is n.

Each power c^{2^i} can be written as $\zeta^j + \zeta^{-j}$ for the unique $j \in \{1, 2, \ldots, n\}$ with $2^i = \pm j$ in $\mathbf{F}_{2n+1}$, since $c^{2^i} = (\zeta + \zeta^{-1})^{2^i} = \zeta^{2^i} + \zeta^{-2^i} = \zeta^j + \zeta^{-j}$. Therefore $(c, c^2, c^{2^2}, \ldots, c^{2^{n-1}})$ is a permutation of $(\zeta + \zeta^{-1}, \zeta^2 + \zeta^{-2}, \zeta^3 + \zeta^{-3}, \ldots, \zeta^n + \zeta^{-n})$.

If a vector $(e_1, e_2, \ldots, e_n) \in \mathbf{F}_2^n$ satisfies $e_1(\zeta + \zeta^{-1}) + e_2(\zeta^2 + \zeta^{-2}) + \cdots + e_n(\zeta^n + \zeta^{-n}) = 0$ then $e_1(\zeta + \zeta^{2n}) + e_2(\zeta^2 + \zeta^{2n-1}) + \cdots + e_n(\zeta^n + \zeta^{n+1}) = 0$, so ζ is a root of the polynomial $p = e_1(1 + z^{2n-1}) + \cdots + e_n(z^{n-1} + z^n) \in \mathbf{F}_2[z]$ of degree at most $2n - 1$. Exchanging ζ and ζ^{-1} shows that ζ^{-1} is also a root of p. If 2 has order $2n$ then the cyclotomic polynomial Φ_{2n+1} is irreducible in $\mathbf{F}_2[z]$, so Φ_{2n+1} divides p. If n is odd and 2 has order n then Φ_{2n+1} factors into the coprime irreducible polynomials $(z - \zeta)(z - \zeta^2) \cdots (z - \zeta^{2^{n-1}})$ and $(z - \zeta^{-1})(z - \zeta^{-2}) \cdots (z - \zeta^{-2^{n-1}})$; each of these polynomials divides p, so again Φ_{2n+1} divides p. Since $\deg(\Phi_{2n+1}) = 2n$ this implies $p = 0$ so $(e_1, e_2, \ldots, e_n) = 0$.

Summary: $(\zeta + \zeta^{-1}, \zeta^2 + \zeta^{-2}, \zeta^3 + \zeta^{-3}, \ldots, \zeta^n + \zeta^{-n})$ is a basis of $\mathbf{F}_{2^n}$, and $(c, c^2, \ldots, c^{2^{n-1}})$ is a normal basis of $\mathbf{F}_{2^n}$. This normal basis is called a "type-II optimal normal basis", and the permutation $(\zeta + \zeta^{-1}, \zeta^2 + \zeta^{-2}, \zeta^3 + \zeta^{-3}, \ldots, \zeta^n + \zeta^{-n})$ is called a "permuted type-II optimal normal basis".

2.2   The Functions N and P

We denote by $N(x)$ the representation of $x \in \mathbf{F}_{2^n}$ with respect to the permuted normal basis $(\zeta + \zeta^{-1}, \zeta^2 + \zeta^{-2}, \zeta^3 + \zeta^{-3}, \ldots, \zeta^n + \zeta^{-n})$. In other words, the vector $N(x) = (N(x)_1, \ldots, N(x)_n) \in \mathbf{F}_2^n$ satisfies $\sum_i N(x)_i(\zeta^i + \zeta^{-i}) = x$.

We denote by $P(x)$ the representation of $x \in \mathbf{F}_{2^n}$ with respect to the polynomial basis $(c, c^2, c^3, \ldots, c^n)$. In other words, the vector $P(x) = (P(x)_1, \ldots, P(x)_n) \in \mathbf{F}_2^n$ satisfies $\sum_i P(x)_i(\zeta + \zeta^{-1})^i = x$. Note that this is not exactly a conventional polynomial basis: the corresponding polynomials have degree $\leq n$ and constant term zero.

2.3   Shokrollahi's Transformation

Shokrollahi extends the normal basis to the redundant generating set $\mathcal{N} = (1, \zeta + \zeta^{-1}, \zeta^2 + \zeta^{-2}, \zeta^3 + \zeta^{-3}, \ldots, \zeta^n + \zeta^{-n})$ and extends the polynomial basis to the redundant generating set $\mathcal{P} = (1, \zeta + \zeta^{-1}, (\zeta + \zeta^{-1})^2, (\zeta + \zeta^{-1})^3, \ldots, (\zeta + \zeta^{-1})^{n--}, (\zeta + \zeta^{-1})^n)$. Shokrollahi recursively defines a transformation S_k from $(1, \zeta + \zeta^{-1}, \zeta^2 + \zeta^{-2}, \zeta^3 + \zeta^{-3}, \ldots, \zeta^{k-1} + \zeta^{1-k})$ to $(1, (\zeta + \zeta^{-1}), (\zeta + \zeta^{-1})^2, (\zeta + \zeta^{-1})^3, \ldots, (\zeta + \zeta^{-1})^{k-1})$, and in particular defines a transformation S_{n+1} from $\mathcal{N}$ to $\mathcal{P}$.

We first describe S_8^{-1} and then generalize. The central observation is that if

$$f_0 + f_1 \left(\zeta + \zeta^{-1}\right) + f_2 \left(\zeta + \zeta^{-1}\right)^2 + f_3 \left(\zeta + \zeta^{-1}\right)^3$$
$$= g_0 + g_1 \left(\zeta + \zeta^{-1}\right) + g_2 \left(\zeta^2 + \zeta^{-2}\right) + g_3 \left(\zeta^3 + \zeta^{-3}\right)$$

and

$$f_4 + f_5 \left(\zeta + \zeta^{-1}\right) + f_6 \left(\zeta + \zeta^{-1}\right)^2 + f_7 \left(\zeta + \zeta^{-1}\right)^3$$
$$= g_4 + g_5 \left(\zeta + \zeta^{-1}\right) + g_6 \left(\zeta^2 + \zeta^{-2}\right) + g_7 \left(\zeta^3 + \zeta^{-3}\right)$$

then

$$f_0 + f_1 \left(\zeta + \zeta^{-1}\right) + f_2 \left(\zeta + \zeta^{-1}\right)^2 + f_3 \left(\zeta + \zeta^{-1}\right)^3$$
$$+ f_4 \left(\zeta + \zeta^{-1}\right)^4 + f_5 \left(\zeta + \zeta^{-1}\right)^5 + f_6 \left(\zeta + \zeta^{-1}\right)^6 + f_7 \left(\zeta + \zeta^{-1}\right)^7$$
$$= g_0 + (g_1 + g_7) \left(\zeta + \zeta^{-1}\right) + (g_2 + g_6) \left(\zeta^2 + \zeta^{-2}\right) + (g_3 + g_5) \left(\zeta^3 + \zeta^{-3}\right)$$
$$+ g_4 \left(\zeta^4 + \zeta^{-4}\right) + g_5 \left(\zeta^5 + \zeta^{-5}\right) + g_6 \left(\zeta^6 + \zeta^{-6}\right) + g_7 \left(\zeta^7 + \zeta^{-7}\right).$$

The conversion S_8^{-1} from coefficients of $1, \zeta + \zeta^{-1}, (\zeta + \zeta^{-1})^2, \ldots, (\zeta + \zeta^{-1})^7$ to coefficients of $1, \zeta + \zeta^{-1}, \zeta^2 + \zeta^{-2}, \zeta^3 + \zeta^{-3}, \ldots, \zeta^7 + \zeta^{-7}$ performs two half-size conversions S_4^{-1} and three additions of bits: first convert f_0, f_1, f_2, f_3 to g_0, g_1, g_2, g_3; separately convert f_4, f_5, f_6, f_7 to g_4, g_5, g_6, g_7; and then add g_7 to g_1, g_6 to g_2, and g_5 to g_3. The inverse S_8, converting from coefficients of $1, \zeta + \zeta^{-1}, \zeta^2 + \zeta^{-2}, \zeta^3 + \zeta^{-3}, \ldots, \zeta^7 + \zeta^{-7}$ to coefficients of $1, \zeta + \zeta^{-1}, (\zeta + \zeta^{-1})^2, \ldots, (\zeta + \zeta^{-1})^7$, has exactly the same cost. These conversions are extremely efficient.

More generally, instead of splitting 8 into $(4, 4)$, one can (and should) split k into $(j, k - j)$, where j is the unique power of 2 satisfying $j + 1 \leq k \leq 2j$. This is exactly what Shokrollahi's transformations S_k and S_k^{-1} do.

2.4   Shokrollahi's Multiplication Algorithm

Shokrollahi expands $N(a)$ and $N(b)$ by inserting a leading 0, obtaining linear combinations of $\mathcal{N}$, and then uses the transformation S_{n+1} to obtain linear combinations of $\mathcal{P}$, which are then interpreted as polynomials of degree at most n. Multiplying these two size-$(n+1)$ polynomials takes $M(n+1)$ bit operations and produces a polynomial of degree at most $2n$. Shokrollahi uses the transformation S_{2n+1}^{-1} to obtain a linear combination of $(1, \zeta + \zeta^{-1}, \zeta^2 + \zeta^{-2}, \zeta^3 + \zeta^{-3}, \ldots, \zeta^{2n} + \zeta^{-2n})$, uses $\zeta^{n+i} + \zeta^{-(n+i)} = \zeta^{n+i-2n-1} + \zeta^{2n+1-(n+i)} = \zeta^{-(n+1-i)} + \zeta^{n+1-i}$ for $1 \leq i \leq n$ to reduce the intermediate result back to $\mathcal{N}$, and finally discards the coefficient of 1 (which is always 0 by [Sho07, Theorem 31]), obtaining $N(ab)$.

2.5   Shokrollahi's Analysis

Shokrollahi shows in [Sho07, Lemma 21] that the cost for a size-2^r transformation is $\eta(r) = 2^{r-1}(r-2) + 1$. He then computes the following upper bound on the cost of his multiplication algorithm:

- two conversions from $\mathcal{N}$ to $\mathcal{P}$, costing $\eta(\lceil \log_2(n+1) \rceil)$ each; plus
- a multiplication of polynomials of degree $\leq n$, costing $M(n+1)$; plus
- one double-length conversion of a polynomial of degree $\leq 2n$, costing $\eta(\lceil \log_2(2n+1) \rceil)$; plus
- n final additions.

See [Sho07, Theorem 32] and [vzGSS07, Theorem 8, first display].

We point out that Shokrollahi's bounds are much higher than the actual costs of his algorithm, often losing a factor of 2 or more outside the $M(n+1)$ term. See Section 5.5 of this paper for an example. The "$+16n \log_2 n$" appearing in [vzGSS07, Section 1], and in more generality in [vzGSS07, Theorem 8, second display], is even more misleading. Those bounds should be disregarded by readers evaluating the performance of normal-basis arithmetic. We present quickly computable formulas for exact operation counts of our algorithms, along with reasonably precise approximations such as $(n/2) \log_2(n/4)$.

3   Streamlined Multiplication in Type-II Normal Basis

This section presents a simpler, smaller, slightly faster algorithm to compute $N(a), N(b) \mapsto N(ab)$. This algorithm is a convenient starting point for the larger speedups discussed in subsequent sections, so we present the algorithm from scratch, but we begin by summarizing the most important differences between the algorithm and Shokrollahi's algorithm.

3.1   Summary of the Simplification

Recall that Shokrollahi's original algorithm extends the basis $\zeta + \zeta^{-1}, \ldots, \zeta^n + \zeta^{-n}$ to $1, \zeta + \zeta^{-1}, \ldots, \zeta^n + \zeta^{-n}$. Note that $1 \neq \zeta^0 + \zeta^{-0}$; evidently 1 plays a special role here.

The algorithm in this section shifts the underlying transformation by one position, avoiding the need to extend the original basis. The new transformation works with only n elements rather than $n+1$, and feeds the multiplier polynomials of size n rather than $n+1$.

3.2   Summary of the Speedup

This multiplication algorithm has overhead approximately $2n \log_2(n/2)$: i.e., it uses approximately $M(n) + 2n \log_2(n/2)$ bit operations. It saves $M(n+1) - M(n)$ bit operations compared to Shokrollahi's original algorithm (according to our analysis of Shokrollahi's algorithm; Shokrollahi's analysis produces a much larger upper bound, as discussed in Section 2.5).

The differences $M(1) - M(0), M(2) - M(1), \ldots, M(n) - M(n-1)$ have sum $M(n) - M(0) = M(n) \in O(n \log_2 n \log_2 \log_2 n)$, so the average difference is bounded by $O(\log_2 n \log_2 \log_2 n)$, which is *asymptotically* not nearly as large as $2n \log_2 n$. However, for *typical* values of n there is a quite significant difference between the best known upper bound on $M(n+1)$ and the best known upper bound on $M(n)$. For example, these differences for $n = 53$, $n = 83$, $n = 89$, $n = 113$, $n = 131$, and $n = 173$ are 67, 121, 73, 81, 154, and 108 respectively.

This section's algorithm makes structurally clear that the polynomials to be multiplied have size only n. An alternate, more complicated, way to save $M(n+1) - M(n)$ is as follows: observe that the coefficient of 1 inside Shokrollahi's algorithm is initialized to 0 and is never modified; conclude that the size-$(n+1)$ polynomials in the algorithm always have constant coefficient 0; speed up the algorithm accordingly. The intermediate conclusion appeared (with a different proof) in [Sho07, Theorem 31, proof, third sentence], but was not exploited in the algorithm.

3.3   The Transformation

For each $k \geq 1$, each vector $e \in \mathbf{F}_2^k$, and each $i \in \{1, 2, \ldots, k\}$, define e_i as the ith component of e. Then $e = (e_1, e_2, \ldots, e_k)$. To support infinite sums over i, as in [Knu97a], we also allow "out-of-range" indices: define $e_i = 0$ for $i \in \mathbf{Z} \setminus \{1, 2, \ldots, k\}$. We also use the notation $[i \neq 0]$ to mean 0 if $i = 0$ and 1 if $i \neq 0$.

For each $k \geq 1$ we define an invertible function $T_k : \mathbf{F}_2^k \rightarrow \mathbf{F}_2^k$ by the following recursion:

- Define $T_1(e) = e$.
- For $k \geq 2$: Define j as the largest power of 2 in $\{1, 2, \ldots, k-1\}$. For each $f \in \mathbf{F}_2^j$ and each $g \in \mathbf{F}_2^{k-j}$ define $T_k(f, g) = (T_j(h), T_{k-j}(g))$ where $h \in \mathbf{F}_2^j$ is defined by $h_i = f_i + [i \neq 0]g_{j-i}$.

To recover f, g from $T_k(f, g) = (T_j(h), T_{k-j}(g))$, first invert T_j and T_{k-j} to obtain h and g, and then compute f from $f_i = h_i + [i \neq 0]g_{j-i}$.

For example:

- $T_2(e_1, e_2) = (e_1, e_2)$. Here $j = 1$, $f = (e_1)$, $g = (e_2)$, and $h = (e_1)$.
- $T_3(e_1, e_2, e_3) = (e_1 + e_3, e_2, e_3)$. Here $j = 2$, $f = (e_1, e_2)$, $g = (e_3)$, and $h = (e_1 + e_3, e_2)$.
- $T_4(e_1, e_2, e_3, e_4) = (e_1 + e_3, e_2, e_3, e_4)$. Here $j = 2$, $f = (e_1, e_2)$, $g = (e_3, e_4)$, and $h = (e_1 + e_3, e_2)$.
- $T_5(e_1, e_2, e_3, e_4, e_5) = (e_1 + e_3 + e_5, e_2, e_3 + e_5, e_4, e_5)$. Here $j = 4$, $f = (e_1, e_2, e_3, e_4)$, $g = (e_5)$, and $h = (e_1, e_2, e_3 + e_5, e_4)$.
- $T_6(e_1, e_2, e_3, e_4, e_5, e_6) = (e_1 + e_3 + e_5, e_2 + e_6, e_3 + e_5, e_4, e_5, e_6)$. Here $j = 4$, $f = (e_1, e_2, e_3, e_4)$, $g = (e_5, e_6)$, and $h = (e_1, e_2 + e_6, e_3 + e_5, e_4)$.

One can visualize the computation of h as folding g onto f in reverse order, but skipping the highest coefficient of f, and skipping the highest coefficient of g if g is as long as f.

Theorem 3.4. *Let k be a positive integer. Let ζ be a nonzero element of a field of characteristic 2. Then $\sum_i (T_k(e))_i (\zeta + \zeta^{-1})^i = \sum_i e_i(\zeta^i + \zeta^{-i})$ for each $e \in \mathbf{F}_2^k$.*

Proof. For $k = 1$: $T_1(e) = e$ so $(T_1(e))_1(\zeta + \zeta^{-1})^1 = e_1(\zeta + \zeta^{-1})$.

For $k \geq 2$: Define j as the largest power of 2 in $\{1, 2, \ldots, k-1\}$. Write e as (f, g) for some $f \in \mathbf{F}_2^j$ and $g \in \mathbf{F}_2^{k-j}$. Then $T_k(e) = (T_j(h), T_{k-j}(g))$ where $h_i = f_i + [i \neq 0]g_{j-i}$.

Both j and $k - j$ are smaller than k, so by induction

$$\sum_i (T_j(h))_i (\zeta + \zeta^{-1})^i = \sum_i h_i(\zeta^i + \zeta^{-i}),$$

$$\sum_i (T_{k-j}(g))_i (\zeta + \zeta^{-1})^i = \sum_i g_i(\zeta^i + \zeta^{-i}).$$

Recall that j is a power of 2. Use $(\zeta + \zeta^{-1})^j = \zeta^j + \zeta^{-j}$ to see that

$$\sum_i (T_k(e))_i (\zeta + \zeta^{-1})^i = \sum_i (T_j(h))_i (\zeta + \zeta^{-1})^i + (\zeta + \zeta^{-1})^j \sum_i (T_{k-j}(g))_i (\zeta + \zeta^{-1})^i$$

$$= \sum_i h_i(\zeta^i + \zeta^{-i}) + (\zeta^j + \zeta^{-j}) \sum_i g_i(\zeta^i + \zeta^{-i})$$

$$= \sum_i (f_i + [i \neq 0]g_{j-i})(\zeta^i + \zeta^{-i}) + \sum_i g_i(\zeta^{i+j} + \zeta^{-i-j} + \zeta^{i-j} + \zeta^{j-i})$$

$$= \sum_i (f_i + g_{j-i})(\zeta^i + \zeta^{-i}) + \sum_i g_i(\zeta^{i+j} + \zeta^{-i-j} + \zeta^{i-j} + \zeta^{j-i})$$

$$= \sum_i f_i(\zeta^i + \zeta^{-i}) + \sum_i g_i(\zeta^{i+j} + \zeta^{-i-j})$$

$$= \sum_i e_i(\zeta^i + \zeta^{-i})$$

as claimed. The replacement of $[i \neq 0]$ by 1 on the fourth line follows from $[i \neq 0](\zeta^i + \zeta^{-i}) = \zeta^i + \zeta^{-i}$; note that $\zeta^0 + \zeta^{-0} = 0$. $\qquad\square$

3.5   Speed of the Transformation

The following in-place algorithm replaces $e \in \mathbf{F}_2^k$ by $T_k(e)$:

- Stop if $k = 1$.
- Define j as the largest power of 2 in $\{1, 2, \ldots, k-1\}$.
- Add e_{2j-i} into e_i for $\max\{1, 2j - k\} \leq i \leq j - 1$. (Now $e = (h, g)$ in the notation of the definition of T_k.)
- Recursively apply T_j to the first j coefficients of e.
- Recursively apply T_{k-j} to the remaining coefficients of e.

Inverting this algorithm is a simple matter of carrying out the same additions in reverse order.

The cost of T_k is $\min\{j - 1, k - j\}$ plus the costs of T_j and T_{k-j}. An easy induction shows that if $k = 2^{k_0} + 2^{k_1} + \cdots$, with $k_0 > k_1 > \ldots$, then the cost of T_k is exactly $\sum_i (2^{k_i-1}(k_i - 2 + 2i) + 1)$.

3.6 The $N \times N \to N$ Multiplication Algorithm

The following algorithm computes $N(ab)$ given $N(a)$ and $N(b)$:

- Compute $A = T_n(N(a))$ and $B = T_n(N(b))$.
- Compute the product $P_2 z^2 + \cdots + P_{2n} z^{2n}$ of the polynomials $A_1 z + \cdots + A_n z^n$ and $B_1 z + \cdots + B_n z^n$ in the polynomial ring $\mathbf{F}_2[z]$.
- Compute $p = T_{2n}^{-1}(0, P_2, \ldots, P_{2n})$.
- Compute $N(ab) = (p_1 + p_{2n}, p_2 + p_{2n-1}, \ldots, p_n + p_{n+1})$.

Recall that $a = \sum_i N(a)_i(\zeta^i + \zeta^{-i})$ by definition of N, so $a = \sum_i A_i(\zeta + \zeta^{-1})^i$ by Theorem 3.4. Similarly $b = \sum_i B_i(\zeta + \zeta^{-1})^i$. Hence $ab = \sum_i P_i(\zeta + \zeta^{-1})^i$, so $ab = \sum_i p_i(\zeta^i + \zeta^{-i})$ by Theorem 3.4, so $ab = (p_1 + p_{2n})(\zeta + \zeta^{-1}) + \cdots + (p_n + p_{n+1})(\zeta^n + \zeta^{-n})$.

The two computations of T_n each cost $\sum_i(2^{n_i - 1}(n_i - 2 + 2i) + 1)$ if $n = 2^{n_0} + 2^{n_1} + \cdots$ with $n_0 > n_1 > \ldots$. The polynomial multiplication costs $M(n)$. The computation of p costs $\sum_i(2^{n_i}(n_i - 1 + 2i) + 1)$. The final computation of $N(ab)$ costs $n = \sum_i 2^{n_i}$.

The total number of bit operations is $M(n) + \sum_i(2^{n_i}(2n_i - 2 + 4i) + 3)$. The overhead term $\sum_i(2^{n_i}(2n_i - 2 + 4i) + 3)$ is approximately $2n \log_2(n/2)$, and a trivial computer calculation shows that it is bounded by $2(n + 2) \log_2(n/2)$ for $4 \le n \le 100000$.

We comment that the 0 component in the T_{2n}^{-1} input allows a subsequent addition of 0 to be eliminated. This speedup might seem too minor to be worth mentioning, and our operation counts in this section do not take it into account, but the underlying idea helps produce much larger savings in subsequent sections.

4   Type-II Polynomial Basis

Let us pause to review the attractive features of the (permuted) type-II normal basis $\zeta + \zeta^{-1}, \zeta^2 + \zeta^{-2}, \ldots, \zeta^n + \zeta^{-n}$. The multiplication overhead, compared to size-n polynomial multiplication, is only about $2n \log_2(n/2)$. Repeated squaring is a very fast permutation, costing no bit operations.

This section presents a multiplication algorithm for the non-traditional polynomial basis $c, c^2, \ldots, c^n$, where $c = \zeta + \zeta^{-1}$. The overhead in the new algorithm is only about $n \log_2 n$. Repeated squaring in this basis is more complicated than a permutation but is still very fast, costing only $n \log_2(n/4)$ bit operations. We refer to this basis as a "type-II optimal polynomial basis" because of its close connection to the type-II optimal normal basis.

For comparison, in a traditional low-weight polynomial basis, the multiplication overhead is typically $2n$ (for trinomials) or $4n$ (for pentanomials), and *single* squarings are fast, but *repeated* squarings such as $a \mapsto a^{2^{\lfloor n/3 \rfloor}}$ are very slow.

We obtain our best results by combining type-II polynomial basis P with type-II normal basis N. This combined system keeps repeated-squaring inputs in

N form, and keeps multiplication inputs in P form. Multiplications are $P \times P \to N$ when the outputs are used for repeated squarings, and $P \times P \to N \to P$ when the outputs are used for repeated squarings *and* for multiplications, but $P \times P \to P$ when the outputs are used solely for multiplications.

4.1   The $N \to P$ and $P \to N$ Conversions

We begin by reinterpreting the transformation T_n in Section 3 as a fast conversion from type-II normal basis to type-II polynomial basis: Theorem 3.4 implies that $T_n(N(a)) = P(a)$. This also means that T_n^{-1} is a fast conversion from type-II polynomial basis to type-II normal basis: $T_n^{-1}(P(a)) = N(a)$. Recall that each of these conversions costs $\sum_i (2^{n_i-1}(n_i - 2 + 2i) + 1) \approx (n/2) \log_2(n/4)$.

For comparison: Shokrollahi in [Sho07, Theorem 28], emphasizing "the most important property" of his multiplier, showed that conversion between a type-II normal basis and the basis $1, c, c^2, \ldots, c^{n-1}$ takes time $O(n \log_2 n)$. We simplify and accelerate the conversion by shifting to the basis $c, c^2, \ldots, c^n$. The speedup is $\Theta(n)$ operations. The simplification is illustrated by the fact that our basis conversion naturally appears as a subroutine in our multiplication algorithm, whereas modifying the multiplication algorithm from [Sho07] to use the basis conversion from [Sho07, Theorem 28] would slow down the multiplication algorithm.

4.2   The $P \times P \to N$ Multiplication Algorithm

We next observe that the $N \times N \to N$ multiplication algorithm of Section 3 factors into two $N \to P$ conversions and a $P \times P \to N$ multiplication algorithm.

Specifically, the first step of the $N(a), N(b) \mapsto N(ab)$ algorithm of Section 3 computes $A = T_n(N(a))$ and $B = T_n(N(b))$; i.e., it computes $A = P(a)$ and $B = P(b)$. The remaining steps make no further use of $N(a)$ and $N(b)$: they start from $A = P(a)$ and $B = P(b)$ and compute $N(ab)$. In other words, the remaining steps are exactly a $P \times P \to N$ multiplication algorithm. This $P \times P \to N$ multiplication algorithm costs $M(n) + \sum_i (2^{n_i}(n_i + 2i) + 1) \approx M(n) + n \log_2 n$.

4.3   The $P \times P \to P$ Multiplication Algorithm

Composing $P \times P \to N$ with a final $N \to P$ conversion would produce a $P \times P \to P$ multiplication algorithm with overhead approximately $n(1.5 \log_2 n - 1)$. This algorithm would feed the size-$2n$ polynomial product through a size-$2n$ transform, then fold the result in half using $\zeta^{2n+1-i} + \zeta^{i-2n-1} = \zeta^i + \zeta^{-i}$, then transform the size-n result from N to P.

We do better by separately handling the two halves of the polynomial product. The point is that

$$T_n(\text{fold}(T_{2n}^{-1}(\text{bottom}, \text{top})))$$
$$= T_n(\text{fold}(T_{2n}^{-1}(\text{bottom}, 0, \ldots, 0))) + T_n(\text{fold}(T_{2n}^{-1}(0, \ldots, 0, \text{top})))$$
$$= \text{bottom} + T_n(\text{fold}(T_{2n}^{-1}(0, \ldots, 0, \text{top}))).$$

Instead of uselessly transforming the bottom half back and forth between P and N, we simply leave it in P and add it at the end. We use transforms as a fast mechanism to reduce the top half from coefficients of $c^{n+1}, \ldots, c^{2n}$ to coefficients of $c^1, \ldots, c^n$.

In the computation of $T_{2n}^{-1}(0, \ldots, 0, \text{top})$, and in the subsequent folding, we systematically eliminate all additions of 0. For any particular n one can do this elimination by hand, keeping track of which intermediate values are 0; or one can generate straight-line code for the entire computation and use standard optimizing-compiler tools.

We have done this optimization for all $n \in \{1, 2, \ldots, 100000\}$ and found that the cost of $\text{fold}(T_{2n}^{-1}(0, \ldots, 0, \text{top}))$ can in every case be computed as follows. Write n as $2^{n_0} + 2^{n_1} + \cdots + 2^{n_r}$ with $n_0 > n_1 > \cdots > n_r$. The cost is then $\sum_i 2^{n_i-1}(n_i + 4i)$ *minus* a nonnegative rebate. The rebate is 1 for each 11 in the binary expansion of n (i.e., each i such that $n_i = n_{i+1} + 1$), plus 2 for each 111 in the binary expansion, plus 4 for each 1111, plus 8 for each 11111, etc.

Examples: If $n = 131$ then n has binary expansion 10000011, so the rebate is 1, and the cost is $2^{7-1}(7) + 2^{1-1}(1+4) + 2^{0-1}(0+8) - 1 = 456$. If $n = 491$ then $n+1$ has binary expansion 111101011, with 4 occurrences of 11, 2 occurrences of 111, and 1 occurrence of 1111, so the rebate is $1+1+1+1+2+2+4 = 12$, and the cost is $2^{8-1}(8) + 2^{7-1}(7+4) + 2^{6-1}(6+8) + 2^{5-1}(5+12) + 2^{3-1}(3+16) + 2^{-1}(1+20) + 2^{0-1}(0+24) - 12 = 2545$.

To summarize, $P \times P \to P$ multiplication involves

- cost $M(n)$ for the polynomial product;
- the cost discussed above, approximately $(n/2) \log_2 n$;
- the cost of T_n, approximately $(n/2) \log_2(n/4)$; and
- cost $n - 1$ for the final addition of the bottom half.

The total cost is approximately $M(n) + n \log_2 n$, similar to the cost of $P \times P \to N$ multiplication. These approximations should not be viewed as equalities: a closer look shows that $P \times P \to P$ multiplication costs about $\sum_i 2^{n_i} i$ more than $P \times P \to N$ multiplication.

4.4 Dynamically Mixing N and P

At this point our basic tools are as follows:

- $N \to P$ conversion: 1 transform, cost $\sum_i (2^{n_i-1}(n_i - 2 + 2i) + 1)$.
- $P \to N$ conversion: 1 transform, cost $\sum_i (2^{n_i-1}(n_i - 2 + 2i) + 1)$.
- $N \to N$ repeated squarings: 0 transforms, cost 0.
- $P \times P \to N$ multiplication: 2 transforms (actually one double-size transform), cost $M(n) + \sum_i (2^{n_i}(n_i + 2i) + 1)$.
- $P \times P \to P$ multiplication: 2 transforms, slightly larger cost as discussed above.

There are several reasonable ways to combine these tools. One extreme is to work everywhere in N, using $N \to N$ repeated squarings (0 transforms) and

$N \times N \rightarrow P \times P \rightarrow N$ multiplications (4 transforms). Another extreme is to work everywhere in P, using $P \rightarrow N \rightarrow N \rightarrow P$ repeated squarings (2 transforms) and $P \times P \rightarrow P$ multiplications (2 transforms).

We take a more fluid approach, mixing the advantages of both extremes. We compute $N(a)$ for variables a that will be used in repeated squarings; we compute $P(a)$ for variables a that will be used in multiplications; we compute both $N(a)$ and $P(a)$ for variables a that will be used in both repeated squarings and multiplications. The overall number of transforms in this approach is 0 for each squaring and between 2 and 4 for each multiplication, depending on the exact pattern of multiplications and repeated squarings. See Section 5 for an illustrative example.

We briefly comment that $P \rightarrow P$ *single* squaring can be sped up by the same idea used in $P \times P \rightarrow P$ multiplication. However, in every application so far where we have tried this approach, we have found a faster solution that uses $N \rightarrow N$ squaring and rearranges the earlier computations.

5   Case Study: ECC2K-130

This section illustrates the use of optimal polynomial bases and optimal normal bases in the ECC2K-130 computation mentioned in Section 1. Specifically, this section shows that the $5B + 5$ multiplications in B iterations of the ECC2K-130 iteration function from [BBB+09] can be carried out with an overhead of only $5851B + 9412$ bit operations. The original Shokrollahi approach, with our improved analysis, would have used $8565B + 8565$ bit operations.

5.1   Review of the Iteration Function

We take the perspective of an implementor faced with the job of implementing the ECC2K-130 iteration function from [BBB+09], the bottleneck in the ECC2K-130 computation. To keep this paper self-contained we now repeat the definition of the iteration function.

The input to an iteration is a pair $(x, y) \in \mathbf{F}_{2^{131}} \times \mathbf{F}_{2^{131}}$ satisfying three conditions: first, $y^2 + xy = x^3 + 1$; second, $x \neq 0$; third, x has trace 0, i.e., $N(x)$ has even Hamming weight. The output of the iteration is the pair (x', y') defined by the equations

$$j = 3 + \left(\frac{\text{weight}(N(x))}{2} \bmod 8 \right), \qquad \lambda = \frac{y + y^{2^j}}{x + x^{2^j}},$$
$$x' = \lambda^2 + \lambda + x + x^{2^j}, \qquad\qquad y' = \lambda(x + x') + x' + y.$$

One can check that $(y')^2 + x'y' = (x')^3 + 1$, that $x' \neq 0$, and that x' has trace 0.

This iteration function can be computed using $3 + (5/B)$ multiplications for a B-way-batched inversion of $x + x^{2^j}$; 1 multiplication of the inverse by $y + y^{2^j}$, producing λ; and 1 multiplication of λ by $x + x'$. All of these stages are discussed in more detail below.

See [EBB+09] for further information on how these iterations are being used to solve the ECC2K-130 challenge. We comment that thousands of CPU and GPU cores (including Core 2 clusters, Cell clusters, and GTX 295 clusters) have already been busy for months computing these iterations, and that many more cores are being added; obviously every speedup in the computation is valuable

5.2   The Main Loop

It is natural to represent the input (x, y) as $(N(x), N(y))$: the first step is to compute the weight of $N(x)$, and both x and y are then fed through repeated squarings. On the other hand, dividing $y + y^{2^j}$ by $x + x^{2^j}$ requires $P(y + y^{2^j})$ and $P(1/(x + x^{2^j}))$. The quotient λ is then used for both a squaring λ^2 and a multiplication $\lambda(x + x')$, so we compute both $N(\lambda)$ and $P(\lambda)$.

Figure 1 shows the resulting data flow between representations of various field elements. There are 4 explicit size-131 transforms, and 2 multiplications $P \times P \to N$ each involving 2 size-131 transforms. Working solely with N, and with an $N \times N \to N$ multiplication subroutine, would require an extra transform for $N(1/d)$. Note that more transforms are saved inside the inversion, as discussed below.

Figure 1 shows computations from $N(y)$ through $N(y^{2^j})$ in parallel with computations from $N(x)$ through $N(x^{2^j})$. To reduce storage requirements, cache misses, etc., the ECC2K-130 software actually delays the $N(y^{2^j})$ computations until after the inversion.

5.3   Batching Inversions

Montgomery in [Mon87, Section 10.3.1] suggested computing $1/d_1$ and $1/d_2$ as $d_2/(d_1d_2)$ and $d_1/(d_1d_2)$. This suggestion eliminates 1 inversion in favor of 3 multiplications. We are not aware of any inversion method for $\mathbf{F}_{2^{131}}$ that can compete with 3 multiplications if the multiplications are performed by state-of-the-art techniques.

A batch of B parallel iterations involves B inversions $1/d_1, 1/d_2, \ldots, 1/d_B$. Merging the first two inversions, then merging with the next, etc., leads to the following standard computation, replacing $B - 1$ inversions with $3(B - 1)$ multiplications: first compute $d_1d_2, d_1d_2d_3, \ldots, d_1d_2 \cdots d_B$ using $B - 1$ multiplications; then compute $1/(d_1d_2 \cdots d_B)$ using a single inversion; then compute $1/d_B = (d_1d_2 \cdots d_{B-1})/(d_1d_2 \cdots d_B)$ and $1/(d_1d_2 \cdots d_{B-1}) = d_B/(d_1d_2 \cdots d_B)$ using 2 multiplications, etc.

The single central inversion begins with squarings, as discussed below, and therefore takes $N(d_1d_2 \cdots d_B)$ as input. However, all of the intermediate products here are used solely for further multiplications, so we represent them in P form. Figure 2 shows the resulting data flow for $B = 4$. Working solely with N, and with an $N \times N \to N$ multiplication subroutine, would double the number of transforms.

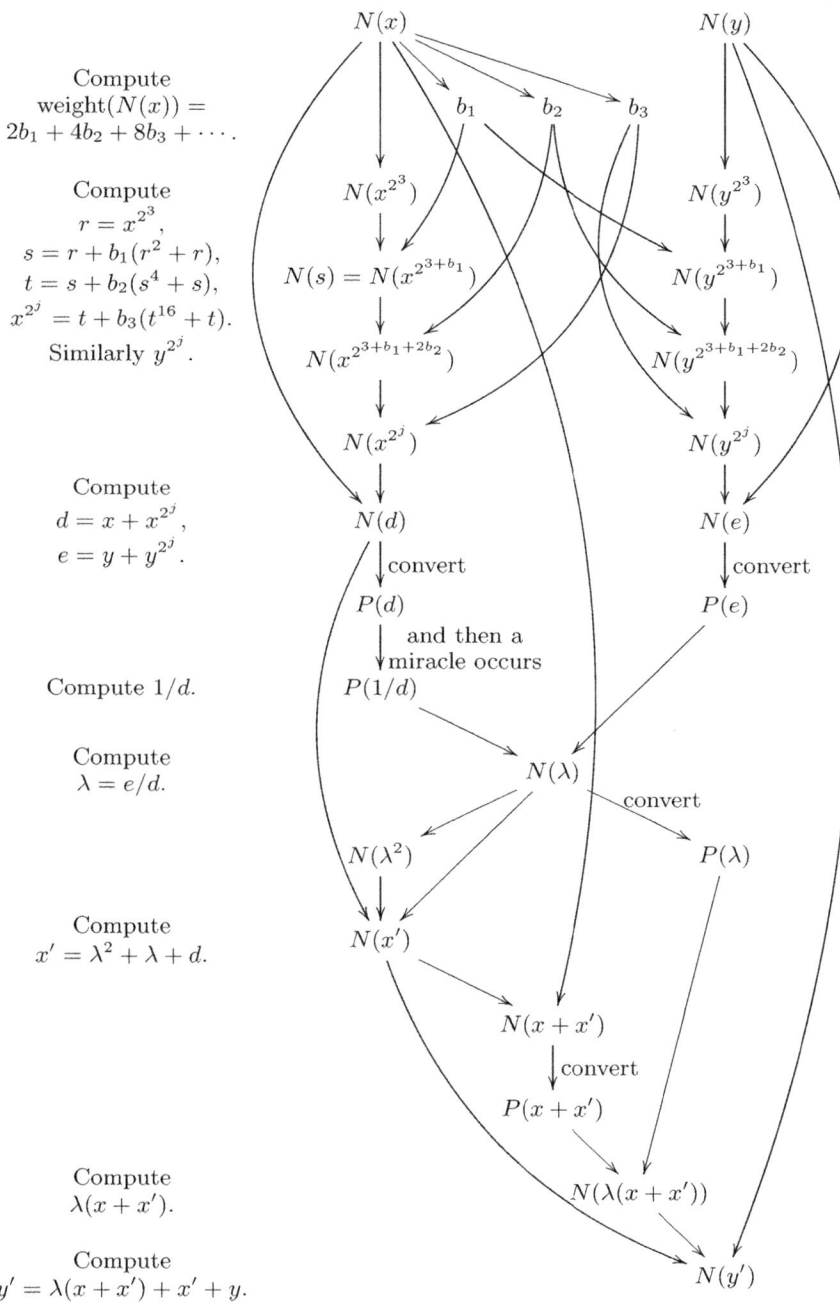

Fig. 1. The ECC2K-130 iteration function

Other merging patterns, such as a balanced tree, reduce latency without changing the number of operations. The same comments regarding P and N apply to arbitrary merging patterns.

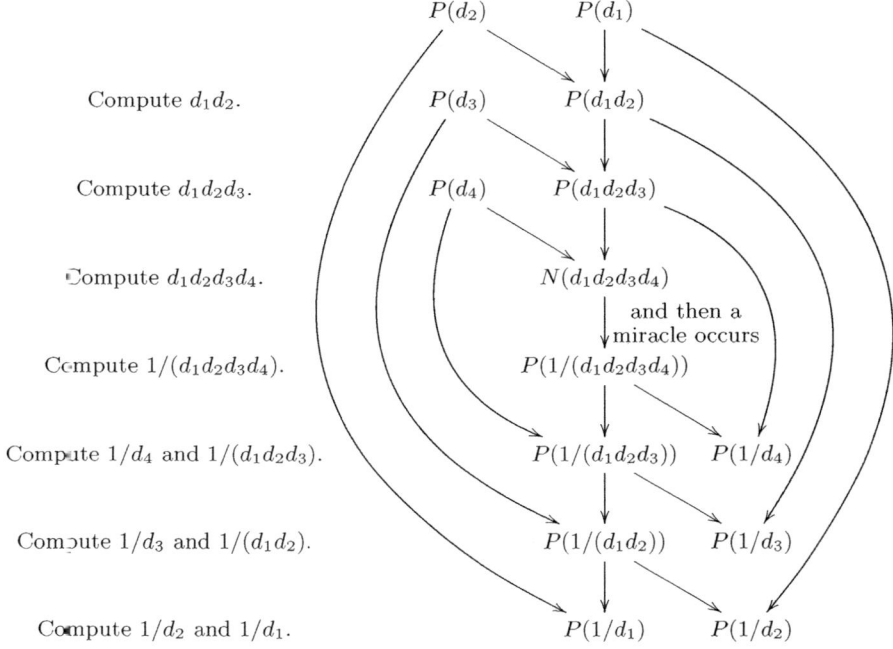

Fig. 2. Batching 4 independent inversions

5.4   Core Inversions

Eventually one has to actually invert something. Inversion time is amortized across a batch of B iterations, but B is often limited by communication costs (such as the cost of copying data between DRAM and limited SRAM), making inversion an important part of the ECC2K-130 computation.

The standard branchless inversion method for $\mathbf{F}_{2^n}$, certainly not the only method, is to compute a $(2^n - 2)$nd power. This inversion method is also important in many other computations, so we describe the details for general n before focusing on $n = 131$.

The standard method of computing a $(2^n - 2)$nd power uses $n - 1$ squarings and just r multiplications, where r is the length of an "ℓ_0 chain" for $n - 1$; an ℓ_0 chain is a particular type of addition chain. The idea is to convert a chain $1 = e_0, e_1, \ldots, e_r = n-1$ into a chain containing $1 = 2^{e_0} - 1, 2^{e_1} - 1, \ldots, 2^{e_r} - 1 = 2^{n-1} - 1$ along with various doublings; i.e., to compute $x^1 = x^{2^{e_0}-1}, \ldots, x^{2^{e_r}-1} = x^{2^{n-1}-1}$ along with various squarings. This powering method was introduced by Brauer in 1939 for the special case of "star chains" and by Hansen in 1959 for all

ℓ_0 chains. See [Bra39], [Han59], and [Knu97b, Section 4.6.3, Theorem G]. Note that the shortest ℓ_0 chains are as short as the shortest addition chains for all integers below 5784689; see [Cli05].

In particular, a simple binary addition chain achieves $r = \lfloor \log_2(n-1) \rfloor +$ weight$(n-1) - 1$, producing an inversion method that takes $n-1$ squarings and $\lfloor \log_2(n-1) \rfloor +$ weight$(n-1) - 1$ multiplications. This inversion method is often credited to the 1988 paper [IT88] by Itoh and Tsujii. For most values of n one can do noticeably better (often more than 1.5× better!) by switching to a standard "windowing" addition chain for $n-1$, producing an inversion method that takes $n-1$ squarings and $(1 + o(1)) \log_2 n$ multiplications. For further discussion of this inversion method see [vzGN99] and [Nöc01].

In the case $n = 131$ we take the length-8 addition chain $1, 2, 4, 8, 16, 32, 64, 65, 130$ for $n-1$. We could compute

$$x, x^2, x^3, x^{12}, x^{15}, x^{240}, x^{255} = x^{2^8-1}, x^{2^{16}-2^8}, x^{2^{16}-1}, x^{2^{32}-2^{16}}, x^{2^{32}-1},$$
$$x^{2^{64}-2^{32}}, x^{2^{64}-1}, x^{2^{65}-2}, x^{2^{65}-1}, x^{2^{130}-2^{65}}, x^{2^{130}-1}, x^{2^{131}-2} = x^{-1}$$

but we eliminate a final transform by moving the final squaring to the beginning:

$$x, x^2, x^4, x^6, x^{24}, x^{30}, x^{480}, x^{510} = x^{2^9-2}, x^{2^{17}-2^9}, x^{2^{17}-2}, x^{2^{33}-2^{17}}, x^{2^{33}-2},$$
$$x^{2^{65}-2^{33}}, x^{2^{65}-2}, x^{2^{66}-4}, x^{2^{66}-2}, x^{2^{131}-2^{66}}, x^{2^{131}-2} = x^{-1}.$$

Figure 3 shows the resulting data flow. Working solely with N, and with an $N \times N \to N$ multiplication subroutine, would require an extra transform for $P(x^2)$, and an extra transform for $P(x^{-1})$.

5.5 Total Overhead

A batch of $B \geq 2$ iterations involves the following multiplications and conversions:

- Inversion (see Figure 3): 8 multiplications $P \times P \to N$ and 15 conversions $N \to P$.
- Batching (see Figure 2 for $B = 4$): 1 multiplication $P \times P \to N$ and $3B - 4$ multiplications $P \times P \to P$. Note that all B inversions together involve $8 + 1 + (3B - 4) = 3B + 5$ multiplications; i.e., $3 + (5/B)$ multiplications per inversion, as mentioned earlier.
- Iteration (B copies of Figure 1): $4B$ conversions $N \to P$ and $2B$ multiplications $P \times P \to N$.

In total there are

- $2B + 9$ multiplications $P \times P \to N$, each having overhead 909;
- $3B - 4$ multiplications $P \times P \to P$, each having overhead 911; and
- $4B + 15$ conversions $N \to P$, each having overhead 325.

The total overhead is $5851B + 9412$, i.e., $5851 + 9412/B$ per iteration. To put this in perspective, the fastest known method for size-131 polynomial multiplication (see [Ber09a]) costs 11961 bit operations, and all of the other operations in the iteration cost 3929 bit operations.

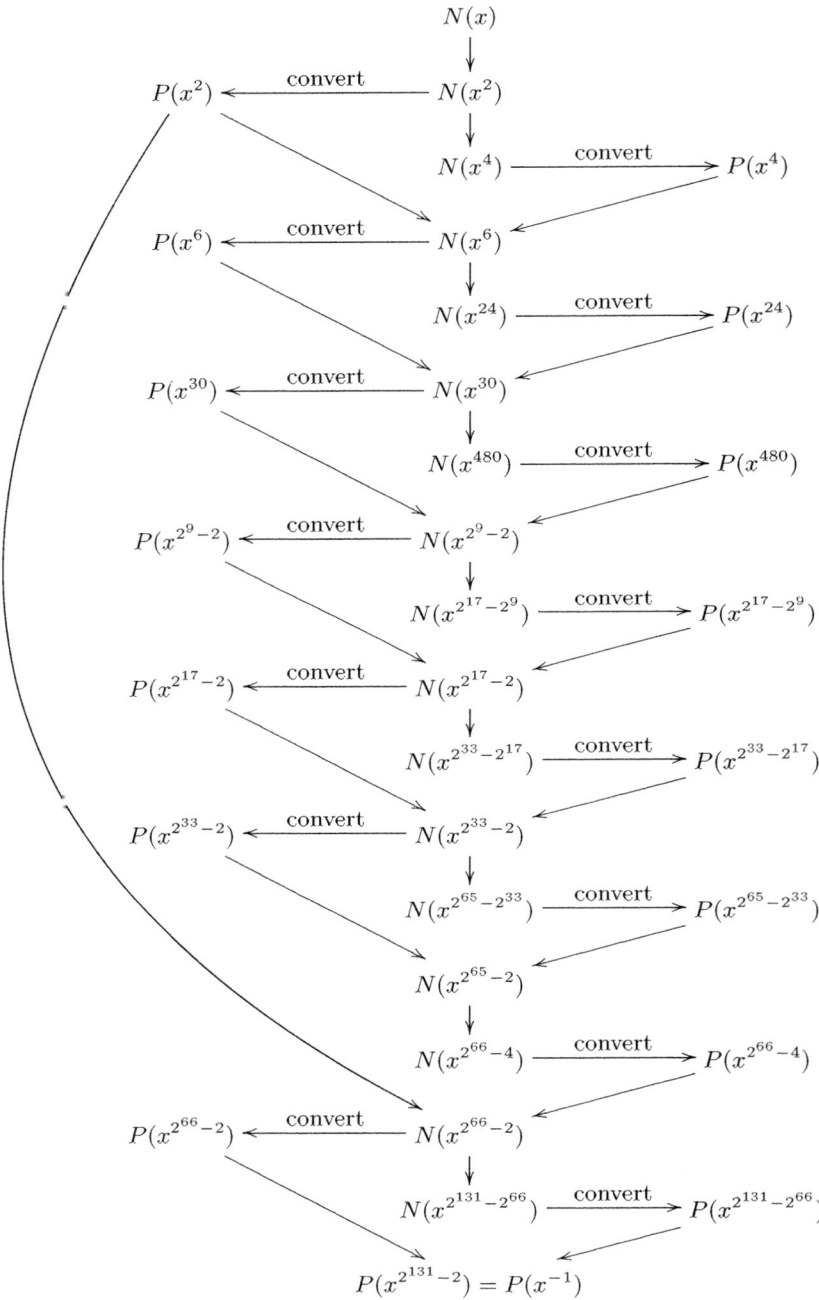

Fig. 3. Core inversion inside the ECC2K-130 iteration function

For comparison, Shokrollahi's original approach would have used $5B + 5$ multiplications $N \times N \rightarrow N$, each costing $M(132) + 1559$. (Shokrollahi's analysis actually says $M(132) + 3462$; $M(132) + 1559$ is the result of our own analysis of Shokrollahi's algorithm, and has been computer-verified.) The fastest known methods for size-132 multiplication involve 154 bit operations more than the fastest known methods for size-131 multiplication; if these methods are used then each $N \times N \rightarrow N$ multiplication has overhead $154 + 1559 = 1713$, for a total overhead of $8565B + 8565$, i.e., 8565 per iteration.

5.6 Comparison to Traditional Low-Weight Polynomial Bases

The current ECC2K-130 attack software uses our techniques. The original ECC2K-130 attack software instead used a low-weight polynomial basis, specifically the basis $1, z, z^2, \ldots, z^{130}$ of $\mathbf{F}_{2^{131}} = \mathbf{F}_2[z]/(z^{131} + z^{13} + z^2 + z + 1)$. There is no trinomial basis for this field.

The obvious approach to multiplication in this polynomial basis has overhead $4 \cdot 130 = 520$: for example, one eliminates the coefficient of z^{260} by adding it to 4 previous coefficients. However, a closer look shows that 65 of these 520 additions can be reused, thanks to the even spacing of $z^2, z, 1$, reducing the multiplication overhead to 455.

Similarly, a *single* squaring costs just 203. The problem is that there are 21 squarings in Figure 1: 10 for x^{2^j} via r^2, s^4, t^{16}; another 10 for y^{2^j}; and another 1 for λ^2. Even worse, one still needs to convert x to $N(x)$ as a stepping-stone to weight($N(x)$).

The total overhead for 5 multiplications and 21 squarings is $5 \cdot 455 + 21 \cdot 203 = 6538$ per iteration. The basis conversion from x to $N(x)$ can be performed in 3380 bit operations as explained in [Ber09b]. We have seen some ideas for slightly reducing these costs, but nothing that could make this low-weight polynomial basis competitive with the approach explained in this paper.

References

[BBB+09] Bailey, D.V., Batina, L., Bernstein, D.J., Birkner, P., Bos, J.W., Chen, H.-C., Cheng, C.-M., van Damme, G., de Meulenaer, G., Dominguez Perez, L.J., Fan, J., Güneysu, T., Gurkaynak, F., Kleinjung, T., Lange, T., Mentens, N., Niederhagen, R., Paar, C., Regazzoni, F., Schwabe, P., Uhsadel, L., Van Herrewege, A., Yang, B.-Y.: Breaking ECC2K-130. Cryptology ePrint Archive, Report 2009/541 (2009), http://eprint.iacr.org/2009/541

[Ber09a] Bernstein, D.J.: Batch binary Edwards. In: Halevi, S. (ed.) CRYPTO 2009. LNCS, vol. 5677, pp. 317–336. Springer, Heidelberg (2009)

[Ber09b] Bernstein, D.J.: Optimizing linear maps modulo 2. In: Workshop Record of SPEED-CC: Software Performance Enhancement for Encryption and Decryption and Cryptographic Compilers, pp. 3–18 (2009), http://cr.yp.to/papers.html#linearmod2

[BG01]   Bolotov, A.A., Gashkov, S.B.: On a quick multiplication in normal bases of finite fields. Discrete Mathematics and Its Applications 11, 327–356 (2001)

[Bra39]   Brauer, A.T.: On addition chains. Bulletin of the American Mathematical Society 45, 736–739 (1939)

[Cer97]   Certicom. Certicom ECC challenge (1997), http://www.certicom.com/images/pdfs/cert_ecc_challenge.pdf

[Cli05]   Clift, N.: Hansen chains do not always produce optimum addition chains. Posting to sci.math (July 31, 2005), http://sci.tech-archive.net/Archive/sci.math/2005-08/msg00447.html

[FBB+10]   Fan, J., Bailey, D.V., Batina, L., Güneysu, T., Paar, C., Verbauwhede, I.: Breaking elliptic curves cryptosystems using reconfigurable hardware (2010)

[FH07]   Fan, H., Hasan, M.A.: Subquadratic computational complexity schemes for extended binary field multiplication using optimal normal bases. IEEE Transactions on Computers 56, 1435–1437 (2007)

[GvzGP95]   Gao, S., von zur Gathen, J., Panario, D.: Gauss periods and fast exponentiation in finite fields (extended abstract). In: Baeza-Yates, R.A., Goles, E., Poblete, P.V. (eds.) LATIN 1995. LNCS, vol. 911, pp. 311–322. Springer, Heidelberg (1995)

[GvzGPS00] Gao, S., von zur Gathen, J., Panario, D., Shoup, V.: Algorithms for exponentiation in finite fields. Journal of Symbolic Computation 29(6), 879–889 (2000)

[Han59]   Hansen, W.: Zum Scholz–Brauerschen Problem. Journal für die reine und angewandte Mathematik (Crelles Journal) 202, 129–136 (1959) (in German)

[IT88]   Itoh, T., Tsujii, S.: A fast algorithm for computing multiplicative inverses in $GF(2^m)$ using normal bases. Information and Computation 78(3), 171–177 (1988)

[IT89]   Itoh, T., Tsujii, S.: Structure of parallel multipliers for a class of fields $GF(2^m)$. Information and Computation 83(1), 21–40 (1989)

[Knu97a]   Knuth, D.E.: The art of computer programming, volume 1: Fundamental algorithms, 3rd edn. Addison-Wesley Series in Computer Science and Information Processing. Addison-Wesley Publishing Company, Reading (1997)

[Knu97b]   Knuth, D.E.: The art of computer programming, volume 2: Seminumerical algorithms, 3rd edn. Addison-Wesley Series in Computer Science and Information Processing. Addison-Wesley Publishing Company, Reading (1997)

[Mon87]   Montgomery, P.L.: Speeding the Pollard and elliptic curve methods of factorization. Mathematics of Computation 48, 243–264 (1987)

[MOVW89]   Mullin, R.C., Onyszchuk, I.M., Vanstone, S.A., Wilson, R.M.: Optimal normal bases in $GF(p^n)$. Discrete Applied Mathematics 22(2), 149–161 (1989)

[Nöc01]   Nöcker, M.: Data structures for parallel exponentiation in finite fields. PhD thesis, Universität Paderborn (2001), http://math-www.uni-paderborn.de/~aggathen/Publications/noc01.ps

[Sho07]   Shokrollahi, J.: Efficient implementation of elliptic curve cryptography on
          FPGAs. PhD thesis, Universität Bonn (2007), `http://hss.ulb.uni-bonn.`
          `de/diss_online/math_nat_fak/2007/shokrollahi_jamshid/0960.pdf`
[vzGN99] von zur Gathen, J., Nöcker, M.: Computing special powers in finite fields
          (extended abstract). In: Proceedings of the 1999 International Symposium
          on Symbolic and Algebraic Computation, ISSAC '99, Vancouver, B.C.,
          Canada, July 29–31, pp. 83–90 (1999)
[vzGSS07] von zur Gathen, J., Shokrollahi, A., Shokrollahi, J.: Efficient multiplication
          using type 2 optimal normal bases. In: Carlet, C., Sunar, B. (eds.) WAIFI
          2007. LNCS, vol. 4547, pp. 55–68. Springer, Heidelberg (2007)

Pseudorandom Vector Sequences Derived from Triangular Polynomial Systems with Constant Multipliers

Alina Ostafe

Institut für Mathematik, Universität Zürich,
Winterthurerstrasse 190 CH-8057, Zürich, Switzerland
alina.ostafe@math.uzh.ch

Abstract. In this paper we study a new class of dynamical systems generated by iterations of a class of multivariate permutation polynomial systems. Using the same techniques studied previously for other generators, we bound exponential sums along the orbits of these dynamical systems and show that they admit stronger estimates than in the general case and thus can be of use for pseudorandom number generation. We also prove a nontrivial bound "on average" over all initial values $\mathbf{v} \in \mathbb{F}_p^m$ on the discrepancy for pseudorandom vectors generated by these iterations.

Keywords: Nonlinear pseudorandom number generators, triangular polynomial systems, exponential sums, discrepancy.

MSC(2010): Primary 65C10; Secondary 11K45; 94A60.

1 Introduction

In the series of papers [14,16,17] the authors considered multivariate polynomial systems $\mathcal{F} = \{F_1, \ldots, F_m\}$ of m polynomials in m variables over a finite field $\mathbb{F}_q$ having the "triangular" form

$$F_1(X_1, \ldots, X_m) = X_1 G_1(X_2, \ldots, X_m) + H_1(X_2, \ldots, X_m),$$

$$\ldots$$

$$F_{m-1}(X_1, \ldots, X_m) = X_{m-1} G_{m-1}(X_m) + H_{m-1}(X_m), \tag{1}$$
$$F_m(X_1, \ldots, X_m) = g_m X_m + h_m,$$

with $G_i, H_i \in \mathbb{F}_q[X_{i+1}, \ldots, X_m]$, $i = 1, \ldots, m-1$, and $g_m, h_m \in \mathbb{F}_q$, $g_m \neq 0$, for which they imposed that the polynomials F_i have unique leading monomial which dominates the other terms in every variable. For this class of polynomials, it has been shown in [16] that the degrees of the iterations of the polynomials F_i, $i = 1, \ldots, m$, grow significantly slower than the exponential growth expected for iterations of a "generic" system of m polynomials in m variables. In turn, this leads to much better estimates of exponential sums, and thus of discrepancy,

M.A. Hasan and T. Helleseth (Eds.): WAIFI 2010, LNCS 6087, pp. 62–72, 2010.
© Springer-Verlag Berlin Heidelberg 2010

for vectors generated by (1) than for those generated originated from arbitrary polynomial systems, see [4,5,15].

Furthermore, it has been shown in [14] that in the case when such a polynomial map generates a permutation of the corresponding vector space, one can get better results "on average" over all initial values. It is also noticed in [14] that in fact one can avoid the use of the *Weil bound* (see [7, Chapter 5]) of exponential sums and achieve a better result with a more elementary argument.

Let p be a prime and $\mathbb{F}_p$ be a finite field with p elements. In this paper we study a special case of the systems (1), namely we consider the polynomials G_i to be constant polynomials. More precisely, we consider systems of $m \geq 2$ polynomials $F_i \in \mathbb{F}_p[X_1, \ldots, X_m]$, $i = 1, \ldots, m$, over $\mathbb{F}_p$ defined in the following way:

$$F_1(X_1, \ldots, X_m) = g_1 X_1 + H_1(X_2, \ldots, X_m),$$
$$\cdots$$
$$F_{m-1}(X_1, \ldots, X_m) = g_{m-1} X_{m-1} + H_{m-1}(X_m),$$
$$F_m(X_1, \ldots, X_m) = g_m X_m + h_m,$$

(2)

where

$$g_i, h_m \in \mathbb{F}_p, \quad g_i \notin \{0, 1\}, \quad H_i \in \mathbb{F}_p[X_{i+1}, \ldots, X_m], \quad i = 1, \ldots, m.$$

We note that in the case when the polynomials H_i, $i = 1, \ldots, m-1$, are constant polynomials, we simply have a system of m independent polynomials. Clearly, iterations of such systems generate vectors of the form $(A_1 g_1^n + B_1, \ldots, A_m g_m^n + B_m)$. Such systems have actually been suggested as pseudorandom number generators, however with very limited progress. In fact, prior the very recent work of Bourgain [1], no interesting results have been known for such systems and corresponding vectors over finite fields. However, for similar systems with constant polynomials H_i, $i = 1, \ldots, m-1$, but defined over a residue ring modulo a prime power p^α for a fixed prime p, one can also use the estimates of [20], which apply to an arbitrary linear recurrence sequence modulo p^α. However, if the polynomials H_i, $i = 1, \ldots, m-1$, are not constant polynomials over a finite field $\mathbb{F}_p$ of p elements (for prime p), this "mixing" increases the length of the orbits and also allows us to use very different methods and thus derive a series of new results. Naturally, the strength of our bounds depends on the multiplicative orders t_i of g_i in $\mathbb{F}_p$, $i = 1, \ldots, m$.

We remark that for the polynomial systems (2) the conditions imposed in [14,16,17] are not satisfied anymore, and thus the results are not applicable for this case.

We follow the same technique as in [14,17] and we exploit the special structure of iterations of the polynomial systems introduced below that allows us to replace the use of the Weil bound (see [7, Chapter 5]) by a more elementary and stronger estimate on the corresponding exponential sums which in turn leads to a better final result on the distribution of the vectors generated by such dynamical systems. In fact, since our construction can easily be extended to polynomials over commutative rings, the new estimate can also be used to study polynomial maps over residue rings (while the Weil bound does not apply there).

Our results expand the class of polynomial dynamical systems which admit good estimates on exponential sums and thus have strong uniform distribution properties of elements in their orbits.

Throughout the paper, the implied constants in the symbols 'O' and '$\ll$' may occasionally, where obvious, depend on the number of variables m, and are absolute otherwise. We recall that the notations $A = O(B)$ and $A \ll B$ are all equivalent to the assertion that the inequality $|A| \leq cB$ holds for some constant $c > 0$.

2 Nonlinear Pseudorandom Number Generators

2.1 Iterations of Triangular Polynomial Systems

For each $i = 1, \ldots, m$ we define the k-th iteration of the polynomials F_i by the recurrence relation

$$F_i^{(0)} = X_i, \quad F_i^{(k)} = F_i(F_1^{(k-1)}, \ldots, F_m^{(k-1)}), \quad k = 1, 2, \ldots. \tag{3}$$

We can describe explicitly the iterations of the polynomials F_i as follows:

Lemma 1. Let $F_1, \ldots, F_m \in \mathbb{F}_p[X_1, \ldots, X_m]$ be as in (2). Then for $i = 1, \ldots, m$ and $k = 0, 1, \ldots$, for the polynomials $F_i^{(k)}$ given by (3) we have

$$F_i^{(k)} = g_i^k X_i + H_{i,k}(X_{i+1}, \ldots, X_m),$$

where $H_{i,k} \in \mathbb{F}_p[X_{i+1}, \ldots, X_m]$ for $i = 1, \ldots, m$.

We note that the system defined above is a permutation system, that is a system of multivariate polynomials in $\mathbb{F}_p[X_1, \ldots, X_m]$ which induces a map that permutes the elements of $\mathbb{F}_p^m$, given by absolutely irreducible polynomials. Moreover, the iterated polynomials $F_i^{(k)}$ have exactly the same form as the polynomials F_i and are also absolutely irreducible polynomials.

2.2 Vector Sequences

Let $\mathcal{F} = \{F_1, \ldots, F_m\}$ be a polynomial system in the ring $\mathbb{F}_p[X_1, \ldots, X_m]$ of the form (2). We consider the m-dimensional multisequence

$$(\mathbf{u}_n) = ((u_{n,1}, \ldots, u_{n,m})) \tag{4}$$

defined by a recurrence relation of the form

$$u_{n+1,i} = F_i(u_{n,1}, \ldots, u_{n,m}), \quad n = 0, 1, \ldots, \quad i = 1, \ldots, m, \tag{5}$$

with some *initial vector* $\mathbf{u}_0 = (u_{0,1}, \ldots, u_{0,m}) \in \mathbb{F}_p^m$.

Using the following vector notation

$$\mathbf{F} = (F_1(X_1, \ldots, X_m), \ldots, F_m(X_1, \ldots, X_m)),$$

we have the recurrence relation

$$\mathbf{u}_{n+1} = \mathbf{F}(\mathbf{u}_n), \quad n = 0, 1, \ldots.$$

In particular, for any $n, k \geq 0$ and $i = 1, \ldots, m$ we have

$$u_{n+k,i} = F_i^{(k)}(\mathbf{u}_n) = F_i^{(k)}(u_{n,1}, \ldots, u_{n,m}) \tag{6}$$

or

$$\mathbf{u}_{n+k} = \mathbf{F}^{(k)}(\mathbf{u}_n).$$

Clearly, as we work over a finite field of p elements, the sequence of vectors $(\mathbf{u}_n)$ is eventually periodic with some period $\tau \leq p^m$. We always assume that the sequence is purely periodic, that is,

$$\mathbf{u}_{n+\tau} = \mathbf{u}_n, \quad n = 0, 1, \ldots.$$

3   Main Results

3.1   Exponential Sums

Assume that the sequence $\{\mathbf{u}_n\}$ generated by (4) and (5) is purely periodic with an arbitrary period τ. For an integer vector $\mathbf{a} = (a_1, \ldots, a_m) \in \mathbb{Z}^m$ we introduce the exponential sum

$$S_{\mathbf{a}}(N) = \sum_{n=0}^{N-1} \mathbf{e}_p \left(\sum_{i=1}^{m} a_i u_{n,i} \right),$$

where

$$\mathbf{e}_p(z) = \exp(2\pi i z / p).$$

Also, as before t_i denotes the multiplicative order of g_i in $\mathbb{F}_p$, $i = 1, \ldots, m$.

As in [14, Lemma 2] we have the following description of the linear combinations of the iterations of the polynomials F_i:

Lemma 2. *Let $\mathcal{F}$ be the polynomial system* (2). *For any two integers $k > l$ and any nonzero integer vector $\mathbf{a} = (a_1, \ldots, a_m) \in \mathbb{F}_p^m$, we define the polynomial*

$$F_{\mathbf{a},k,l} = \sum_{i=1}^{m} a_i \left(F_i^{(k)} - F_i^{(l)} \right),$$

where the polynomials $F_i^{(k)}$ are given by (3). *If $g_s^k \not\equiv g_s^l \pmod{t_s}$ where $1 \leq s \leq m$ is the smallest integer such that $a_s \neq 0$, then we have*

$$\sum_{x_1, \ldots, x_m = 1}^{p} \mathbf{e}_p \left(F_{\mathbf{a},k,l}(x_1, \ldots, x_m) \right) = 0.$$

Proof. By Lemma 1 we have

$$F_{\mathbf{a},k,l}(x_1,\ldots,x_m)$$
$$= \sum_{i=s}^{m} a_i \left((g_i^k - g_i^l)x_i + (H_{i,k}(x_{i+1},\ldots,x_m) - H_{i,l}(x_{i+1},\ldots,x_m)) \right)$$
$$= a_s(g_s^k - g_s^l)x_s + \Psi_{\mathbf{a},k,l}(x_{s+1},\ldots,x_m),$$

where

$$\Psi_{\mathbf{a},k,l}(x_{s+1},\ldots,x_m) = \sum_{i=s+1}^{m} \left(a_i(g_i^k - g_i^l)x_i \right) +$$
$$\sum_{i=s}^{m} a_i \left(H_{i,k}(x_{s+1},\ldots,x_m) - H_{i,l}(x_{s+1},\ldots,x_m) \right).$$

Therefore,

$$\sum_{x_1,\ldots,x_m=1}^{p} \mathbf{e}_p\left(F_{\mathbf{a},k,l}(x_1,\ldots,x_m) \right) = p^s \sum_{x_{s+1},\ldots,x_m=1}^{p} \mathbf{e}_p\left(\Psi_{\mathbf{a},k,l}(x_{s+1},\ldots,x_m) \right)$$
$$\cdot \sum_{x_s=1}^{p} \mathbf{e}_p\left(a_s(g_s^k - g_s^l)x_s \right).$$

Recalling the identity

$$\sum_{u=1}^{p} \mathbf{e}_p(cu) = \begin{cases} p, & \text{if } c \equiv 0 \pmod{p}, \\ 0, & \text{if } c \not\equiv 0 \pmod{p}, \end{cases}$$

see [8, Equation (5.9)], we get the desired result.   □

Following the same technique as in [16,17] we obtain now the following estimate for the exponential sum $S_{\mathbf{a}}(N)$:

Theorem 1. *Let the sequence $\{\mathbf{u}_n\}$ be generated by (4) and (5), where the system of $m \geq 2$ polynomials $\mathcal{F} = \{F_1,\ldots,F_m\} \in \mathbb{F}_p[X_1,\ldots,X_m]$ is of the form (2). Assume that $\{\mathbf{u}_n\}$ is purely periodic with period τ. Then for any positive integer $N \leq \tau$ and any nonzero vector $\mathbf{a} \in \mathbb{F}_p^m$ we have the bound*

$$S_{\mathbf{a}}(N) \ll N^{1/2}t_s^{-1/2}p^{m/2},$$

where $1 \leq s \leq m$ is the smallest integer such that $a_s \neq 0$ and t_s is the order of g_s in $\mathbb{F}_p$

Proof. We follow the same argument as in the proof of [17, Theorem 4].
    In particular, as in [17], we obtain that for any integer $K \geq 1$,

$$K|S_{\mathbf{a}}(N)| \leq W + K^2, \tag{7}$$

where

$$W = \left| \sum_{n=0}^{N-1} \sum_{k=1}^{K} \mathbf{e}_p \left(\sum_{i=1}^{m} a_i u_{n+k,i} \right) \right|.$$

Using the Cauchy-Schwarz inequality we derive (again exactly the same way as in [16,17])

$$W^2 \le N \sum_{k,l=1}^{K} \sum_{x_1,\dots,x_m \in \mathbb{F}_p^m} \mathbf{e}_p(F_{\mathbf{a},k,l}(x_1,\dots,x_m)).$$

Because $g_s^k - g_s^l \equiv 0 \pmod{p}$ if and only if we have $k \equiv l \pmod{t_s}$, for $O(K(Kt_s^{-1} +1))$ elements k, l such that $k \equiv l \pmod{t_s}$ we estimate the sum trivially by p^m. Furthermore, for $k \not\equiv l \pmod{t_s}$, using Lemma 2 we see that the sum simply vanishes. We obtain the estimate

$$W^2 \ll NK(Kt_s^{-1} + 1)p^m.$$

Choosing now $K = t_s$ and inserting the above bound in (7) we obtain the desired result.   □

3.2 Discrepancy

Given a sequence Γ of N points

$$\Gamma = \left\{ (\gamma_{n,1}, \dots, \gamma_{n,m})_{n=0}^{N-1} \right\} \tag{8}$$

in the m-dimensional unit cube $[0,1)^m$ it is natural to measure the level of its statistical uniformity in terms of the *discrepancy* $D_N(\Gamma)$. More precisely,

$$D_N(\Gamma) = \sup_{B \subseteq [0,1)^m} \left| \frac{T_\Gamma(B)}{N} - |B| \right|,$$

where $T_\Gamma(B)$ is the number of points of Γ inside the box

$$B = [\alpha_1, \beta_1) \times \dots \times [\alpha_m, \beta_m) \subseteq [0,1)^m$$

and the supremum is taken over all such boxes, see [3,6].

We recall that the discrepancy is a widely accepted quantitative measure of uniformity of distribution of sequences, and thus good pseudorandom sequences should (after an appropriate scaling) have a small discrepancy, see [11,12].

Typically the bounds on the discrepancy of a sequence are derived from bounds of exponential sums with elements of this sequence. The relation is made explicit in the celebrated *Erdős–Turán–Koksma inequality*, see [3, Theorem 1.21], which we present it in the following form.

Lemma 3. *For any integer $H > 1$ and any sequence Γ of N points (8) the discrepancy $D_N(\Gamma)$ satisfies the following bound:*

$$D_N(\Gamma) = O\left(\frac{1}{H} + \frac{1}{N} \sum_{0 < |\mathbf{h}| \le H} \prod_{j=1}^{m} \frac{1}{|h_j| + 1} \left| \sum_{n=0}^{N-1} \exp\left(2\pi i \sum_{j=1}^{m} h_j \gamma_{n,j} \right) \right| \right),$$

where the sum is taken over all integer vectors $\mathbf{h} = (h_1, \ldots, h_m) \in \mathbb{Z}^m$ with $|\mathbf{h}| = \max_{j=1,\ldots,m} |h_j| < H$.

Using now Theorem 1 and Lemma 3 we obtain the following estimate on the discrepancy of the sequence of vectors generated by polynomial systems of the form (2).

Theorem 2. *Let the sequence* $\{\mathbf{u}_n\}$ *be generated by* (4) *and* (5), *where the system of* $m \geq 2$ *polynomials* $\mathcal{F} = \{F_1, \ldots, F_m\} \in \mathbb{F}_p[X_1, \ldots, X_m]$ *is of the form* (2). *Assume that* $\{\mathbf{u}_n\}$ *is purely periodic with period* τ. *Then for any positive integer* $N \leq \tau$, *the discrepancy* $D_N(\Gamma)$ *of the sequence*

$$\Gamma = \left\{ \left(\frac{u_{n,1}}{p}, \ldots, \frac{u_{n,m}}{p} \right), \qquad n = 0, \ldots, N-1 \right\},$$

satisfies the bound

$$D_N(\Gamma) = O(N^{-1/2} t^{-1/2} p^{m/2} (\log p)^m)$$

where $t = \min\{t_s | s = 1, \ldots, m\}$.

We note that both Theorem 1 and Theorem 2 are nontrivial if $\tau \geq N \geq t^{-1}(\log p)^{2m} p^m$.

3.3   Average Case over All Initial Values

We follow the scheme previously introduced in [13] for estimating the discrepancy on average of the sequence generated by (4) and (5).

For a vector $\mathbf{a} = (a_1, \ldots, a_m) \in \mathbb{F}_p^m$ and integers c, M, N with $M \geq 1$ and $N \geq 1$, we introduce

$$V_{\mathbf{a},c}(M, N) = \sum_{v_1,\ldots,v_m \in \mathbb{F}_p} \left| \sum_{n=0}^{N-1} \mathbf{e}_p \left(\sum_{j=1}^{m} a_j F_j^{(n)}(v_1, \ldots, v_m) \right) \mathbf{e}_M(cn) \right|^2.$$

Theorem 3. *Let the polynomial system of* m *polynomials*

$$\mathcal{F} = \{F_1, \ldots, F_m\} \in \mathbb{F}_p[X_1, \ldots, X_m], \qquad m \geq 2,$$

of the form (2). *Then for any positive integers* c, M, N *and any nonzero vector* $\mathbf{a} = (a_1 \ldots, a_m) \in \mathbb{F}_p^m$ *we have*

$$V_{\mathbf{a},c}(M, N) \ll A(N, p),$$

where

$$A(N, p) = \begin{cases} Np^m & \text{if } N \leq t_s, \\ N^2 t_s^{-1} p^m & \text{if } N > t_s, \end{cases}$$

and $s \leq m$ *is the smallest integer such that* $a_s \neq 0$.

Proof. We have

$$V_{\mathbf{a},c}(M,N) = \sum_{k,l=0}^{N-1} \mathbf{e}_M\left(c(k-l)\right)$$

$$\sum_{\mathbf{v}\in\mathbb{F}_p^m} \mathbf{e}_p\left(\sum_{j=1}^{m} a_j\left(F_j^{(k)}(\mathbf{v}) - F_j^{(l)}(\mathbf{v})\right)\right)$$

$$\leq \sum_{k,l=0}^{N-1}\left|\sum_{\mathbf{v}\in\mathbb{F}_p^m} \mathbf{e}_p\left(\sum_{j=1}^{m} a_j\left(F_j^{(k)}(\mathbf{v}) - F_j^{(l)}(\mathbf{v})\right)\right)\right|.$$

As in Theorem 1, for $O(N(Nt_s^{-1}+1))$ elements k,l such that $k \equiv l \pmod{t_s}$, we estimate the inner sum trivially by p^m. Furthermore, for $k \not\equiv l \pmod{t_s}$, using Lemma 2 we see that the sum simply vanishes.

Hence,

$$V_{\mathbf{a},c}(M,N) \ll N(Nt_s^{-1}+1)p^m. \tag{9}$$

Because $\mathcal{F}$ is a permutation polynomial system and using (6), for any integer L we obtain

$$\sum_{\mathbf{v}\in\mathbb{F}_p^m}\left|\sum_{n=L}^{L+N-1} \mathbf{e}_p\left(\sum_{j=1}^{m} a_j F_j^{(n)}(\mathbf{v})\right) \mathbf{e}_M(cn)\right|^2$$

$$= \sum_{\mathbf{v}\in\mathbb{F}_p^m}\left|\sum_{n=0}^{N-1} \mathbf{e}_p\left(\sum_{j=1}^{m} a_j F_j^{(n)}\left(F_1^{(L)}(\mathbf{v}),\ldots,F_m^{(L)}(\mathbf{v})\right)\right) \mathbf{e}_M(cn)\right|^2$$

$$= \sum_{\mathbf{v}\in\mathbb{F}_p^m}\left|\sum_{n=0}^{N-1} \mathbf{e}_p\left(\sum_{j=1}^{m} a_j F_j^{(n)}(\mathbf{v})\right) \mathbf{e}_M(cn)\right|^2 = V_{\mathbf{a},c}(M,N).$$

Therefore, for any positive integer $K \leq N$, separating the inner sum into at most $N/K+1$ subsums of length at most K, and using (9), we derive

$$V_{\mathbf{a},c}(M,N) \ll K(Kt_s^{-1}+1)p^m N^2 K^{-2} = N^2 t_s^{-1} p^m + K^{-1}N^2 p^m.$$

Thus, selecting $K = \min\{N, t_s\}$ we obtain the desired result.   □

Now, exactly as in [13,17], combining Lemma 3 with the bound obtained in Theorem 3 we obtain stronger estimates for the discrepancy "on average" over all initial values.

Theorem 4. *Let $0 < \varepsilon < 1$ and let the sequence $\{\mathbf{u}_n\}$ be generated by (4) and (5), where the system of $m \geq 2$ polynomials*

$$\mathcal{F} = \{F_1,\ldots,F_m\} \in \mathbb{F}_p[X_1,\ldots,X_m]$$

is of the form (2). *Then for all initial values* $\mathbf{v} \in \mathbb{F}_p^m$ *except at most* $O(\varepsilon p^m)$ *of them, and any positive integer* $N \leq p^m$, *the discrepancy* $D_N(\Gamma(\mathbf{v}))$ *of the sequence*

$$\Gamma(\mathbf{v}) = \left\{ \left(\frac{u_{n,1}}{p}, \ldots, \frac{u_{n,m}}{p} \right), \qquad n = 0, \ldots, N-1 \right\},$$

satisfies the bound

$$D_N(\Gamma(\mathbf{v})) \leq \varepsilon^{-1} B(N, p),$$

where

$$B(N, p) = \begin{cases} N^{-1/2}(\log N)^{m+1} \log p & \text{if } N \leq t, \\ t^{-1/2}(\log N)^{m+1} \log p & \text{if } N > t, \end{cases}$$

and $t = \min\{t_s | s = 1, \ldots, m\}$.

We note that Theorem 4 is nontrivial if $N \geq (\log p)^{2+\epsilon}$ for some $\epsilon > 0$.

4 Remarks and Questions

We remark that our bounds of exponential sums can be immediately extended to arbitrary finite fields. Furthermore, our approach also applies to the same polynomial systems over residue rings and also leads to similar results.

Although low discrepancy is a very important requirement on any pseudorandom number generator, this is not the only one. For example, the notion of linear complexity also plays an important role in this area. We recall that the *linear complexity* L of an N-element sequence $s_0, \ldots, s_{N-1}$ in a ring $\mathcal{R}$ is defined as the smallest L such that

$$s_{u+L} = c_{L-1}s_{u+L-1} + \ldots + c_0 s_u, \qquad 0 \leq u \leq N - L - 1,$$

for some $c_0, \ldots, c_{L-1} \in \mathcal{R}$, see [2,9,10,21,22].

We remark that the degree argument which has been used in [18] to prove the linear complexity bounds for the polynomial systems (1) cannot be applied here anymore, so more ideas are needed in order to be able to give nontrivial estimates in the case of the polynomial systems considered in this paper.

Furthermore, in the case of vector sequences it is also natural to consider linear relations with vector coefficients. Namely, it would be interesting to give nontrivial estimates for the smallest L such that for some m-dimensional vectors $\mathbf{c}_0, \ldots, \mathbf{c}_L$ over $\mathbb{F}_q$ where $\mathbf{c}_L$ is a non-zero vector, we have

$$\sum_{h=0}^{L} \mathbf{c}_h \cdot \mathbf{u}_{n+h} = 0$$

for all $h = 0, \ldots, N - L - 1$, where $\mathbf{c} \cdot \mathbf{u}$ denotes the scalar product. This can be extended to sequences over arbitrary finite fields.

It will be also very interesting to investigate character sums using polynomial systems of the form (2). For some results involving the systems (1) see [19].

Acknowledgment

The author would like to thank Igor Shparlinski and Arne Winterhof for many valuable discussions and for a careful reading of this work. During the preparation of this paper, the author was supported in part by the Swiss National Science Foundation Grant-121874.

References

1. Bourgain, J.: Mordell's exponential sum estimate revisited. J. Amer. Math. Soc. 18, 477–499 (2005)
2. Cusick, T.W., Ding, C., Renvall, A.: Stream ciphers and number theory. Elsevier, Amsterdam (2003)
3. Drmota, M., Tichy, R.: Sequences, discrepancies and applications. Springer, Berlin (1997)
4. Griffin, F., Niederreiter, H., Shparlinski, I.E.: On the distribution of nonlinear recursive congruential pseudorandom numbers of higher orders. In: Fossorier, M.P.C., Imai, H., Lin, S., Poli, A. (eds.) AAECC 1999. LNCS, vol. 1719, pp. 87–93. Springer, Heidelberg (1999)
5. Gutierrez, J., Gomez-Perez, D.: Iterations of multivariate polynomials and discrepancy of pseudorandom numbers. In: Bozta, S., Sphparlinski, I. (eds.) AAECC 2001. LNCS, vol. 2227, pp. 192–199. Springer, Heidelberg (2001)
6. Kuipers, L., Niederreiter, H.: Uniform distribution of sequences. Wiley-Intersci., New York (1974)
7. Lidl, R., Niederreiter, H.: On orthogonal systems and permutation polynomials in several variables. Acta Arith. 22, 257–265 (1973)
8. Lidl, R., Niederreiter, H.: Finite fields. Cambridge Univ. Press, Cambridge (1997)
9. Menezes, A.J., van Oorschot, P.C., Vanstone, S.A.: Handbook of Applied Cryptography. CRC Press, Boca Raton (1997)
10. Niederreiter, H.: Linear complexity and related complexity measures for sequences. In: Johansson, T., Maitra, S. (eds.) INDOCRYPT 2003. LNCS, vol. 2904, pp. 1–17. Springer, Heidelberg (2003)
11. Niederreiter, H.: Quasi-Monte Carlo methods and pseudo-random numbers. Bull. Amer. Math. Soc. 84, 957–1041 (1978)
12. Niederreiter, H.: Random number generation and Quasi–Monte Carlo methods. SIAM Press, Philadelphia (1992)
13. Niederreiter, H., Shparlinski, I.E.: On the average distribution of inversive pseudorandom numbers. Finite Fields and Their Appl. 8, 491–503 (2002)
14. Ostafe, A.: Multivariate permutation polynomial systems and pseudorandom number generators. Finite Fields and Their Appl. (to appear, 2009)
15. Ostafe, A., Pelican, E., Shparlinski, I.E.: On pseudorandom numbers from multivariate polynomial systems. Preprint (2010)
16. Ostafe, A., Shparlinski, I.E.: On the degree growth in some polynomial dynamical systems and nonlinear pseudorandom number generators. Math. Comp. 79, 501–511 (2010)
17. Ostafe, A., Shparlinski, I.E.: Pseudorandom numbers and hash functions from iterations of multivariate polynomials. Cryptography and Communications 2, 49–67 (2010)

18. Ostafe, A., Shparlinski, I.E., Winterhof, A.: On the generalized joint linear complexity profile of a class of nonlinear pseudorandom multisequences. Adv. Math. Comm. (2010)
19. Ostafe, A., Shparlinski, I.E., Winterhof, A.: Multiplicative character sums of a class of nonlinear recurrence vector sequences (preprint) (2009)
20. Shparlinski, I.E.: Bounds for exponential sums with recurrence sequences and their applications. In: Proc. Voronezh State Pedagogical Inst., vol. 197, pp. 74–85 (1978)
21. Topuzoğlu, A., Winterhof, A.: Pseudorandom sequences. In: Topics in Geometry, Coding Theory and Cryptography, pp. 135–166. Springer, Berlin (2006)
22. Winterhof, A.: Linear complexity and related complexity measures. In: Selected Topics in Information and Coding Theory, pp. 3–40. World Scientific, Singapore (2010)

Structure of Pseudorandom Numbers Derived from Fermat Quotients

Zhixiong Chen[1], Alina Ostafe[2], and Arne Winterhof[3]

[1] Department of Mathematics, Putian University, Fujian 351100, China
ptczx@126.com
[2] Institut für Mathematik, Universität Zürich, Winterthurerstr. 190,
CH-8057 Zürich, Switzerland
alina.ostafe@math.uzh.ch
[3] Johann Radon Institute for Computational and Applied Mathematics, Austrian
Academy of Sciences, Altenberger Straße 69, A-4040 Linz, Austria
arne.winterhof@oeaw.ac.at

Abstract. We study the distribution of s-dimensional points of Fermat quotients modulo p with arbitrary lags. If no lags coincide modulo p the same technique as in [21] works. However, there are some interesting twists in the other case. We prove a discrepancy bound which is unconditional for $s = 2$ and needs restrictions on the lags for $s > 2$. We apply this bound to derive results on the pseudorandomness of the binary threshold sequence derived from Fermat quotients in terms of bounds on the well-distribution measure and the correlation measure of order 2, both introduced by Mauduit and Sárközy. We also prove a lower bound on its linear complexity profile. The proofs are based on bounds on exponential sums and earlier relations between discrepancy and both measures above shown by Mauduit, Niederreiter and Sárközy. Moreover, we analyze the lattice structure of Fermat quotients modulo p with arbitrary lags.

Keywords: Fermat quotients, finite fields, pseudorandom sequences, exponential sums, discrepancy, well-distribution measure, correlation measure, linear complexity, lattice test.

MSC(2010): Primary 11T23; Secondary 65C10, 94A55, 94A60.

1 Introduction

For a prime p and an integer u with $\gcd(u, p) = 1$ the *Fermat quotient $q_p(u)$ modulo p* is defined as the unique integer with

$$q_p(u) \equiv \frac{u^{p-1} - 1}{p} \pmod{p}, \qquad 0 \le q_p(u) \le p - 1,$$

and we also define

$$q_p(kp) = 0, \qquad k \in \mathbb{Z}.$$

We note that $(q_p(u))$ is a p^2-periodic sequence modulo p for $u \ge 1$. There are several results which involve the distribution and structure of Fermat quotients

M.A. Hasan and T. Helleseth (Eds.): WAIFI 2010, LNCS 6087, pp. 73–85, 2010.

$q_p(u)$ modulo p and it has numerous applications in computational and algebraic number theory, see e.g. [8,10,11,21] and references therein.

In particular, Heath-Brown [11] presented a nontrivial upper bound on exponential sums with $q_p(u)$, $u = d_0 + 1, \ldots, d_0 + N$, for any integers d_0 and $N \geq p^{1/2+\varepsilon}$ for any fixed ε. (The result in [11] is weaker but using the Burgess bound instead of the Polya-Vinogradov bound in the proof one can easily obtain it, see [21, Lemma 2].) Furthermore, the second author and Shparlinski [21] have recently presented a nontrivial bound of exponential sums with linear combinations of $s \geq 1$ consecutive values $(q_p(u), \ldots, q_p(u+s-1))$, $u = d_0+1, \ldots, d_0+N$ for longer intervals of length $N \geq p^{1+\varepsilon}$.

Here we first study the distribution of the points

$$\Gamma(D, N, s) = \left\{ \left(\frac{q_p(u + d_0)}{p}, \ldots, \frac{q_p(u + d_{s-1})}{p} \right) : u = 1, \ldots, N \right\} \qquad (1)$$

in the s-dimensional unit interval for any lags $D = (d_0, \ldots, d_{s-1})$ with $0 \leq d_0 < \cdots < d_{s-1} < p^2$. More precisely, we prove an exponential sum bound (which implies a discrepancy bound using the Erdős-Turan-Koksma inequality) which is nontrivial for $s = 2$ and arbitrary lags $0 \leq d_0 < d_1 < p^2$ and for $s > 2$ if no three lags are equivalent modulo p. We note that in the case when $d_i \not\equiv d_j$ (mod p) for all $0 \leq i < j < s$, the proof is exactly the same as in [21, Theorem 11]. However, the other case brings interesting twists and will be discussed in Theorem 1 below. We also indicate that the exponential sums can be trivial for $s > 2$ if there exist three equivalent lags modulo p.

As applications we use some results of [14] to derive bounds on the *well-distribution measure* $W(E_{p^2})$ and the *correlation measure* $C_2(E_{p^2})$ *of order* 2 (see Section 3 below for the definitions) of the binary sequence $E_{p^2} = \{e_1, e_2, \ldots, e_{p^2}\} \in \{0, 1\}^{p^2}$ defined by

$$e_u = \begin{cases} 0, & \text{if } 0 \leq q_p(u)/p < \frac{1}{2}, \\ 1, & \text{if } \frac{1}{2} \leq q_p(u)/p < 1, \end{cases} \quad 1 \leq u \leq p^2. \qquad (2)$$

Note that for such applications a discrepancy bound with arbitrary lags is needed. Most known discrepancy bounds on nonlinear pseudorandom numbers found in the literature consider only equidistant lags. In many cases the analysis of the discrepancy becomes much more intricate for arbitrary lags, see for example [18].

It was shown in [2] that for a "truly random" sequence $E_T \in \{0,1\}^T$ both pseudorandomness measures $W(E_T)$ and $C_2(E_T)$ are "small". More precisely, a sequence E_T can be considered as a "good" pseudorandom sequence if both $W(E_T)$ and $C_2(E_T)$ are small and are ideally greater than $T^{1/2}$ only by at most a power of $\log T$. We prove bounds on well-distribution measure and correlation measure of order 2 for the binary threshold sequence E_{p^2} derived from Fermat quotients modulo p of the desired order of magnitude.

Moreover, we use the bounds on exponential sums of Fermat quotients to derive a bound on the linear complexity profile (see Section 4 for the definition) of the sequence E_{p^2} defined by (2).

Finally we study the lattice structure of the sequence $(q_p(u))$. The following lattice test was introduced in [20]. Let (w_u), $u = 1, 2, \ldots$, be a T-periodic sequence over the finite field $\mathbb{F}_p$ of p elements. For given integers $s \geq 1$, $0 \leq d_0 < d_1 < \ldots < d_{s-1} < T$, and $N \geq 2$, we say that (w_u) passes the s-dimensional N-lattice test with lags $d_0, \ldots, d_{s-1}$ if the vectors $\{\mathbf{w}_u - \mathbf{w}_1 : 1 \leq u \leq N\}$ span $\mathbb{F}_p^s$, where

$$\mathbf{w}_u = (w_{u+d_0}, w_{u+d_1}, \ldots, w_{u+d_{s-1}}), \quad 1 \leq u \leq N.$$

In the case $d_i = i$ for $0 \leq i < s$, this test coincides essentially with the lattice test introduced in [5] and further analyzed in [3,4,5,6,9,22]. The latter lattice test is closely related to the concept of the linear complexity profile, see [5,6,19]. If additionally $N \geq T$, this special lattice test was proposed by Marsaglia [13].

We note that in the case $d_i \not\equiv d_j \pmod{p}$ for all $0 \leq i < j < s$, the lattice test can be analyzed essentially along the same lines as in the proof of the linear complexity bounds in [21, Theorems 13,14].

The implied constants in the symbols 'O', and '$\ll$' are absolute. We recall that the notations $U = O(V)$ and $U \ll V$ are both equivalent to the assertion that the inequality $|U| \leq cV$ holds for some constant $c > 0$.

2 Distribution of Fermat Quotients

Given a sequence Γ of N points

$$\Gamma = \left\{ (\gamma_{n,0}, \ldots, \gamma_{n,s-1})_{n=1}^{N} \right\} \tag{3}$$

in the s-dimensional unit cube $[0,1)^s$ it is natural to measure the level of its statistical uniformity in terms of the *discrepancy* $\Delta(\Gamma)$ defined by

$$\Delta(\Gamma) = \sup_{B \subseteq [0,1)^s} \left| \frac{T_\Gamma(B)}{N} - |B| \right|,$$

where $T_\Gamma(B)$ is the number of points of Γ inside the box

$$B = [\alpha_0, \beta_0) \times \cdots \times [\alpha_{s-1}, \beta_{s-1}) \subseteq [0,1)^s$$

and the supremum is taken over all such boxes, see [7,17].

Typically the bounds on the discrepancy of a sequence are derived from bounds of exponential sums with elements of this sequence. The relation is made explicit in the celebrated *Erdős-Turan-Koksma inequality*, see [7, Theorem 1.21], which we present in the following form.

Lemma 1. *For any integer $H > 1$ and any sequence Γ of N points (3) the discrepancy $\Delta(\Gamma)$ satisfies*

$$\Delta(\Gamma) \leq \left(\frac{3}{2} \right)^s \left(\frac{2}{H+1} + \frac{1}{N} \sum_{0 < |\mathbf{a}| \leq H} \prod_{j=0}^{s-1} \frac{1}{\max\{|a_j|, 1\}} \left| \Sigma_N^{(s)}(\Gamma, \mathbf{a}) \right| \right),$$

where

$$\Sigma_N^{(s)}(\Gamma, \mathbf{a}) = \sum_{n=1}^{N} \exp\left(2\pi i \sum_{j=0}^{s-1} a_j \gamma_{n,j}\right)$$

and the outer sum is taken over all integer vectors $\mathbf{a} = (a_0, \ldots, a_{s-1}) \in \mathbb{Z}^s \setminus \{\mathbf{0}\}$ *with* $|\mathbf{a}| = \max\limits_{j=0,\ldots,s-1} |a_j| \le H$.

Our results are based on the following well-known property of Fermat quotients. For any integers k and u with $\gcd(u,p) = 1$ we have

$$q_p(u + kp) \equiv q_p(u) - ku^{-1} \pmod{p}, \tag{4}$$

see, for example, [8, (2)].

Let $\psi(z) = \exp(2\pi i z/p)$ denote the *additive canonical character* of $\mathbb{F}_p$. For integers $N \ge 1$, $s \ge 1$ and $\mathbf{a} = (a_0, \ldots, a_{s-1}) \in \mathbb{Z}^s$ we consider the exponential sums

$$\Sigma_N^{(s)}(D; \mathbf{a}) = \sum_{u=1}^{N} \psi\left(\sum_{j=0}^{s-1} a_j q_p(u + d_j)\right),$$

for any integer vector $D = (d_0, d_1, \ldots, d_{s-1})$ with $0 \le d_0 < d_1 < \cdots < d_{s-1} < p^2$.

Theorem 1. *For* $s \ge 1$ *and* $D = (d_0, d_1, \ldots, d_{s-1})$ *with* $0 \le d_0 < d_1 < \cdots < d_{s-1} < p^2$ *such that no triple* (d_l, d_h, d_t) *satisfies* $d_l \equiv d_h \equiv d_t \pmod{p}$ *for* $0 \le l < h < t < s$, *we have*

$$\max_{\gcd(a_0,\ldots,a_{s-1},p)=1} \left|\Sigma_N^{(s)}(D; \mathbf{a})\right| \ll s \max\{p\log p, N p^{-1/2}\} \quad \text{for } 1 \le N \le p^2.$$

If $s = 2$ *or* $d_{s-1} < p$, *the stronger bound* $sp\log p$ *holds.*

Proof. For $s = 1$ the result follows from [11] and we assume $s \ge 2$. Select any $\mathbf{a} = (a_0, \ldots, a_{s-1}) \in \mathbb{Z}^s$ with $\gcd(a_0, \ldots, a_{s-1}, p) = 1$. Let denote by l the smallest index such that $\gcd(a_l, p) = 1$. For $d_l \not\equiv d_j \pmod{p}$ for all $l < j < s$, we can obtain the desired result by following the proof path of [21, Theorem 11].

Now we suppose that there exists h with $l < h < s$ such that $d_l \equiv d_h \pmod{p}$ but $d_l \not\equiv d_j \pmod{p}$ for all $j \ne h$ with $l < j < s$ by our assumption. Let $d_h = d_l + k_0 p$ for some integer $1 \le k_0 < p$. Take $K = \lceil N/p \rceil$ and note that $K \le p$. Using (4) we get

$$\Sigma_N^{(s)}(D; \mathbf{a})$$

$$= \sum_{u=1}^{Kp} \psi\left(\sum_{j=0}^{s-1} a_j q_p(u + d_j)\right) + O(p)$$

$$= \sum_{u=1}^{Kp} \psi\left(a_l q_p(u + d_l) + a_h q_p(u + d_l + k_0 p) + \sum_{\substack{j=l+1 \\ j \ne h}}^{s-1} a_j q_p(u + d_j)\right) + O(p)$$

$$
= \sum_{\substack{u=1 \\ u \not\equiv -d_l \pmod p}}^{Kp} \psi\Big(-k_0 a_h (u+d_l)^{-1} + (a_l + a_h) q_p(u+d_l)
$$

$$
+ \sum_{\substack{j=l+1 \\ j \neq h}}^{s-1} a_j q_p(u+d_j)\Big) + O(p)
$$

$$
= O(p) + \sum_{\substack{v=1 \\ v \not\equiv -d_l \pmod p}}^{p} \psi\left(-k_0 a_h (v+d_l)^{-1}\right)
$$

$$
\cdot \sum_{k=0}^{K-1} \psi\left((a_l + a_h) q_p(v+d_l+kp) + \sum_{\substack{j=l+1 \\ j \neq h}}^{s-1} a_j q_p(v+d_j+kp)\right),
$$

where we substituted $u = v + kp$ in the last step.

If $a_l + a_h \not\equiv 0 \pmod p$ we get the result following the proof of [21, Theorem 11]. Let $\mathcal{V}$ be the set of $1 \le v \le p$ with $v \not\equiv -d_j \pmod p$ for $l \le j < s$. Then we have

$$
\left|\Sigma_N^{(s)}(D; \mathbf{a})\right|
$$

$$
\le \sum_{v \in \mathcal{V}} \left|\sum_{k=0}^{K-1} \psi\left((a_l + a_h) q_p(v+d_l+kp) + \sum_{\substack{j=l+1 \\ j \neq h}}^{s-1} a_j q_p(v+d_j+kp)\right)\right| + O(sp)
$$

$$
= \sum_{v \in \mathcal{V}} \left|\sum_{k=0}^{K-1} \psi\left(k\left((a_l + a_h)(v+d_l)^{-1} + \sum_{\substack{j=l+1 \\ j \neq h}}^{s-1} a_j(v+d_j)^{-1}\right)\right)\right| + O(sp)
$$

$$
\ll sp \log p,
$$

where we used [21, Lemma 3] in the last step and the fact that

$$
F(X) = \frac{a_l + a_h}{X + d_l} + \sum_{\substack{j=l+1 \\ j \neq h}}^{s-1} \frac{a_j}{X + d_j}
$$

is a nonconstant rational function of degree $O(s)$. (Note that $-d_l$ is a single pole of $F(X)$.)

If $a_l \equiv -a_h \pmod p$ and there is a $j \neq h$ with $l < j < s$ such that $\gcd(a_j, p) = 1$ and d_j is either not equivalent to any other lag d_k or $a_j \not\equiv -a_k$ we see that $F(X)$ is not constant again and derive the bound $sp \log p$ in the same way.

In the last case all lags d_j with $\gcd(a_j, p) = 1$ appear in pairs $d_j, d_{h(j)}$ with $d_{h(j)} \equiv d_j + k_j p \pmod p$ for some $1 \le k_j < p$ such that $a_j \equiv -a_{h(j)} \pmod p$. In this case we get

$$\Sigma_N^{(s)}(D;\mathbf{a}) = \sum_{u=1}^{N} \psi\left(\sum_j a_j k_j (u + d_j)^{-1}\right)$$

and get the bound

$$sp^{1/2}\left(\frac{N}{p} + \log p\right)$$

using the standard method for reducing incomplete exponential sums to complete ones, see [12, Chapter 12], and the bound of Moreno and Moreno [16]. (Note that we have $\lfloor N/p \rfloor$ complete sums and one incomplete sum.) For $s = 2$ the sum over j contains only one summand and we can obtain the better bound

$$\frac{N}{p} + p^{1/2}\log p \ll p$$

and the result follows. $\qquad\square$

Together with Lemma 1, Theorem 1 implies an upper bound on the discrepancy of points (1).

Corollary 1. *For $s \geq 1$ and $D = (d_0, d_1, \ldots, d_{s-1})$ with $0 \leq d_0 < d_1 < \cdots < d_{s-1} < p^2$ such that no triple (d_l, d_h, d_t) satisfies $d_l \equiv d_h \equiv d_t \pmod{p}$, $0 \leq l < h < t < s$, the discrepancy of points $\Gamma(D, N, s)$ defined by (1) satisfies*

$$\Delta(\Gamma(D, N, s)) = O\left(\left(\frac{3}{2}\right)^s s \max\{N^{-1}p\log p, p^{-1/2}\}(\log p)^s\right) \text{ for } 1 \leq N \leq p^2.$$

If $s = 2$ or $d_{s-1} < p$, we have $\Delta(\Gamma(D, N, s)) = O((3/2)^s s N^{-1}p(\log p)^{s+1})$.

However, Theorem 1, hence Corollary 1, are not extendable if there exist at least three lags congruent modulo p, as the following example shows.

Example. For $D = (d_0, d_1, d_2)$ with $0 \leq d_0 < d_1 < d_2 < p^2$ and $d_0 \equiv d_1 \equiv d_2 \pmod{p}$ let $d_1 = d_0 + k_1 p$ and $d_2 = d_0 + k_2 p$ for some integers $1 \leq k_1 < k_2 < p$, then we have

$$\Sigma_N^{(3)}(D;\mathbf{a})$$

$$= \sum_{u=1}^{N} \psi\left(\sum_{j=0}^{2} a_j q_p(u + d_j)\right)$$

$$= \sum_{u=1}^{N} \psi\left(\sum_{j=0}^{2} a_j q_p(u + d_0) - a_1 k_1 (u + d_0)^{-1} - a_2 k_2 (u + d_0)^{-1}\right)$$

$$= \sum_{u=1}^{N} \psi\left(\sum_{j=0}^{2} a_j q_p(u + d_0) - (a_1 k_1 + a_2 k_2)(u + d_0)^{-1}\right).$$

We get a trivial bound on $\Sigma_N^{(3)}(D;\mathbf{a})$ if $a_0 + a_1 + a_2 \equiv 0 \pmod{p}$ and $a_1 k_1 + a_2 k_2 \equiv 0 \pmod{p}$. In fact, for example, one can select $a_0 = 1, a_1 = -2, a_2 = 1$ if we take $k_1 = 1$ and $k_2 = 2$.

3 Pseudorandom Measures of the Binary Threshold Sequence

In a series of papers starting from [15], Mauduit and Sárközy (partly with further coauthors) introduced certain measures of pseudorandomness and studied finite binary pseudorandom sequences. For a finite binary sequence of length T

$$E_T = \{e_1, \ldots, e_T\} \in \{0, 1\}^T,$$

the *well-distribution measure* of E_T is defined as

$$W(E_T) = \max_{a,b,t} \left| \sum_{j=0}^{t-1} (-1)^{e_{a+bj}} \right|,$$

where the maximum is taken over all $a, b, t \in \mathbb{N}$ such that $1 \le a \le a+b(t-1) \le T$, and the *correlation measure of order s* of E_T is defined as

$$C_s(E_T) = \max_{M,D} C_s(E_T, M, D),$$

where the maximum is taken over all integer vectors $D = (d_0, \ldots, d_{s-1})$ and $M > 0$ such that $0 \le d_0 < d_1 < \ldots < d_{s-1} \le T - M$, and

$$C_s(E_T, M, D) = \left| \sum_{n=1}^{M} (-1)^{e_{n+d_0} + e_{n+d_1} + \cdots + e_{n+d_{s-1}}} \right|.$$

(Note that [15] actually deals with the sequences $E'_T = \{e'_1, \ldots, e'_T\} \in \{-1, 1\}^T$ defined by $e'_n = (-1)^{e_n}$, $1 \le n \le T$, and the corresponding definitions of $W(E'_T) = W(E_T)$ and $C_s(E'_T) = C_s(E_T)$.)

In this section, we estimate the well-distribution measure and the correlation measure of order 2 for the binary sequence E_{p^2} defined as in (2). For estimates on the correlation measure of higher order using the same method we would need a discrepancy bound without restrictions on the lags. However, it seems to fail according to the example in Section 2.

Theorem 2. *For the binary threshold sequence E_{p^2} defined as in (2), we have*

$$W(E_{p^2}) \ll p(\log p)^2.$$

Proof. For any integers $a, b, t \in \mathbb{N}$ with $1 \le a \le a + b(t-1) \le p^2$, we have by [14, Theorem 2]

$$\left| \sum_{j=0}^{t-1} (-1)^{e_{a+jb}} \right| \ll t\Delta(\Gamma(t)),$$

where

$$\Gamma(t) = \left\{ \frac{q_p(a)}{p}, \frac{q_p(a+b)}{p}, \ldots, \frac{q_p(a+b(t-1))}{p} \right\}.$$

Note that
$$q_p(ub) \equiv q_p(u) + q_p(b) \pmod{p}, \quad \gcd(ub, p) = 1.$$
Since otherwise the result is trivial we may assume $b < p$. Hence we have
$$\left| \sum_{j=0}^{t-1} \psi(q_p(a + jb)) \right| = \left| \sum_{j=0}^{t-1} \psi(q_p(ab^{-1} + j)) \right| + O(p) = O(p \log p),$$
where b^{-1} denotes the inverse of b modulo p^2. Finally the Erdős-Turan-Koksma inequality gives
$$\Delta(\Gamma(t)) \ll t^{-1} p (\log p)^2,$$
which implies the result. □

Theorem 3. *For the binary threshold sequence E_{p^2} defined as in (2), and $D = (d_0, d_1)$ with $0 \le d_0 < d_1 < p^2$ we have*
$$C_2(E_{p^2}) \ll p (\log p)^3.$$

Proof. With assumptions on D and M such that $M + d_1 \le p^2$, by [14, Theorem 1] and Corollary 1, we have
$$\left| \sum_{n=1}^{M} (-1)^{e_{n+d_0} + e_{n+d_1}} \right| \ll M \Delta(\Gamma(D, M, 2)) \ll p (\log p)^3$$
and the result follows. □

4   Linear Complexity Profile of the Binary Threshold Sequence

We recall that the *linear complexity profile* $L(E_T, N)$ is the least order L of a linear recurrence relation over $\{0, 1\}$
$$e_{n+L} = c_0 e_n + c_1 e_{n+1} + \cdots + c_{L-1} e_{n+L-1} \quad \text{for } 1 \le n \le N - L$$
which is satisfied by the first N terms of E_T.

Theorem 4. *For the binary threshold sequence E_{p^2} defined as in (2), we have*
$$L(E_{p^2}, N) \gg \frac{\log(N/p)}{\log \log p} \quad \text{for } 2 \le N \le p^2.$$

Proof. The proof is reminiscent to that of [1, Theorem 1]. Since otherwise the bound is trivial we may assume $L = L(E_{p^2}, N) < \log p$. Choose $c_0, c_1, \ldots, c_{L-1} \in \{0, 1\}$ such that
$$e_{n+L} = c_{L-1} e_{n+L-1} + \cdots + c_0 e_n, \quad 1 \le n \le N - L.$$

Putting $c_L = -1$, we get

$$N - L = \sum_{n=0}^{N-L-1} (-1)^{\sum_{i=0}^{L} c_i e_{n+i}}.$$

The sum on the right hand side can be estimated by

$$\max_{\substack{D \\ 1 \le s \le L+1}} C_s(E_{p^2}, N - d_{s-1}, D),$$

where the maximum is taken over all $D = (d_0, d_1, \ldots, d_{s-1})$ with $0 \le d_0 < d_1 < \cdots < d_{s-1} \le L < p$. We note that in this case $d_i \not\equiv d_j \pmod{p}$ for all $0 \le i < j < s$. For all such D, by Corollary 1 we have

$$\max \Delta(\Gamma(D, N, s)) \ll \left(\frac{3}{2}\right)^s N^{-1} sp(\log p)^{s+1}.$$

So by [14, Theorem 1] we get

$$C_s(E_{p^2}, N - d_{s-1}, D) \le 2^s N \max \Delta(\Gamma(D, N, s)) \ll 3^s sp(\log p)^{s+1}$$
$$\le Lp(3 \log p)^{L+1},$$

which leads to

$$L \gg N - p(3 \log p)^{L+2}$$

and the result follows. □

5 Lattice Tests

In this section we study the behavior of the sequence $(q_p(u))$, $u = 1, 2, \ldots$ under the lattice test.

We denote by

$$S((w_u), N, D) = \max\{s : \langle(w_{u+d_0} - w_{1+d_0}, \ldots, w_{u+d_{s-1}} - w_{1+d_{s-1}}),$$
$$1 \le u \le N\rangle = \mathbb{F}_p^s\}$$

the greatest dimension s such that (w_u) satisfies the s-dimensional N-lattice test for the lags $D = (d_0, \ldots, d_{s-1})$ with $0 \le d_0 < \cdots < d_{s-1} < p^2$.

As we mentioned before, in the case $d_i \not\equiv d_j \pmod{p}$ for all $0 \le i < j < s$, we can essentially proceed as in the proof of [21, Theorem 13].

Theorem 5. For $N \ge 2$ and $D = (d_0, d_1, \ldots, d_{s-1})$ with $0 \le d_0 < d_1 < \cdots < d_{s-1} < p^2$ such that no triple (d_l, d_h, d_t) satisfies $d_l \equiv d_h \equiv d_t \pmod{p}$, $0 \le l < h < t < s$, we have

$$S((q_p(u)), N, D) \ge \min\left\{\frac{p-1}{2}, \frac{N-p-1}{2}\right\}.$$

Proof. We assume that the sequence $(q_p(u))$ does not pass the s-dimensional N-lattice test for some lags $0 \le d_0 < d_1 < \ldots < d_{s-1} < p^2$. Put

$$\mathbf{w}_u = (q_p(u + d_0), q_p(u + d_1), \ldots, q_p(u + d_{s-1})), \quad \text{for } u = 1, \ldots, N,$$

and let V be the subspace of $\mathbb{F}_p^s$ spanned by all $\mathbf{w}_u - \mathbf{w}_1$ for $1 \le u \le N$. Let denote by $V^\perp = \{\mathbf{u} \in \mathbb{F}_p^s : \mathbf{u} \cdot \mathbf{v} = 0 \text{ for all } \mathbf{v} \in V\}$ the *orthogonal space* of V, where $\cdot$ denotes the usual inner product. Then $\dim(V) < s$ and $\dim(V^\perp) \ge 1$. Take $\mathbf{0} \ne \alpha \in V^\perp$, then

$$\alpha \cdot (\mathbf{w}_u - \mathbf{w}_1) = 0 \quad \text{for } 1 \le u \le N.$$

We denote

$$\delta = \alpha \cdot \mathbf{w}_u = \alpha \cdot \mathbf{w}_1 \quad \text{for } 1 \le u \le N.$$

If $\alpha = (\alpha_0, \alpha_1, \ldots, \alpha_{s-1})$, then let j be the smallest index with $\alpha_j \ne 0$ (so $0 \le j < s$). Then we get

$$\sum_{i=j}^{s-1} \alpha_i q_p(u + d_i) \equiv \delta \pmod{p} \quad \text{for } 1 \le u \le N. \tag{5}$$

Let $R = \min(p, N - p)$. We see from (5) that for $1 \le u \le R$ we have

$$\sum_{i=j}^{s-1} \alpha_i q_p(u + p + d_i) \equiv \delta \pmod{p}. \tag{6}$$

Recalling (4) and using (5) again, we now see that for any integer u with $u + d_i \not\equiv 0 \pmod{p}$, $i = j, \ldots, s - 1$, we have

$$\sum_{i=j}^{s-1} \alpha_i q_p(u + p + d_i) \equiv \sum_{i=j}^{s-1} \alpha_i \left(q_p(u + d_i) - (u + d_i)^{-1} \right)$$

$$\equiv \delta - \sum_{i=j}^{s-1} \alpha_i (u + d_i)^{-1} \pmod{p}. \tag{7}$$

Comparing (6) and (7) we see that

$$\sum_{i=j}^{s-1} \alpha_i (u + d_i)^{-1} \equiv 0 \pmod{p} \tag{8}$$

for at least $R - s + j$ values of u with

$$1 \le u \le R, \quad u + d_i \not\equiv 0 \pmod{p}, \quad i = j, \ldots, s - 1.$$

We consider first the case where $d_j \not\equiv d_h \pmod{p}$, for all $j < h < s$. Clearing the denominators of (8), we obtain a nontrivial polynomial congruence

$$\sum_{i=j}^{s-1} \alpha_i \prod_{\substack{e=j \\ e \ne i}}^{s-1} (u + d_e) \equiv 0 \pmod{p}$$

of degree $s - j - 1 \leq s$, which has at least $R - s + j$ solutions (to see that it is nontrivial it is enough to substitute $u \equiv -d_j \pmod{p}$ in the polynomial on the left hand side). Therefore $s - j - 1 \geq R - s + j$ and the result follows.

In the case $d_j \equiv d_h \pmod{p}$, for some $j < h < s$, taking $d_h = k_0 p + d_j$ for some $k_0 \geq 1$ and proceeding in the same way as above (but recalling that $u + d_j \equiv u + d_h \pmod{p}$), we get

$$(\alpha_j + \alpha_h) \prod_{\substack{e=j+1 \\ e \neq h}}^{s-1} (u + d_e) + (u + d_j) \sum_{\substack{i=j+1 \\ i \neq h}}^{s-1} \alpha_i \prod_{\substack{e=j \\ e \neq i}}^{s-1} (u + d_e) \equiv 0 \pmod{p}. \qquad (9)$$

If $\alpha_j + \alpha_h \not\equiv 0 \pmod{p}$ then the nontriviality of this polynomial equation is obvious again.

In the case of $\alpha_j + \alpha_h \equiv 0 \pmod{p}$, we have reduced the s-dimensional lattice test to the $(s - 2)$-dimensional one. If we are in a case where no two lags are equivalent or there are some equivalent lags $d_{j'}, d_{h'}$ with corresponding $\alpha_{j'} + \alpha_{h'} \not\equiv 0 \pmod{p}$ we easily see that (9) is nontrivial.

Hence, we are left with the case that there are only pairs d_i, $d_{h(i)} = d_i + k_i p$ of equivalent lags such that the sum of the corresponding coefficients $\alpha_i + \alpha_{h(i)}$ vanishes modulo p. However, in this case we get

$$\delta \equiv \sum_i (\alpha_i + \alpha_h(i)) q_p(u + d_i) + \sum_i \alpha_i k_i (u + d_i)^{-1} \equiv \sum_i \alpha_i k_i (u + d_i)^{-1}.$$

Since we can assume $\alpha_i \not\equiv 0 \pmod{p}$ for some i and the remaining d_i are pairwise distinct modulo p now, we have a nontrivial polynomial equation from which we obtain our result.   □

However, in the case when there exist three lags $d_l, d_h, d_t, 0 \leq l < h < t < s$, such that $d_l \equiv d_h \equiv d_t \pmod{p}$, the lattice test fails as the next result shows.

Theorem 6. *For $N \geq 2$ and $D = (d_0, d_1, \ldots, d_{s-1})$ with $0 \leq d_0 < d_1 < \cdots < d_{s-1} < p^2$ such that there exists a triple (d_l, d_h, d_t) satisfying $d_l \equiv d_h \equiv d_t \pmod{p}$, $0 \leq l < h < t < s$, we have*

$$S((q_p(u)), N, D) = 2.$$

Proof. To prove this result it is sufficient to consider the case $s = 3$ and to see that for $d_0 \equiv d_1 \equiv d_2 \pmod{p}$ the 3-dimensional test fails. For this let $\alpha = (\alpha_0, \alpha_1, \alpha_2)$ be an orthogonal vector on each $\mathbf{w}_u$, $u = 1, 2, \ldots$. Then we have

$$0 \equiv \alpha \cdot (\mathbf{w}_u - \mathbf{w}_1) \equiv \alpha \cdot \mathbf{w}_u$$
$$\equiv \alpha_0 q_p(u + d_0) + \alpha_1 q_p(u + d_1) + \alpha_2 q_p(u + d_2) \pmod{p}.$$

This congruence is trivially satisfied for all u with $u + d_0 \equiv 0 \pmod{p}$. For u with $u + d_0 \not\equiv 0 \pmod{p}$ we get

$$0 \equiv (\alpha_0 + \alpha_1 + \alpha_2) q_p(u + d_0) + \left(\alpha_1 \frac{d_1 - d_0}{p} + \alpha_2 \frac{d_2 - d_0}{p} \right) (u + d_0)^{-1} \pmod{p},$$

which gives the system of equations

$$\alpha_0 + \alpha_1 + \alpha_2 \equiv \alpha_1((d_1 - d_0)/p) + \alpha_2((d_2 - d_0)/p) \equiv 0 \pmod{p}.$$

It is clear that this system has a nontrivial solution α and then we easily verify that for all $u = 1, 2, \ldots$ we have

$$(\alpha_0, \alpha_1, \alpha_2) \cdot (\mathbf{w}_u - \mathbf{w}_1) = 0.$$

Hence, the orthogonal space is nontrivial and the lattice test is failed for $s = 3$, and thus for every $s > 3$. □

As in [20], the greatest dimension s such that (w_u) satisfies the s-dimensional N-lattice test for all lags $D = (d_0, \ldots, d_{s-1})$ is denoted by $S((w_u), N)$, i.e.,

$$S((w_u), N) = \max_D S((w_u), N, D) = \max \left\{ s : \forall\, 0 \le d_0 < \cdots < d_{s-1} < T : \right.$$
$$\left\langle \left(w_{u+d_0} - w_{1+d_0}, \ldots, w_{u+d_{s-1}} - w_{1+d_{s-1}} \right), 1 \le u \le N \right\rangle = \mathbb{F}_p^s \right\}.$$

Corollary 2. *For $N \ge 2$, we have*

$$S((q_p(u)), N) = 2.$$

Acknowledgement

This paper was written during a pleasant visit of Z.X.C. and A.O. to Linz. They wish to thank the Austrian Academy of Sciences for hospitality and support.

Z.X.C. was partially supported by the HAIXI Project of Fujian Province under grant 2008HX03 and the Natural Science Foundation of Fujian Province of China under grant 2007F3086. A.O. was partially supported by the Swiss National Science Foundation Grant 121874.

The authors thank Wilfried Meidl and Igor Shparlinski for useful discussions.

References

1. Brandstätter, N., Winterhof, A.: Linear complexity profile of binary sequences with small correlation measure. Periodica Mathematica Hungarica 52(2), 1–8 (2006)
2. Cassaigne, J., Mauduit, C., Sárközy, A.: On finite pseudorandom binary sequences, VII: the measures of pseudorandomness. Acta Arithmetica 103, 97–118 (2002)
3. Dorfer, G.: Lattice profile and linear complexity profile of pseudorandom number sequences. In: Mullen, G.L., Poli, A., Stichtenoth, H. (eds.) Fq7 2003. LNCS, vol. 2948, pp. 69–78. Springer, Heidelberg (2004)
4. Dorfer, G., Meidl, W., Winterhof, A.: Counting functions and expected values for the lattice profile at n. Finite Fields Appl. 10, 636–652 (2004)
5. Dorfer, G., Winterhof, A.: Lattice structure and linear complexity profile of nonlinear pseudorandom number generators. Appl. Algebra Engrg. Comm. Comput. 13, 499–508 (2003)

6. Dorfer, G., Winterhof, A.: Lattice structure of nonlinear pseudorandom number generators in parts of the period. In: Niederreiter, H. (ed.) Monte Carlo and Quasi-Monte Carlo Methods 2002, pp. 199–211. Springer, Berlin (2004)
7. Drmota, M., Tichy, R.: Sequences, Discrepancies and Applications. Springer, Berlin (1997)
8. Ernvall, R., Metsänkylä, T.: On the p-divisibility of Fermat quotients. Math. Comp. 66, 1353–1365 (1997)
9. Fu, F.-W., Niederreiter, H.: On the counting function of the lattice profile of periodic sequences. J. Complexity 23, 423–435 (2007)
10. Granville, A.: Some conjectures related to Fermat's Last Theorem. In: Number Theory W. de Gruyter, NY, pp. 177–192 (1990)
11. Heath-Brown, R.: An estimate for Heilbronn's exponential sum. In: Analytic Number Theory: Proc. Conf. in honor of Heini Halberstam, Birkhäuser, Boston, pp. 451–463 (1996)
12. Iwaniec, H., Kowalski, E.: Analytic number theory. American Mathematical Society Colloquium Publications, vol. 53. American Mathematical Society, Providence (2004)
13. Marsaglia, G.: The structure of linear congruential sequences. In: Zaremba, S.K. (ed.) Applications of Number Theory to Numerical Analysis, pp. 249–285. Academic Press, New York (1972)
14. Mauduit, C., Niederreiter, H., Sárközy, A.: On pseudorandom (0,1) and binary sequences. Publ. Math. Debrecen 71(3-4), 305–324 (2007)
15. Mauduit, C., Sárközy, A.: On finite pseudorandom binary sequences I: measures of pseudorandomness, the Legendre symbol. Acta Arith. 82, 365–377 (1997)
16. Moreno, C.J., Moreno, O.: Exponential sums and Goppa codes. I. Proc. Amer. Math. Soc. 111, 523–531 (1991)
17. Niederreiter, H.: Random Number Generation and Quasi-Monte Carlo Methods. SIAM, Philadelphia (1992)
18. Niederreiter, H., Rivat, J.: On the correlation of pseudorandom numbers generated by inversive methods. Monatshefte für Mathematik 153(3), 251–264 (2008)
19. Niederreiter, H., Winterhof, A.: Lattice structure and linear complexity of nonlinear pseudorandom numbers. Appl. Algebra Engrg. Comm. Comput. 13, 319–326 (2002)
20. Niederreiter, H., Winterhof, A.: On the structure of inversive pseudorandom number generators. In: Boztaş, S., Lu, H.-F. (eds.) AAECC 2007. LNCS, vol. 4851, pp. 208–216. Springer, Heidelberg (2007)
21. Ostafe, A., Shparlinski, I.E.: Pseudorandomness and dynamics of Fermat quotients, 1–26 (preprint) (2009)
22. Wang, L.-P., Niederreiter, H.: Successive minima profile, lattice profile, and joint linear complexity profile of pseudorandom multisequences. J. Complexity 24, 144–153 (2008)

Distribution of Boolean Functions
According to the Second-Order Nonlinearity*

Stéphanie Dib

Institut de Mathématiques de Luminy, Marseille, France

Abstract. The nonlinearity of a Boolean function is the minimum number of substitutions required in its truth table to change it into an affine function. Hence, in a cryptographic context, it is used to measure the strength of cryptosystems when facing linear attacks. As for the nonlinearity of order r of a Boolean function, which equals the least number of substitutions needed to change it into a function of degree at most r, it is examined when dealing with low-degree approximation attacks [7,14].

Many studies aimed at the distribution of Boolean functions according to the r-th order nonlinearity. Asymptotically, a lower bound is established in the higher order cases for almost all boolean functions, whereas a concentration point is shown in the (first order) nonlinearity case. We present a more accurate distribution by proving a concentration point in the second-order nonlinearity case.

Keywords: Boolean functions, nonlinearity, Reed-Muller code.

1   Introduction

We shall denote by $\mathcal{B}_n$ the set of all Boolean functions of n variables.
A Boolean function $f : \mathbb{F}_2^n \longrightarrow \mathbb{F}_2$ is often represented by its Algebraic Normal Form, that is the unique n-variable polynomial over $\mathbb{F}_2$ of the form

$$f(x_1, ..., x_n) = \sum_{u \in \mathbb{F}_2^n} a_u \Big(\prod_{i=1}^{n} x_i^{u_i} \Big),$$

where $c_u \in \mathbb{F}_2$.
Its degree, denoted by $deg(f)$, is called the algebraic degree of the function.
An affine function is a Boolean function that consists of a linear transformation over the vector space $\mathbb{F}_2^n$ followed by a translation, and thus a function whose algebraic degree is at most 1.
The Hamming weight $w_H(f)$ of a Boolean function f equals the cardinality of its support, which is the set $\{x \in \mathbb{F}_2^n \mid f(x) = 1\}$. The Hamming distance $d_H(f, g)$ between two functions f and g equals the cardinality of the set $\{x \in \mathbb{F}_2^n \mid f(x) \neq g(x)\}$.

---

* This work has been done with the support of the Région Provence-Alpes-Côte d'Azur.

M.A. Hasan and T. Helleseth (Eds.): WAIFI 2010, LNCS 6087, pp. 86–96, 2010.

The nonlinearity $NL(f)$ of a Boolean function f is its Hamming distance to the set of affine functions, that is

$$NL(f) = \min_{g \text{ affine}} d_H(f,g).$$

The nonlinearity of order r generalizes the usual nonlinearity. For a given function f, it is its Hamming distance to the set of all functions whose algebraic degrees do not exceed r. Namely, for every $0 \le r \le n$, the Reed-Muller code of order r can be considered as the set of all the vectors of values taken by each function f of degree at most r, when its argument x ranges over $\mathbb{F}_2^n$. Let $NL_r(f)$ denote the r-th order nonlinearity of f, we have

$$NL_r(f) = \min_{g \in \text{RM}(r,n)} d_H(f,g).$$

Let us briefly recall the main results about the nonlinearity of Boolean functions. The nonlinearity of a Boolean function f of n variables is bounded from above by $2^{n-1} - 2^{n/2-1}$. This upper bound is reached if n is even by the so-called bent functions. When n tends to infinity, it was shown by C. Carlet [2], D. Olejàr and M. Stanek [10], that almost all Boolean functions have nonlinearities greater than $2^{n-1} - c \cdot 2^{\frac{n}{2}-1}\sqrt{2n \ln 2}$, where c is a real number greater than 1. However, the distribution of Boolean functions between the lower and upper bound remained unknown until F. Rodier [11,12,13] proved that the nonlinearity of almost all Boolean functions lies in the neighbourhood of $2^{n-1} - 2^{n/2-1}\sqrt{2n \log 2}$. This result has been proven as well later by S. Litsyn and A. Shpunt with a different approach [8].

Before we present our result, we describe the distribution of Boolean functions in the second-order nonlinearity case. It has been shown, by C. Carlet and S. Mesnager [4], that the covering radius of $\text{RM}(2,n)$, which coincides with the maximum possible nonlinearity of order 2 of Boolean functions, is bounded from above by $2^{n-1} - \sqrt{15} \times 2^{n/2-1} + O(1)$. Asymptotically, C. Carlet [3] proved that almost all Boolean functions have high r-th order nonlinearities. In fact, he showed that the density of the set of functions satisfying

$$NL_r(f) > 2^{n-1} - c \ \sqrt{\binom{n}{r} \ln 2} \ 2^{\frac{n-1}{2}},$$

tends to 1 when n tends to infinity, if c is a real number greater than 1.

In what follows, we will focus on the distribution of Boolean functions for large n in the case of the second-order nonlinearity. By applying mainly fundamental combinatorics and probability theory, we were enabled to establish a concentration point for almost all Boolean functions, whereas Rodier's approach involves harmonic analysis. Our method is indeed convenient in the first order nonlinearity case.

To attain this goal, we investigate the subset of Boolean functions whose second-nonlinearities are greater than $\delta = 2^{n-1} - c \ \sqrt{\binom{n}{2} \ln 2} \ 2^{\frac{n-1}{2}}$ where c is a strictly positive real number less than 1, δ being slightly greater than the lower bound given for almost all Boolean functions. By definition, functions whose second-nonlinearities are greater than a given δ are those for which the Hamming distance to $RM(2, n)$ exceeds δ. A geometrical viewpoint is the way to go. Therefore, we consider the Hamming closed balls of radius δ centered at all codewords of $RM(2, n)$, defined by $B_\delta[g] = \{f \in \mathcal{B}_n \mid d_H(f, g) \leq \delta\}$, where g ranges over $RM(2, n)$. Notice that such functions are strictly those for which the indicator function of any of the above-mentioned Hamming balls equals 0, thus the sum of all indicator functions of these balls as well. Then, we assign probabilities to the subsets of $\mathcal{B}_n$ of equiprobable Boolean functions. Hence, the problem of determining the probability of the event $\{NL_2(f) > \delta\}$ is equivalent to that of the event $\{\eta = 0\}$, where $\eta = \sum_{g \in RM(2,n)} 1_{B_\delta[g]}$ is an integer-valued function defined on $\mathcal{B}_n$. This is followed by application of the Chebyshev's inequality [1, chapter IV]

$$P(\eta = 0) \leq P\big(|\eta - E(\eta)| \geq E(\eta)\big) \leq \frac{E(\eta^2) - E(\eta)^2}{E(\eta)^2}.$$

This upper bound allows us to prove a concentration point around $2^{n-1} - \sqrt{\binom{n}{2} \ln 2}\, 2^{\frac{n-1}{2}}$, when n tends to infinity.
This work was done with François Rodier's guidance.

2   Expected Value of the Random Variable η

In this section, we obtain a lower bound of the expected value of the random variable η. Note that the mean of any random variable $1_{B_g[\delta]}$, which is equal to the probability of $B_g[\delta]$, does not depend on the choice of g for a simple translation reason. Thus, $E(\eta)$ equals to $2^k P(B_0[\delta])$, where $k = \sum_{j=0}^{2} \binom{n}{j}$ is the dimension of $RM(2, n)$.

We present in the following proposition a lower bound of the density of the subset $B_0[\delta]$. But in order to avoid later interruptions, we introduce first a function of great importance that will occur very often in our proof . The function is defined by

$$\mathcal{H}(y) = \int_{-\infty}^{y} e^{-u^2} du$$

and the double inequality

$$(1 - \frac{1}{2y^2}) \frac{e^{-y^2}}{-2y} < \mathcal{H}(y) < \frac{e^{-y^2}}{-2y} \tag{1}$$

holds for every $y < 0$ [5, page 175].

Proposition 1. *Let* $\rho = 2^{n-1} + \alpha(n) 2^{\frac{n-1}{2}}$, *where* α *is a function of* n *such that* $-\frac{\alpha(n)}{n}$ *tends to a positive constant. The density in* $\mathcal{B}_n$ *of the subset* $B_0[\rho]$ *greater than*

$$\frac{1}{\sqrt{\pi}}\mathcal{H}\big(\alpha(n)\big)\big(1+o(1)\big)$$

when n tends to infinity.

Proof: The number of Boolean functions f whose Hamming distance to 0 is bounded from above by some number ρ equals

$$\sum_{0\leq i\leq\rho}\binom{2^n}{i}$$

In order to obtain a lower bound of the binomial coefficient, we follow Feller's [5, chapter VII] method for the normal approximation to the binomial distribution and do likewise. Let q be a positive integer, $0\leq p\leq q$. We shall take $q=2\nu$ even to avoid indefiniteness. We denote by a_t the probability of getting exactly $\nu+t$ successes in q trials with success probability equals $\frac{1}{2}$, where t runs from $-\nu$ to ν. In other words, we consider the terms of the symmetric binomial distribution by their distance to the central term which is a_0. Since $a_{-t}=a_t$, we shall consider only $t\geq 0$.

If ν sufficiently large and t is restricted to values $0<t<T_q$ such that $T_q^3/\nu^2\to 0$, we have from Feller's book [5, page 180]

$$a_t>a_0e^{-\frac{t^2}{\nu}}e^{-\frac{T_q^3}{\nu^2}}.$$

Under these circumstances,

$$a_t>a_0e^{-\frac{t^2}{\nu}}\Big(1-\frac{T_q^3}{\nu^2}\Big).$$

When the binomial coefficient is expressed using the double inequality derived from Stirling's formula bounding the factorial, we get

$$a_0>\frac{1}{\sqrt{\pi\nu}}\big(1+o(1)\big).$$

Adding the last two inequalities and by definiton of a_t, we obtain

$$\binom{q}{p}>\frac{2^q e^{-\frac{(p-q/2)^2}{q/2}}}{\sqrt{\pi\frac{q}{2}}}\big(1+o(1)\big)$$

provided that $\nu\to\infty$ and $0<p-q/2<T_q$ such that $\frac{T_q^3}{(q/2)^2}\to 0$. It is valid as well when $0<q-p-q/2<T_q$ because of symmetry, thus when $q/2-T_q<p<q/2+T_q$.

Returning to the summation of the binomial coefficients $\binom{2^n}{i}$ over all integers between 0 and ρ, the relation above holds when $n\to\infty$, for values of i greater than $2^{n-1}-T_{2^n}$ for some T_{2^n} such that $\frac{T_{2^n}^3}{2^{2(n-1)}}\to 0$, say $\frac{T_{2^n}^3}{2^{2(n-1)}}=\frac{1}{n}$. We don't need to concern ourselves with other values of i since we are looking for a lower bound, and adding that the exponential function involved is increasing, we have therefore

$$\sum_{0 \le i \le \rho} \binom{2^n}{i} > \frac{2^{2^n}}{\sqrt{\pi 2^{n-1}}} (1 + o(1)) \int_{2^{n-1}-T_{2^n}-1}^{\rho} e^{-\frac{(i-2^{n-1})^2}{2^{n-1}}} di$$

$$> \frac{2^{2^n}}{\sqrt{\pi 2^{n-1}}} (1 + o(1)) \left(\int_0^{\rho} e^{-\frac{(i-2^{n-1})^2}{2^{n-1}}} di - \int_0^{2^{n-1}-T_{2^n}-1} e^{-\frac{(i-2^{n-1})^2}{2^{n-1}}} di \right)$$

$$> \frac{2^{2^n}}{\sqrt{\pi}} (1 + o(1)) \left(\int_{-\infty}^{\alpha(n)} e^{-u^2} du - \int_{-\infty}^{-\frac{T_{2^n}-1}{\sqrt{2^{n-1}}}} e^{-u^2} du \right).$$

Noticing that $-\frac{T_{2^n}-1}{\sqrt{2^{n-1}}}$ is much less than $\alpha(n)$, the proposition is complete using (1).

3   Expected Value of η^2

We recall that $\eta = \sum_{g \in RM(2,n)} 1_{B_\delta[g]}$, $\delta = 2^{n-1} - c \sqrt{\binom{n}{2} \ln 2} \, 2^{\frac{n-1}{2}}$, where $0 < c < 1$ and k is the dimension of $RM(2,n)$. We put $\alpha = \frac{\delta - 2^{n-1}}{\sqrt{2^{n-1}}}$.
The random variable η^2, defined on the set $\mathcal{B}_n$ of all Boolean functions, is equal to

$$\eta + \sum_{\substack{(g_1,g_2) \in RM(2,n) \times RM(2,n) \\ g_1 \ne g_2}} 1_{B_\delta[g_1] \cap B_\delta[g_2]}$$

3.1   Number of Points in the Intersection of Two Balls

We need to establish an upper bound of the expected value of η^2. For this purpose, we will deal first with the probability of intersection of two Hamming balls of radius δ, centered at g_1 and g_2. Note that the number of points in the intersection of these balls depends only on the Hamming distance between their centers. Thus, we can always suppose without loss of generality that the centers are 0 and g such that $w_H(g) = d_H(g_1, g_2)$. This means that we have to consider the weight distribution of the second-order Reed-Muller code.
Let A_w be the number of Boolean functions of weight w in $RM(2,n)$. Then $A_w = 0$ unless $w = 2^{n-1}$ or $w = 2^{n-1}(1 \pm 2^{-h})$ for some h, $0 \le h \le \lfloor \frac{n}{2} \rfloor$. We have [9, Chapter XV], [6]

$$A_{2^{n-1} \pm 2^{n-1-h}} = 2^{2h(n-h)+h} \frac{\prod_{i=0}^{2h-1}(1 - 2^{i-n})}{\prod_{i=1}^{h}(1 - 4^{-i})}$$

where $\frac{\prod_{i=0}^{2h-1}(1-2^{i-n})}{\prod_{i=1}^{h}(1-4^{-i})} < \frac{3}{2}$, and $A_{2^{n-1}} = 2^{1+n+\binom{n}{2}} - \sum_{w \ne 2^{n-1}} A_w$.

Accordingly, the expected value of η^2 is equal to

$$E(\eta) + 2^k \sum_{w \ne 0} A_w \cdot P(B_\delta[0] \cap B_\delta[g]).$$

We emphasize that the sum just mentioned is taken over all weights in $RM(2, n)$ and that g is any codeword of weight w.

Proposition 2. *Let g be a Boolean function in n variables of algebraic degree at most 2, and w its Hamming weight. The density in $\mathcal{B}_n$ of the subset $B_\delta[0] \cap B_\delta[g]$ is less than*

$$\frac{1}{\pi}(1 + o(1)) \int_{-\infty}^{\alpha} \int_{-\infty}^{\alpha} e^{-\frac{u^2 + v^2 + 2uvt}{1 - t^2}} \, du \, dv$$

where t is such that $w = 2^{n-1}(1 + t)$ and n tends to infinity.

Proof: We start by considering the intersection of two Hamming spheres $S_{\delta_1}[0]$ et $S_{\delta_2}[g]$, centered at 0 and g, of radius δ_1 and δ_2, where δ_1 and δ_2 take values from $0, ..., \delta$. Let f be a function that belongs to this intersection. Due to this belonging, f has $\frac{\delta_1 - \delta_2 + w}{2}$ elements of its support for which g equals 1 and $\frac{\delta_1 + \delta_2 - w}{2}$ elements of its support for which g equals 0. Thus the size of the subset $S_{\delta_1}[0] \cap S_{\delta_2}[g]$ is

$$\binom{w}{\frac{\delta_1 - \delta_2 + w}{2}} \binom{2^n - w}{\frac{\delta_1 + \delta_2 - w}{2}}$$

As a result, the size of the subset $B_\delta[0] \cap B_\delta[g]$ is

$$\sum_{(\delta_1, \delta_2)} \binom{w}{\frac{\delta_1 - \delta_2 + w}{2}} \binom{2^n - w}{\frac{\delta_1 + \delta_2 - w}{2}}$$

where (δ_1, δ_2) ranges over $D = \{(\delta_1, \delta_2) \mid 0 \leq \delta_1 \leq \delta, 0 \leq \delta_2 \leq \delta, 0 \leq \frac{\delta_1 - \delta_2 + w}{2} \leq w, 0 \leq \frac{\delta_1 + \delta_2 - w}{2} \leq 2^n - w\}$, and have same parity since the Hamming weight w of $g \in RM(2, n)$ is even.

As in the preceding section, we can prove that

$$\binom{q}{p} < \frac{2^q e^{-\frac{(p - q/2)^2}{q/2}}}{\sqrt{\pi \frac{q}{2}}} (1 + o(1))$$

for $\frac{(p - \frac{q}{2})^3}{(q/2)^2} \to 0$ and $q \to \infty$.

One can prove that the sum taken over all (δ_1, δ_2) that fail these conditions is negligible. Also, it can be shown that $\sum_{(\delta_1, \delta_2)} \binom{w}{\frac{\delta_1 - \delta_2 + w}{2}} \binom{2^n - w}{\frac{\delta_1 + \delta_2 - w}{2}}$ is less than

$$\frac{2^{2^n}}{\pi \sqrt{w(2^n - w)}} (1 + o(1)) \int_0^\delta \int_{\delta_2 - w}^\delta e^{-\frac{(\delta_1 - w)^2 + (\delta_2 - w)^2}{2^{1-n} w(2^n - w)} - \frac{w(2^n - 2w)(2^n w - 2\delta_1 \delta_2)}{2w(2^n - w)}} \, d\delta_1 \, d\delta_2.$$

The following change of variables $u = \frac{\delta_1 - 2^{n-1}}{\sqrt{2^{n-1}}}, v = \frac{\delta_2 - 2^{n-1}}{\sqrt{2^{n-1}}}$ and $w = 2^{n-1}(1 + t)$ leads us to this upper bound

$$\frac{2^{2^n}}{\pi} (1 + o(1)) \int_{-\infty}^{\alpha} \int_{-\infty}^{\alpha} e^{-\frac{u^2 + v^2 + 2uvt}{1 - t^2}} \, du \, dv$$

when n tends to infinity. And that ends the proof.

Let us see what this proposition turns into when codewords have weights close to 2^{n-1} Recall that t is either equal to $\pm 2^{-h}$ for some h such that $0 \le h \le \lfloor \frac{n}{2} \rfloor$ or to 0. We shall consider $t \le 0$, the other case being similar. In this case, we have

$$-(u^2 + v^2 + 2uvt) \le -(1+t)(u^2 + v^2).$$

Thus,

$$\int_{-\infty}^{\alpha} \int_{-\infty}^{\alpha} e^{-\frac{u^2+v^2+2uvt}{1-t^2}} \, du \, dv \le \left(\int_{-\infty}^{\alpha} e^{-\frac{u^2}{1-t}} \, du \right)^2 = \left(\sqrt{1-t} \mathcal{H}(\frac{\alpha}{\sqrt{1-t}}) \right)^2 < \left((1-t) \frac{e^{-\alpha^2/(1-t)}}{-2\alpha} \right)^2.$$

For $\alpha^2 t$ sufficiently small, we have

$$e^{-\alpha^2/(1-t)} = e^{-\alpha^2} - \alpha^2 t e^{-\alpha^2} + O(\alpha^4 t^2) e^{-\alpha^2}.$$

Using the left-hand inequality of (1) we find

$$\int_{-\infty}^{\alpha} e^{-\frac{u^2}{1-t}} \, du < \mathcal{H}(\alpha) - x^2 t \mathcal{H}(\alpha) + O(\alpha^4 t^2) \mathcal{H}(\alpha).$$

We have thus proved that

$$\int_{-\infty}^{\alpha} \int_{-\infty}^{\alpha} e^{-\frac{u^2+v^2+2uvt}{1-t^2}} \, du \, dv < \mathcal{H}(\alpha)^2 \left(1 + o(1) \right), \qquad (2)$$

provided that $\alpha^2 t$ is small. We shall use this relation when $t = \pm 2^{-h}$ so that $2^{-h} \le \frac{1}{\sqrt[3]{3}}$, thus for weights corresponding to $h \ge 3 \log_2 n$ and for $t = 0$.
This calls for splitting the sum $\sum_{w \ne 0} A_w \cdot P(B_\delta[0] \cap B_\delta[g])$ taken over w into two parts according to whether the weight is in the interval of length $\frac{2}{n^3}$ centered at 2^{n-1} or not.
Let g_0 be any codeword of weight 2^{n-1}. We get the following:

$$S_1(w) = \sum_{\substack{w \\ 3\log_2 n \le h \le \lfloor \frac{n}{2} \rfloor}} A_w \cdot P(B_\delta[0] \cap B_\delta[g]) + A_{2^{n-1}} \cdot P(B_\delta[0] \cap B_\delta[g_0])$$

and

$$S_2(w) = \sum_{\substack{w \ne 0 \\ 0 \le h < 3\log_2 n}} A_w \cdot P(B_\delta[0] \cap B_\delta[g]).$$

3.2   The Sum over the Weights Close to 2^{n-1}: The Main Part

We need to determine an upper bound of $S_1(w)$. In view of (2), we have

$$S_1(w) < \sum_{\substack{w \\ 3\log_2 n \le h \le \lfloor \frac{n}{2} \rfloor}} A_w \frac{\mathcal{H}(\alpha)^2}{\pi} \left(1 + o(1) \right) + A_{2^{n-1}} \frac{\mathcal{H}(\alpha)^2}{\pi} \left(1 + o(1) \right)$$

$$< \left(2^{1+n+\binom{n}{2}} - \sum_{\substack{w \\ 0 \le h < 3\log_2 n}} A_w \right) \frac{\mathcal{H}(\alpha)^2}{\pi} \left(1 + o(1) \right).$$

What we are about to find out is that almost all codewords of the second-order Reed-Muller code have weights setting in a small neighbourhood of 2^{n-1}. Knowing the weight distribution of the code, this property can be easily found. We proceed as follows:

$$\sum_{\substack{w \\ 0 \le h < 3 \log_2 n}} A_w < \frac{3}{2} \sum_{0 \le h < 3 \log_2 n} e^{-2 \ln 2 \left((h - \frac{2^{n+1}}{4})^2 - \frac{(2n+1)^2}{16} \right)}$$

$$< \frac{3}{2} e^{\frac{(2n+1)^2}{8}} \ln 2 \int_0^{3 \log_2 n + 1} e^{-2 \ln 2 (h - \frac{2n+1}{4})^2} dh$$

$$< \frac{3}{2\sqrt{2 \ln 2}} e^{\frac{(2n+1)^2}{8}} \ln 2 \, \mathcal{H}\left(\sqrt{2 \ln 2}(3 \log_2 n - \frac{2n+1}{4} + 1) \right)$$

$$< \frac{3}{2\sqrt{2 \ln 2}} e^{\frac{(2n+1)^2}{8}} \ln 2 \, \frac{e^{-2 \ln 2 (3 \log_2 n - \frac{2n+1}{4} + 1)^2}}{-2\sqrt{2 \ln 2}(3 \log_2 n - \frac{2n+1}{4} + 1)}$$

$$< e^{3 \ln 2 (2n+1) \log_2 n}$$

$$= o(2^{1+n+\binom{n}{2}}),$$

which is the required result.
Finally,

$$\mathcal{S}_1(w) < 2^{1+n+\binom{n}{2}} \frac{\mathcal{H}(\alpha)^2}{\pi} \big(1 + o(1)\big).$$

3.3   The Sum over the Weights Far from 2^{n-1}

One can see that the number of intersections between two Hamming balls increases when they get closer and decreases otherwise. This yields

$$\sum_{\substack{w \ne 0 \\ 0 \le h < 3 \log_2 n}} A_w \cdot P(B_\delta[0] \cap B_\delta[g]) \le P(B_\delta[0] \cap B_\delta[g_1]) \sum_{\substack{w \\ 0 \le h < 3 \log_2 n}} A_w$$

where g_1 has the smallest possible non-zero weight, that is to say $2^{n-1}(1 - \frac{1}{2})$. According to proposition 2,

$$P(B_\delta[0] \cap B_\delta[g_1]) < \frac{1}{\pi}\big(1 + o(1)\big) \int_{-\infty}^{\alpha} \int_{-\infty}^{\alpha} e^{-\frac{u^2 + v^2 - uv}{3/4}} du\, dv.$$

We have

$$\int_{-\infty}^{\alpha} \int_{-\infty}^{\alpha} e^{-\frac{u^2 + v^2 - uv}{3/4}} du\, dv = \frac{\sqrt{3}}{2} \int_{-\infty}^{\alpha} e^{-v^2} \int_{-\infty}^{\frac{2}{\sqrt{3}}(\alpha - \frac{v}{2})} e^{-u^2} du\, dv.$$

Since the sign of $\alpha - \frac{v}{2}$ varies with v, a separation is in order

$$\int_{-\infty}^{\alpha} \int_{-\infty}^{\alpha} e^{-\frac{u^2 + v^2 - uv}{3/4}} du\, dv < \frac{\sqrt{3}}{2} \int_{\frac{3\alpha}{2}}^{\alpha} e^{-v^2} \int_{-\infty}^{\frac{2}{\sqrt{3}}(\alpha - \frac{v}{2})} e^{-u^2} du\, dv + \frac{\sqrt{3}}{2} \int_{-\infty}^{\frac{3\alpha}{2}} e^{-v^2} dv \int_{-\infty}^{+\infty} e^{-u^2} du.$$

Using (1), we get

$$\int_{-\infty}^{\alpha} \int_{-\infty}^{\alpha} e^{-\frac{u^2+v^2-uv}{3/4}} du dv < \frac{\sqrt{3}}{2} \int_{\frac{3\alpha}{2}}^{\alpha} e^{-v^2} \frac{e^{-\left(\frac{2}{\sqrt{3}}(\alpha-\frac{v}{2})\right)^2}}{\frac{2}{\sqrt{3}}(v-2\alpha)} dv + \frac{\sqrt{3\pi}}{2} \frac{e^{-\frac{9}{4}\alpha^2}}{-3\alpha}.$$

Rearranging gives

$$\int_{-\infty}^{\alpha} \int_{-\infty}^{\alpha} e^{-\frac{u^2+v^2-uv}{3/4}} du dv < \frac{3}{4} \int_{\frac{3\alpha}{2}}^{\alpha} \frac{e^{\left(-\alpha^2-\left(\frac{2}{\sqrt{3}}(v-\frac{x}{2})\right)^2\right)}}{\frac{3}{2}\alpha - 2\alpha} dv + \frac{\sqrt{3\pi}}{2} \frac{e^{-\frac{9}{4}\alpha^2}}{-3\alpha}.$$

A little manipulation with the use of (1) yields

$$\int_{-\infty}^{\alpha} \int_{-\infty}^{\alpha} e^{-\frac{u^2+v^2-uv}{3/4}} du dv < \frac{3\sqrt{3}}{4\alpha^2} e^{-\frac{4}{3}\alpha^2} + \frac{\sqrt{3\pi}}{2} \frac{e^{-\frac{9}{4}\alpha^2}}{-3\alpha}$$
$$< e^{-\frac{4}{3}\alpha^2}\left(1 + o(1)\right).$$

Putting all together, we have

$$S_2(w) < \frac{1}{\pi} e^{-\frac{4}{3}\alpha^2 + 3\ln 2(2n+1)\log_2 n}\left(1 + o(1)\right).$$

Our final step to the evaluation of the mean of η^2 is to show that $S_2(w)$ is too weak to participate in the related sum. We have

$$S_2(w) \cdot 2^{1+n+\binom{n}{2}} \frac{\mathcal{H}(\alpha)^2}{\pi})^{-1} < e^{-\frac{4}{3}\alpha^2 + 3\ln 2(2n+1)\log_2 n} \cdot e^{-\ln 2\left(1+n+\binom{n}{2}\right)} \cdot 4\alpha^2 \cdot e^{2\alpha^2}\left(1+o(1)\right),$$

which tends to 0 if $|\alpha| < n\frac{\sqrt{3}}{2}\sqrt{\ln 2}$ and this condition is satisfied given that $c < 1$. This gives the necessary result.

4   Final Result

Theorem 1. Let $\delta = 2^{n-1} - c\sqrt{\binom{n}{2}\ln 2}\, 2^{\frac{n-1}{2}}$, where $0 < c < 1$. The density of the subset $\{f \in \mathcal{B}_n | NL_2(f) > \delta\}$ tends to 0 when n tends to infinity.

Proof: We have

$$P\left(NL_2(f) > \delta\right) \leq \frac{E(\eta^2) - E(\eta)^2}{E(\eta)^2}$$
$$\leq \frac{2^{2k}\frac{\mathcal{H}(\alpha)^2}{\pi}\left(1 + o(1)\right) + E(\eta) - \left(2^k \frac{\mathcal{H}(\alpha)}{\sqrt{\pi}}\left(1 + o(1)\right)\right)^2}{E(\eta)^2}$$
$$\leq \frac{1}{E(\eta)}\left(1 + o(1)\right),$$

and

$$E(\eta) > 2^k \frac{\mathcal{H}(\alpha)}{\sqrt{\pi}}\left(1 + o(1)\right).$$

Since $0 < c < 1$,

$$2^k \frac{\mathcal{H}(\alpha)}{\sqrt{\pi}} \left(1 + o(1)\right) \sim 2^k \frac{1}{\sqrt{\pi}} \frac{e^{-\binom{n}{2} \ln 2}}{2\sqrt{\binom{n}{2} \ln 2}} \frac{e^{\binom{n}{2} \ln 2 (1 - c^2)}}{c}$$

tends to infinity with n. This concludes the proof.

5   Conclusion

We have proved that $2^{n-1} - \sqrt{\binom{n}{2} \ln 2} \, 2^{\frac{n-1}{2}}$ represents a "concentration" point of the second-order nonlinearity of Boolean functions, when n tends to infinity. Unfortunately, the situation gets critical when dealing with Reed-Muller codes of higher orders because of the little knowledge we have about their weight distribution. Nevertheless, one can prove that almost all codewords have weights lying in a small neighborhood of 2^{n-1} as pointed out by Carlet. But the problem arises regarding the sum taken over the weights far from 2^{n-1} that we are unable yet to enhance in order to give the needed precisions.

References

1. Alon Noga, N., Spencer, J.: The probabilistic method. With an appendix on the life and work of Paul Erdös, 3rd edn. Wiley-Interscience Series in Discrete Mathematics and Optimization. John Wiley & Sons, Inc., Hoboken (2008)
2. Carlet, C.: On cryptographic complexity of Boolean functions. In: Mullen, G.L., Stichtenoth, H., Tapia-Recillas, H. (eds.) Proceedings of the Sixth Conference on Finite Fields with Applications to Coding Theory, Cryptography and Related Areas, pp. 53–69. Springer, Heidelberg (2002)
3. Carlet, C.: The complexity of Boolean functions from cryptographic viewpoint (2006), http://drops.dagstuhl.de/volltexte/2006/604
4. Carlet, C., Mesnager, S.: Improving the upper bounds on the covering radii of binary Reed-Muller codes. IEEE Trans. Inform. Theory 53(1), 162–173 (2007)
5. Feller, W.: An introduction to probability theory and its applications, 3rd edn., vol. I. John Wiley & Sons, Inc., New York (1968)
6. Helleseth, T., Klove, T., Levenshtein, V.: Error-correction capability of binary linear codes. IEEE Trans. Inform. Theory 51(4), 1408–1423 (2005)
7. Knudsen, L., Robshaw, M.: Non-linear approximations in linear cryptanalysis. In: Maurer, U.M. (ed.) EUROCRYPT 1996. LNCS, vol. 1070, pp. 224–236. Springer, Heidelberg (1996)
8. Litsyn, S., Shpunt, A.: On the distribution of Boolean function nonlinearity. SIAM J. Discrete Math. 23(1), 79–95 (2008/2009)
9. MacWilliams, F.J., Sloane, N.J.A.: The theory of error-correcting codes. I. North-Holland Mathematical Library, vol. 16. North-Holland Publishing Co., Amsterdam (1977)
10. Olejár, D., Stanek, M.: On cryptographic properties of random Boolean functions. J. UCS 4(8), 705–717 (1998)

11. Rodier, F.: Sur la non-linéarité des fonctions booléennes. In: Acta Arithmetica, vol. _15, pp. 1–22 (2004); prétirage de l'IML n° 2002-07, disponible sur ArXiv: math.NT/0306395
12. Rodier, F.: On the nonlinearity of Boolean functions. In: de Augot, D., Charpin, P., Kabatianski, G. (eds.) Proceedings of WCC 2003, Workshop on coding and cryptography 2003, sous la direction. INRIA, pp. 397–405 (2003)
13. Rodier, F.: Asymptotic nonlinearity of Boolean functions. Designs, Codes and Cryptography 40, 1 (2006)
14. Shimoyama 1, T., Kaneko, T.: Quadratic Relation of S-box and Its Application to the Linear Attack of Full Round DES. In: Krawczyk, H. (ed.) CRYPTO 1998. LNCS, vol. 1462, pp. 129–147. Springer, Heidelberg (1998)

Hyper-bent Boolean Functions with Multiple Trace Terms

Sihem Mesnager

LAGA (Laboratoire Analyse, Géometrie et Applications), UMR 7539, CNRS,
Department of Mathematics, University of Paris XIII and University of Paris VIII, 2
rue de la liberté, 93526 Saint-Denis Cedex, France
mesnager@math.jussieu.fr

Abstract. Bent functions are maximally nonlinear Boolean functions with an even number of variables. These combinatorial objects, with fascinating properties, are rare. The class of bent functions contains a subclass of functions the so-called hyper-bent functions whose properties are still stronger and whose elements are still rarer. In fact, hyper-bent functions seem still more difficult to generate at random than bent functions and many problems related to the class of hyper-bent functions remain open. (Hyper)-bent functions are not classified. A complete classification of these functions is elusive and looks hopeless.

In this paper, we contribute to the knowledge of the class of hyper-bent functions on finite fields $\mathbb{F}_{2^n}$ (where n is even) by studying a subclass $\mathfrak{F}_n$ of the so-called Partial Spreads class PS^- (such functions are not yet classified, even in the monomial case). Functions of $\mathfrak{F}_n$ have a general form with multiple trace terms. We describe the hyper-bent functions of $\mathfrak{F}_n$ and we show that the bentness of those functions is related to the Dickson polynomials. In particular, the link between the Dillon monomial hyper-bent functions of $\mathfrak{F}_n$ and the zeros of some Kloosterman sums has been generalized to a link between hyper-bent functions of $\mathfrak{F}_n$ and some exponential sums where Dickson polynomials are involved. Moreover, we provide a possibly new infinite family of hyper-bent functions. Our study extends recent works of the author and is a complement of a recent work of Charpin and Gong on this topic.

Keywords: Boolean function, Bent functions, Hyper-bent functions, Maximum nonlinearity, Walsh-Hadamard transformation, Kloosterman sums, Cubic sums, Dickson polynomials.

1 Introduction

Bent functions are those Boolean functions whose Hamming distance to the set of all affine functions equals $2^{n-1} \pm 2^{\frac{n}{2}-1}$ (where the number n of variables is even). They were introduced by Rothaus [18] and have attracted a lot of research, specially in the last 15 years for their own sake as interesting combinatorial objects but also because of their applications in cryptography (design of stream ciphers) and their relations to coding theory. Despite their simple and natural definition,

M.A. Hasan and T. Helleseth (Eds.): WAIFI 2010, LNCS 6087, pp. 97–113, 2010.
© Springer-Verlag Berlin Heidelberg 2010

bent functions have turned out to admit a very complicated structure in general. Currently, some algebraic properties of bent functions are well known but the general structure of bent functions on $\mathbb{F}_{2^n}$ is not yet clear. In particular a complete classification of bent functions is elusive and looks hopeless. On the other hand many special explicit constructions are known. Some infinite classes of bent functions have been obtained, thanks to the identification between the vectorspace $\mathbb{F}_2^n$ and the Galois field $\mathbb{F}_{2^n}$. A non exhaustive list of references devoted to the description of classes of bent functions, expressed by means of trace-functions is [1, 5, 7–11, 13–16, 20]. Current results on the known properties and general constructions of bent functions can be found in [2] (pages 77-109). The class of bent functions contains a subclass of functions introduced by Youssef and Gong in [19], the so-called *hyper-bent functions*, those Boolean functions over $\mathbb{F}_{2^n}$ (n even) whose Hamming distances to all functions $Tr_1^n(ax^i) \oplus \epsilon$ ($a \in \mathbb{F}_{2^n}$, $\epsilon \in \mathbb{F}_2$) where Tr_1^n is the trace function from $\mathbb{F}_{2^n}$ to $\mathbb{F}_2$ and where i is co-prime with $2^n - 1$, equals $2^{n-1} \pm 2^{\frac{n}{2}-1}$. The classification of hyperbent functions and many related problems remain open. In particular, it seems difficult to define precisely an infinite class of hyperbent functions, as indicated by the number of open problems proposed by Charpin and Gong in [5].

In [5] the authors have studied the bentness of the class of Boolean functions f defined on $\mathbb{F}_{2^n}$ by $f(x) := \sum_{r \in R} Tr_1^n(\beta_r x^{r(2^m-1)})$, $\beta_r \in \mathbb{F}_{2^n}$, where $n := 2m$ and R is a subset of a set of representatives of the cyclotomic cosets modulo $2^m + 1$ for which each coset has the full size $n = 2m$. When r is co-prime with $2^m + 1$, the functions f are the sums of several Dillon monomial functions. A new tool by means of Dickson polynomials to describe hyper-bent functions f has been introduced in [5]. In fact, Charpin and Gong have shown that the bentness of those functions is related to the Dickson polynomials under some restriction on the coefficients β_r. Thanks to this new approach, a characterization of a new class of binomial hyper-bent functions has been given: $Tr_1^n \left(a \left(x^{(2^r-1)(2^m-1)} + x^{(2^r+1)(2^m-1)} \right) \right)$, where $a \in \mathbb{F}_{2^m}^*$ and r is an integer such that $0 < r < m, \{2^r - 1, 2^r + 1\} \subset R$. Continuing their interesting approach, Gologlu [12] has proved recently that the following functions defined on $\mathbb{F}_{2^n}$ ($n = 2m$), are hyper-bent:

- $f(x) := \sum_{i=1}^{2^{m-1}-1} Tr_1^n \left(\beta x^{i(2^m-1)} \right)$; $\beta \in \mathbb{F}_{2^m} \setminus \mathbb{F}_2$.
- $f(x) := \sum_{i=1}^{2^{m-2}-1} Tr_1^n \left(\beta x^{i(2^m-1)} \right)$; m odd and $\beta^{(2^m-4)^{-1}} \in \{x \in \mathbb{F}_{2^m}^* \mid Tr_1^r(x) = 0\}$.

Recently, two new infinite families of hyper-bent Boolean functions in polynomial forms defined on $\mathbb{F}_{2^n}$ ($n = 2m$) have been exhibited and studied in [15, 17]:

- $f_{a,t}(x) := Tr_1^n(ax^{r(2^m-1)}) + Tr_1^2(bx^{\frac{2^n-1}{3}})$; m odd, $\gcd(r, 2^m+1) = 1$, $a \in \mathbb{F}_{2^n}^*$ and $b \in \mathbb{F}_4^*$ ([15]).
- $g_{a,t}(x) := Tr_1^n(ax^{3(2^m-1)}) + Tr_1^2(bx^{\frac{2^n-1}{3}})$; m odd, $a \in \mathbb{F}_{2^n}^*$ and $b \in \mathbb{F}_4^*$ ([17]).

In particular, an explicit characterization of the hyper-bent functions of those families $f_{a,b}$ and $g_{a,b}$ by means of the Kloosterman sums, has been given.

In the line of the recent works [5, 15, 17], this paper is devoted to the study of a subclass $\mathfrak{F}_n$ of the so-called class PS^-. Functions of $\mathfrak{F}_n$ are of the form: $x \in \mathbb{F}_{2^n} \mapsto \sum_{r \in R} Tr_1^n(a_r x^{r(2^m-1)}) + Tr_1^2(bx^{\frac{2^n-1}{3}})$ where R is a set of representatives of the cyclotomic cosets modulo $2^n - 1$ of maximal size $n := 2m$, $\{a_r, r \in R\}$ is a collection of elements of $\mathbb{F}_{2^m}$ and b is an element of $\mathbb{F}_4$. The set of the functions $\mathfrak{F}_n$ includes the functions studied in [15] and in [17].

The paper is organized as follows. Section 2, we fix our main notation and recall the necessary background. Next, in Section 3, we show that hyper-bent functions of $\mathfrak{F}_n$ can be described by means of exponential sums involving Dickson polynomials (Theorem 13 and Theorem 15). In particular, when b is a primitive element of $\mathbb{F}_4$, we provide a way to transfer the characterization of hyperbentness of an element of $\mathfrak{F}_n$ to the evaluation of the Hamming weight of some Boolean functions. To illustrate our results, we show in the Sub-section 3.2.3 that the results presented in [15] and in [17] can be deduced. Finally, in the end of section 3, we provide a possibly new infinite family of hyper-bent functions provided that some set is not empty (Conjecture 1).

2   Notation and Preliminaries

For any set E, $E^\star = E \setminus \{0\}$ and $|E|$ will denote the cardinality of E.

- *Boolean functions and polynomial forms*:

Let n be a positive integer. A Boolean function f on $\mathbb{F}_{2^n}$ is an $\mathbb{F}_2$-valued function on the Galois field $\mathbb{F}_{2^n}$ of order 2^n. The *weight* of f, denoted by $\mathrm{wt}(f)$, is the *Hamming weight* of the image vector of f i.e. the cardinality of its support $supp(f) := \{x \in \mathbb{F}_{2^n} \mid f(x) = 1\}$.

For any positive integer k, and r dividing k, the trace function from $\mathbb{F}_{2^k}$ to $\mathbb{F}_{2^r}$, denoted by Tr_r^k, is the mapping defined as: $Tr_r^k(x) := \sum_{i=0}^{\frac{k}{r}-1} x^{2^{ir}}$. In particular, we denote the *absolute trace* over $\mathbb{F}_2$ of an element $x \in \mathbb{F}_{2^n}$ by $Tr_1^n(x) = \sum_{i=0}^{n-1} x^{2^i}$. Recall that, the absolute trace satisfies $(Tr_1^n(x))^2 = Tr_1^n(x) = Tr_1^n(x^2)$ for every $x \in \mathbb{F}_{2^n}$ and that, for every integer r dividing k, the trace function Tr_r^k satisfies the transitivity property, that is, $Tr_1^k = Tr_1^r \circ Tr_r^k$.

Every non-zero Boolean function f defined on $\mathbb{F}_{2^n}$ has a (unique) trace expansion of the form:

$$\forall x \in \mathbb{F}_{2^n}, \quad f(x) = \sum_{j \in \Gamma_n} Tr_1^{o(j)}(a_j x^j) + \epsilon(1 + x^{2^n-1})$$

called its polynomial form, where Γ_n is the set of integers obtained by choosing one element in each cyclotomic class $\{j \times 2^i \pmod{2^n - 1}; \quad i \in \mathbb{N}\}$ of 2 modulo $2^n - 1$, $o(j)$ is the size of the cyclotomic coset of 2 modulo $2^n - 1$ containing j, $a_j \in \mathbb{F}_{2^{o(j)}}$ and, $\epsilon = \mathrm{wt}(f)$ modulo 2.

- *Walsh transform, bent and hyper-bent functions*:

Let f be a Boolean function on $\mathbb{F}_{2^n}$. Its *"sign" function* is the integer-valued function $\chi(f) := (-1)^f$. The *Walsh Hadamard transform* of f is the discrete Fourier transform of χ_f, whose value at $\omega \in \mathbb{F}_{2^n}$ is defined as follows:

$$\forall \omega \in \mathbb{F}_{2^n}, \quad \widehat{\chi_f}(\omega) = \sum_{x \in \mathbb{F}_{2^n}} (-1)^{f(x) + Tr_1^n(\omega x)}.$$

Bent functions can be defined as follows:

Definition 1. A Boolean function $f : \mathbb{F}_{2^n} \to \mathbb{F}_2$ (n even) is said to be bent if $\widehat{\chi_f}(\omega) = \pm 2^{\frac{n}{2}}$, for all $\omega \in \mathbb{F}_{2^n}$.

Hyper-bent functions have properties still stronger than bent functions. More precisely, they can be defined as follows:

Definition 2. A Boolean function $f : \mathbb{F}_{2^n} \to \mathbb{F}_2$ (n even) is said to be hyper-bent if the function $x \mapsto f(x^i)$ is bent, for every integer i co-prime with $2^n - 1$.

Note that bent and hyper-bent functions defined on $\mathbb{F}_{2^n}$ exist only for even n. Moreover, it is well known that their Hamming weight is even. Therefore, their polynomial form is

$$\forall x \in \mathbb{F}_{2^n}, \quad f(x) = \sum_{j \in \Gamma_n} Tr_1^{o(j)}(a_j x^j) \tag{1}$$

where Γ_n, $o(j)$ are defined as above and $a_j \in \mathbb{F}_{2^{o(j)}}$.

- *Some results on bent and hyper-bent Boolean fucntions*:

Recall the following well-known result which includes the definition of the Partial Spreads class $\mathcal{PS}^-$ introduced by Dillon.

Theorem 1. *[8] Let E_i, $i = 1, 2, \cdots, N$, be N subspaces of $\mathbb{F}_{2^n}$ of dimension m satisfying $E_i \cap E_j = \{0\}$ for all $i, j \in \{1, 2, \cdots, N\}$ with $i \neq j$. Let f be a Boolean function over $\mathbb{F}_{2^n}$ ($n = 2m$). Assume that the support of f can be written as*
$$supp(f) = \bigcup_{i=1}^{N} E_i^\star, \quad \text{where } E_i^\star := E_i \setminus \{0\}.$$
Then f is bent if and only if $N = 2^{m-1}$. In this case f is said to be in the $\mathcal{PS}^-$ class.

Youssef and Gong have shown that hyper-bent functions exist. They partially state this main result of [19] in terms of sequences. The following proposition is an easy translation of their result stated using only the terminology of Boolean functions (see [3])

Proposition 2. *[19] Let $n = 2m$ be an even integer. Let α be a primitive element of $\mathbb{F}_{2^n}$. Let f be a Boolean function defined on $\mathbb{F}_{2^n}$ such that $f(\alpha^{2^m+1}x) = f(x)$ for every $x \in \mathbb{F}_{2^n}$ and $f(0) = 0$. Then, f is a hyper-bent function if and only if the weight of the vector $(f(1), f(\alpha), f(\alpha^2), \cdots, f(\alpha^{2^m}))$ equals 2^{m-1}.*

Charpin and Gong [5] have derived a slightly different version of the preceding Proposition.

Proposition 3. *[5] Let $n = 2m$ be an even integer. Let α be a primitive element of $\mathbb{F}_{2^n}$. Let f be a Boolean function defined on $\mathbb{F}_{2^n}$ such that $f(\alpha^{2^m+1}x) = f(x)$ for every $x \in \mathbb{F}_{2^n}$ and $f(0) = 0$. Denote by G the cyclic subgroup of $\mathbb{F}_{2^n}^\star$ of order $2^m + 1$. Let ζ be a generator of G. Then, f is a hyper-bent function if and only if the cardinality of the set $\{i \mid f(\zeta^i) = 1, 0 \le i \le 2^m\}$ equals 2^{m-1}.*

Remark 1. It is important to point out that bent Boolean functions f defined on $\mathbb{F}_{2^n}$ such that $f(\alpha^{2^m+1}x) = f(x)$ for every $x \in \mathbb{F}_{2^n}$ (where α is a primitive element of $\mathbb{F}_{2^n}$) and $f(0) = 0$ are hyper-bent (see proof of Theorem 2 in [5] or observe that the support $supp(f)$ of such Boolean functions f can be decomposed as $supp(f) = \bigcup_{i \in S} \alpha^i \mathbb{F}_{2^m}^\star$, where $S = \{i \mid f(\alpha^i) = 1\}$, that is, thanks to Theorem 1, functions f are bent if and only if $|S| = 2^{m-1}$, proving that these bent functions are hyper-bent functions, according to Proposition 2).

Dillon exhibits a subclass of $\mathcal{PS}^-$, denoted by $\mathcal{PS}_{ap}$, whose elements are defined in an explicit form ($\mathbb{F}_{2^{2m}}$ is a $\mathbb{F}_{2^m}$-vectorspace of dimension 2; every element $z \in \mathbb{F}_{2^{2m}}$ can be decomposed as $z = x + wy$ with $(x, y) \in \mathbb{F}_{2^m} \times \mathbb{F}_{2^m}$ where $\{1, w\}$ stands for a basis of the $\mathbb{F}_{2^m}$-vectorspace $\mathbb{F}_{2^{2m}}$):

Definition 3. Let $n = 2m$. The Partial Spreads class $\mathcal{PS}_{ap}$ consists of all functions f defined as follows: let g be a balanced Boolean function from $\mathbb{F}_{2^m}$ to $\mathbb{F}_2$ such that $g(0) = 0$. Define a Boolean function f from $\mathbb{F}_{2^m} \times \mathbb{F}_{2^m}$ to $\mathbb{F}_2$ as $f(x, y) = g(xy^{2^m-2})$ for every $(x, y) \in \mathbb{F}_{2^m} \times \mathbb{F}_{2^m}$.

It is well-known (see *e.g* [3]) that, all the functions of the class $\mathcal{PS}_{ap}$ are hyper-bent. Carlet and Gaborit have proved in [3] the following more precise statement of Proposition 2.

Proposition 4. *[3] Boolean functions of Proposition 2 such that $f(1) = 0$ are elements of the class $\mathcal{PS}_{ap}$. Those such that $f(1) = 1$ are the functions of the form $f(x) = g(\delta x)$ for some $g \in \mathcal{PS}_{ap}$ and $\delta \in \mathbb{F}_{2^n} \setminus \{1\}$ such that $g(\delta) = 1$.*

- *Some classical binary exponential sums:*

Recall two classical binary exponential sums on $\mathbb{F}_{2^n}$ (where n is a positive integer):

Definition 4. The binary Kloosterman sums on $\mathbb{F}_{2^n}$ are:

$$K_n(a) := \sum_{x \in \mathbb{F}_{2^n}} \chi\left(Tr_1^n\left(ax + \frac{1}{x}\right)\right), \quad a \in \mathbb{F}_{2^n}$$

Recall the following result

Proposition 5. *[13] The Kloosterman sums K_n on $\mathbb{F}_{2^n}$ takes integer values in the range $[-2^{(n+2)/2} + 1, 2^{(n+2)/2} + 1]$.*

Definition 5. The binary cubic sums on $\mathbb{F}_{2^n}$ are:

$$C_n(a, b) := \sum_{x \in \mathbb{F}_{2^n}} \chi\left(Tr_1^n(ax^3 + bx)\right), \quad a \in \mathbb{F}_{2^n}^\star, b \in \mathbb{F}_{2^n}$$

The exact values of the cubic sums $C_m(a, a)$ on $\mathbb{F}_{2^m}$ can be computed thanks to Carlitz's result [4] by means of the Jacobi symbol. Recall that the Jacobi symbol $\left(\frac{2}{m}\right)$ is a generalization of the Legendre symbol (which is defined when m is an odd prime). For m odd, $\left(\frac{2}{m}\right) = (-1)^{\frac{(m^2-1)}{8}}$.

Proposition 6. *[4] Let m be an odd integer. Let $a \in \mathbb{F}_{2^m}^\star$ and $c \in \mathbb{F}_{2^m}$, Then*

1. $C_m(1, 1) = \left(\frac{2}{m}\right) 2^{\frac{m+1}{2}}$ *where* $\left(\frac{2}{m}\right)$ *is the Jacobi symbol.*
2. *If* $Tr_1^m(c) = 0$, *then* $C_m(1, c) = 0$.
3. *If* $Tr_1^m(c) = 1$ *(with $c \neq 1$), then* $C_m(1, c) = \chi(Tr_1^m(\gamma^3 + \gamma)) \left(\frac{2}{m}\right) 2^{\frac{m+1}{2}}$ *where* $c = \gamma^4 + \gamma + 1$ *for some* $\gamma \in \mathbb{F}_{2^m}$.

• *Dickson Polynomials:*

Recall that the family of Dickson polynomials $D_r(X) \in \mathbb{F}_2[X]$ is defined by

$$D_r(X) = \sum_{i=0}^{\frac{r}{2}} \frac{r}{r-i} \binom{r-i}{i} X^{r-2i}, \quad r = 2, 3, \cdots$$

Moreover, the family of Dickson polynomials $D_r(X) \in \mathbb{F}_2[X]$ can also be defined by the following recurrence relation:

$$D_{i+2}(X) = X D_{i+1}(X) + D_i(X)$$

with initial values

$$D_0(X) = 0, \quad D_1(X) = X.$$

Now, recall the following properties which we use in the sequel. For any non-zero positive integers r and p, Dickson polynomials satisfy:

1. $\deg(D_r(X)) = r$,
2. $D_{rp}(X) = D_r(D_p(X))$,
3. $D_r(x + x^{-1}) = x^r + x^{-r}$.

3   Hyper-bent Functions Whose Expression is the Sum of Multiple Trace Terms

In the sequel, n is an even positive integer, $m = \frac{n}{2}$ is an odd integer and E is a set of representatives of the cyclotomic classes modulo $2^n - 1$ for which each class has the full size n. We denote by $\mathfrak{F}_n$ the set of Boolean functions f_b, $(b \in \mathbb{F}_4)$ defined on $\mathbb{F}_{2^n}$ whose polynomial forms are:

$$f_b(x) := \sum_{r \in R} Tr_1^n(a_r x^{r(2^m - 1)}) + Tr_1^2(bx^{\frac{2^n-1}{3}}). \tag{2}$$

where $R \subseteq E$ and all the coefficients a_r are in $\mathbb{F}_{2^m}$.

Note that the size of the cyclotomic coset of 2 modulo $2^n - 1$ containing $\frac{2^n-1}{3}$ is equal to 2 (i.e. $o(\frac{2^n-1}{3}) = 2$) and that, the function f_b does not belong to the class considered by Charpin and Gong in [5].

For m odd, $2^m + 1$ is a multiple of 3 and thus all exponents for x in (2) are multiples of $2^m - 1$. Therefore, every Boolean function f_b in $\mathfrak{F}_n$ satisfies

$$\forall x \in \mathbb{F}_{2^n}, \quad f_b(\alpha^{2^m+1}x) = f_b(x).$$

where α denotes any primitive element of $\mathbb{F}_{2^n}$. Furthermore, since every Boolean f_b of $\mathfrak{F}_n$ vanishes at 0, one can then apply Proposition 3 to get the following characterization of hyper-bentness for an element of $\mathfrak{F}_n$.

Proposition 7. *Let $f_b \in \mathfrak{F}_n$. Set $\Lambda(f_b) := \sum_{u \in U} \chi(f_b(u))$ where U is the group of $(2^m + 1)$-st roots of unity, that is, $U = \{x \in \mathbb{F}_{2^n} \mid x^{2^m+1} = 1\}$. Then, f_b is hyper-bent if and only if $\Lambda(f_b) = 1$. Moreover, a hyper-bent function f_b is in the Partial Spreads class PS_{ap} if and only if $b \in \mathbb{F}_2$.*

Proof. The Boolean function f_b satisfies the assumptions of Proposition 3. Therefore f_b is hyper-bent if and only if its restriction to U has Hamming weight 2^{m-1} according to Proposition 3. Now, one has $\Lambda(f_b) = 2^m + 1 - 2|\{u \in U \mid f_b(u) = 1\}|$. Therefore, the Hamming weight of the restriction of f_b to U equals 2^{m-1} if and only if $\Lambda(f_b) = 1$. The second part of the proposition is a direct application of Proposition 4. Indeed, note that $f_b(1) = \sum_{r \in R} Tr_1^n(a_r) + Tr_1^2(b) = Tr_1^2(b)$ (since $Tr_1^n(a_r) = 0$ for every $r \in R$ because $a_r \in \mathbb{F}_{2^m}$) and it is clear that the elements b of $\mathbb{F}_4$ whose trace over $\mathbb{F}_4$ equals 0, are the elements of $\mathbb{F}_2$.

3.1   The Case $b = 0$

Charpin and Gong [5] have studied the functions of $\mathfrak{F}_n$ in the case where $b = 0$ and provide the following characterization of the hyper-bentness in terms of Dickson polynomials.

Theorem 8. *[5] Let $n = 2m$. Let E' be a set of representatives of the cyclotomic cosets modulo $2^m + 1$ for which each coset has the full size n. Let f be the function defined on $\mathbb{F}_{2^n}$ by $f(x) = \sum_{r \in R} Tr_1^n(a_r x^{r(2^m-1)})$, $a_r \in \mathbb{F}_{2^m}$ where $R \subseteq E'$. Let g be the Boolean function defined on $\mathbb{F}_{2^m}$ by $g(x) = \sum_{r \in R} Tr_1^m(a_r D_r(x))$. Then f is hyper-bent if and only if*

$$\sum_{x \in \mathbb{F}_{2^m}} \chi(Tr_1^m(x^{-1}) + g(x)) = 2^m - 2\,\mathrm{wt}(g).$$

Remark 2. The bentness of monomial functions of $\mathfrak{F}_n$ has been studied. More precisely, the exponent $2^m - 1$ has been considered by Dillon in [8] as an example of bent functions belonging to $\mathcal{PS}^-$ (Theorem 1). Using results from coding theory, Dillon has proved in [8] that the function $x \in \mathbb{F}_{2^n} \mapsto Tr_1^n(ax^{2^m-1})$ is (hyper)-bent if and only if the Kloosterman sum K_m on $\mathbb{F}_{2^m}$ satisfies $K_m(a) = 0$. Further, the exponent $r(2^m - 1)$ where r is co-prime with $2^m + 1$, has been considered firstly by Leander [14] and next by Charpin and Gong [5] (in fact Leander has found another proof of Dillon's result which gives more insight; a small error in his proof has been corrected in [5]). It has been proved that the function $x \in \mathbb{F}_{2^n} \mapsto Tr_1^n(a_r x^{r(2^m-1)})$ where $\gcd(r, 2^m + 1) = 1$, is hyper-bent if and only if a is a zero of the Kloosterman sum K_m on $\mathbb{F}_{2^m}$.

3.2   The Case Where $b \in \mathbb{F}_4^\star$

We are interested in characterizing the hyper-bentness of the Boolean function of the form (2) where $b \neq 0$. To this end, we begin by introducing some additional notation while underlining some facts.

Let β be a primitive element of $\mathbb{F}_4$. Suppose that $\beta = \alpha^{\frac{2^n-1}{3}}$ for some primitive element α of $\mathbb{F}_{2^n}$. Set $\xi := \alpha^{2^m-1}$ so that ξ is a generator of the cyclic group $U := \{u \in \mathbb{F}_{2^n} \mid u^{2^m+1} = 1\}$. Note that U can be decomposed as : $U = \bigcup_{i=0}^{2} \xi^i V$ where $V := \{u^3, u \in U\}$. Next, let introduce the sums

$$S_i := \sum_{v \in V} \chi(f_0(\xi^i v)), \quad \forall i \in \{0, 1, 2\} \tag{3}$$

First of all, note that

$$S_0 + S_1 + S_2 = \sum_{u \in U} \chi(f_0(u)). \tag{4}$$

Next, one has

Lemma 9. $S_1 = S_2$.

Proof. Since the trace map is invariant under the Frobenius automorphism $x \mapsto x^2$, we get applying m times the Frobenius automorphism : $\forall x \in \mathbb{F}_{2^n}$,

$$f_0(x) = \sum_{r \in R} Tr_1^n \left(a_r^{2^m} x^{2^m r(2^m-1)} \right) = \sum_{r \in R} Tr_1^n \left(a_r x^{2^m r(2^m-1)} \right) = f_0(x^{2^m})$$

because all the coefficients a_r are in $\mathbb{F}_{2^m}$. Hence, $S_1 = \sum_{v \in V} \chi(f_0(\xi^{2^m} v^{2^m})) = \sum_{v \in V} \chi(f_0(\xi^2(\xi^{2^m-2} v^{2^m})))$. Now, since m is odd, 3 divides $2^m + 1$ and then divides $2^m - 2$. Hence, ξ^{2^m-2} is a cube of U and the mapping $v \mapsto \xi^{(2^m-2)} v^{2^m}$ is a permutation of V. Consequently, $S_1 = \sum_{v \in V} \chi(f_0(\xi^2 v)) = S_2$. ☐

Now, for $b \in \mathbb{F}_4^\star$, we establish expressions for $\Lambda(f_b) := \sum_{u \in U} \chi(f_b(u))$ (where U is the group of $(2^m + 1)$-st roots of unity) involving the sums S_i.

Proposition 10. $\Lambda(f_\beta) = \Lambda(f_{\beta^2}) = -S_0$ and $\Lambda(f_1) = S_0 - 2S_1$.

Proof. Introduce for every element c of $\mathbb{F}_4$ $T(c) := \sum_{b \in \mathbb{F}_4} \Lambda(f_b) \chi(Tr_1^2(bc))$. Recall that one has

$$\Lambda(f_b) = \frac{1}{4} \sum_{c \in \mathbb{F}_4} T(c) \chi(Tr_1^2(bc)). \tag{5}$$

Indeed

$$\sum_{c \in \mathbb{F}_4} T(c) \chi(Tr_1^2(bc))$$

$$= \sum_{c \in \mathbb{F}_4} \sum_{d \in \mathbb{F}_4} \Lambda(f_d) \chi(Tr_1^2(dc)) \chi(Tr_1^2(bc))$$

$$= \sum_{d \in \mathbb{F}_4} \Lambda(f_d) \sum_{c \in \mathbb{F}_4} \chi(Tr_1^2(c(d + b)))$$

But $\sum_{c\in\mathbb{F}_4}\chi(Tr_1^2(c(d+b)))=4$ if $d=b$ (i.e $b+d=0$) and 0 otherwise. Then, one gets $\sum_{c\in\mathbb{F}_4}T(c)\chi(Tr_1^2(bc))=4\Lambda(f_b)$.

Now, note that $T(c)=\sum_{u\in U}\chi(f_0(u))\sum_{b\in\mathbb{F}_4}\chi\left(Tr_1^2\left(b\left(c+u^{\frac{2^n-1}{3}}\right)\right)\right)$. Furthermore, one has

$$\sum_{b\in\mathbb{F}_4}\chi\left(Tr_1^2\left(b\left(c+u^{\frac{2^n-1}{3}}\right)\right)\right)=0 \text{ if } u^{\frac{2^n-1}{3}}\neq c \text{ and } 4 \text{ otherwise.}$$

Since, $u^{\frac{2^n-1}{3}}\neq 0$ for every $u\in U$, $T(0)=0$. Since β is a primitive element of $\mathbb{F}_4$, let suppose from now that $c=\beta^i$, $i\in\{0,1,2\}$. Recall that $\beta=\alpha^{\frac{2^n-1}{3}}$ and $\xi=\alpha^{2^m-1}$ for some primitive element α of $\mathbb{F}_{2^n}$. Then $\beta^i=\xi^{i\frac{2^m+1}{3}}$. Hence, $T(\beta^i)=4\sum_{u\in U,\ u^{\frac{2^n-1}{3}}=\beta^i=\xi^{i\frac{2^m+1}{3}}}\chi(f_0(u))$. Now,

$$u^{\frac{2^n-1}{3}}=\xi^{i\frac{2^m+1}{3}} \iff \left(u^{-2}\xi^{-i}\right)^{\frac{2^m+1}{3}}=1 \iff u^{-2}\in\xi^i V.$$

That follows from the fact that the only elements x of U such that $x^{\frac{2^m+1}{3}}=1$ are the elements of V. Next, noting that the map $x\mapsto x^{2^{m-1}}$ is one-to-one from $\xi^i V$ to $\xi^i V$ (because $\xi^{i(2^{m-1}-1)}$ is a cube since $2^{m-1}-1\equiv 0\pmod 3$ for m odd), one gets that $u^{\frac{2^n-1}{3}}=\xi^{i\frac{2^m+1}{3}} \iff u\in\xi^i V$. Therefore

$$T(\beta^i)=4\sum_{v\in V}\chi(f_0(\xi^i v))=4S_i.$$

Finally, by the inversion formula (5), one gets $\Lambda(f_b)=\frac{1}{4}\sum_{c\in\mathbb{F}_4}T(c)\chi(Tr_1^2(bc))$ that is,

$$\Lambda(f_1)=S_0\chi(Tr_1^2(1))+S_1\chi(Tr_1^2(\beta))+S_2\chi(Tr_1^2(\beta^2)),$$
$$\Lambda(f_\beta)=S_0\chi(Tr_1^2(\beta))+S_1\chi(Tr_1^2(\beta^2))+S_2\chi(Tr_1^2(1)),$$
$$\Lambda(f_{\beta^2})=S_0\chi(Tr_1^2(\beta^2))+S_1\chi(Tr_1^2(1))+S_2\chi(Tr_1^2(\beta)).$$

The result follows then from Lemma 9 and from the fact that $Tr_1^2(1)=0$ and $Tr_1^2(\beta)=Tr_1^2(\beta^2)=1$.

From Proposition 7, Proposition 10, Lemma 9 and (4), one straight-forwardly deduces the following statement.

Lemma 11. *Let $n=2m$ be an even integer with m odd. For $b\in\mathbb{F}_4$, let f_b be a function defined by (2). Let β be a primitive element of $\mathbb{F}_4$. Let U be the cyclic group of (2^m+1)-st roots of unity and V be the set of the cube of U. Then,*

1. *f_β is hyper-bent if and only if $\sum_{v\in V}\chi(f_0(v))=-1$.*
2. *f_β is hyper-bent if and only if f_{β^2} is hyper-bent.*
3. *f_1 is hyperbent if and only if $2\sum_{v\in V}\chi(f_0(v))-\sum_{u\in U}\chi(f_0(u))=1$.*

The case where b is a primitive element of $\mathbb{F}_4$. According to assertion (2) of Lemma 11, we can suppose that $b = \beta$ without loss of generality. As in the case where $b = 0$ (Theorem 8), one can establish a characterization of the hyper-bentness of f_β involving the Dickson polynomials. To this end, we begin with proving the following result.

Lemma 12. *Let f_0 be the function defined on $\mathbb{F}_{2^n}$ by (2) with $b = 0$. Let g be the related function defined on $\mathbb{F}_{2^m}$ by $g(x) = \sum_{r \in R} Tr_1^m(a_r D_r(x))$, where $D_r(x)$ is the Dickson polynomial of degree r. Let U be the cyclic group of $(2^m + 1)$-st roots of unity. Then, for any positive integer p, we have*

$$\sum_{u \in U} \chi\left(f_0(u^p)\right) = 1 + 2 \sum_{c \in \mathbb{F}_{2^m}^\star, Tr_1^m(c^{-1})=1} \chi\left(g(D_p(c))\right).$$

Proof. Using the transitivity rule $Tr_1^n = Tr_1^m \circ Tr_m^n$, the fact that the coefficients a_r are in the subfield $\mathbb{F}_{2^m}$ of $\mathbb{F}_{2^n}$ and the fact that the mapping $u \mapsto u^{2^m-1}$ is a permutation of U, one has

$$\sum_{u \in U} \chi\left(f_0(u^p)\right) = \sum_{u \in U} \chi\left(\sum_{r \in R} Tr_1^m\left(a_r(u^{(2^m-1)rp} + u^{2^m(2^m-1)rp})\right)\right)$$

$$= \sum_{u \in U} \chi\left(\sum_{r \in R} Tr_1^m\left(a_r(u^{rp} + u^{-rp})\right)\right) = \sum_{u \in U} \chi\left(\sum_{r \in R} Tr_1^m\left(a_r D_{rp}(u + u^{-1})\right)\right)$$

since $u^p + u^{-p} = D_p(u + u^{-1})$. Recall now that every element $1/c$ where $c \in \mathbb{F}_{2^m}^\star$ with $T r_1^m(c) = 1$ can be uniquely represented as $u + u^{2^m} = u + u^{-1}$ with $u \in U$. Thus

$$\sum_{u \in U} \chi\left(f_0(u^p)\right) = 1 + \sum_{u \in U \setminus \{1\}} \chi\left(\sum_{r \in R} Tr_1^m\left(a_r D_{rp}(u + u^{-1})\right)\right)$$

$$= 1 + 2 \sum_{c \in \mathbb{F}_{2^m}^\star, Tr_1^m(c)=1} \chi\left(\sum_{r \in R} Tr_1^m\left(a_r D_{rp}(1/c)\right)\right)$$

$$= 1 + 2 \sum_{c \in \mathbb{F}_{2^m}^\star, Tr_1^m(c^{-1})=1} \chi\left(\sum_{r \in R} Tr_1^m\left(a_r D_{rp}(c)\right)\right).$$

In the last equality, we use the fact that the map $c \mapsto 1/c$ is a permutation on $\mathbb{F}_{2^m}$. Now, since $D_{rp} = D_r \circ D_p$, one gets

$$\sum_{u \in U} \chi\left(f_0(u^p)\right) = 1 + 2 \sum_{c \in \mathbb{F}_{2^m}^\star, Tr_1^m(c^{-1})=1} \chi\left(g(D_p(c))\right).$$

From Lemma 11 and Lemma 12, one deduce the following statement.

Theorem 13. *Let $n = 2m$ be an even integer with m odd. Let β be a primitive element of $\mathbb{F}_4$. Let f_β be the function defined on $\mathbb{F}_{2^n}$ by (2). Let g be the related function defined on $\mathbb{F}_{2^m}$ by $g(x) = \sum_{r \in R} Tr_1^m(a_r D_r(x))$, where $D_r(x)$ is the Dickson polynomial of degree r. Then, the three assertions are equivalent*

1. f_β is hyper-bent.

2. $\displaystyle\sum_{x\in\mathbb{F}_{2^m}^\star,Tr_1^m(x^{-1})=1} \chi\Big(g(D_3(x))\Big) = -2.$

3. $\displaystyle\sum_{x\in\mathbb{F}_{2^m}^\star} \chi\Big(Tr_1^m(x^{-1}) + g(D_3(x))\Big) = 2^m - 2\,\mathrm{wt}(g\circ D_3) + 4.$

Proof. According to Lemma 12, we have

$$S_0 = \sum_{v\in V}\chi\Big(f_0(v)\Big) = \frac{1}{3}\sum_{u\in U}\chi\Big(f_0(u^3)\Big) = \frac{1}{3}\left(1 + 2\sum_{x\in\mathbb{F}_{2^m}^\star,Tr_1^m(x^{-1})=1}\chi\Big(g(D_3(x))\Big)\right).$$

The equivalence between assertions (1) and (2) follows then from assertion (1) of Lemma 11.

Now, note that the indicator of the set $\{x \in \mathbb{F}_{2^m}^\star \mid Tr_1^m(x^{-1}) = 1\}$ can be written as $\frac{1}{2}\left(1 - \chi(Tr_1^m(x^{-1}))\right)$. Therefore,

$$\sum_{x\in\mathbb{F}_{2^m}^\star,Tr_1^m(x^{-1})=1}\chi\Big(g(D_3(x))\Big)$$

$$= \frac{1}{2}\left(\sum_{x\in\mathbb{F}_{2^m}^\star}\chi\Big(g(D_3(x))\Big) - \sum_{x\in\mathbb{F}_{2^m}^\star}\chi\Big(Tr_1^m(x^{-1}) + g(D_3(x)))\Big)\right)$$

$$= \frac{1}{2}\left(\sum_{x\in\mathbb{F}_{2^m}}\chi\Big(g(D_3(x))\Big) - \sum_{x\in\mathbb{F}_{2^m}}\chi\Big(Tr_1^m(x^{-1}) + g(D_3(x)))\Big)\right).$$

Now, f_β is hyper-bent if and only if $\sum_{x\in\mathbb{F}_{2^m}^\star,Tr_1^m(x^{-1})=1}\chi\Big(g(D_3(x))\Big) = -2$. Therefore, using the fact that, for a Boolean function h defined on $\mathbb{F}_{2^n}$, $\sum_{x\in\mathbb{F}_{2^n}}\chi(h(x)) = 2^n - 2\,\mathrm{wt}(h)$, we get that f_β is hyper-bent if and only if

$$\sum_{x\in\mathbb{F}_{2^m}}\chi\Big(Tr_1^m(x^{-1}) + g(D_3(x))\Big) = 4 + 2^m - 2\,\mathrm{wt}(g\circ D_3).$$

One also has

Proposition 14. *Let $n = 2m$ be an even integer with m odd. Let d be a positive integer. Suppose that d and $\frac{2^m+1}{3}$ are co-prime. Let β be a primitive element of $\mathbb{F}_4$. Let f_β be the function defined by (2) and h_β be the function whose expression is*

$$\sum_{r\in R}Tr_1^n(a_r x^{dr(2^m-1)}) + Tr_1^2(\beta x^{\frac{2^n-1}{3}})$$

where $a_r \in \mathbb{F}_{2^m}$. Then, f_β is hyper-bent if and only if h_β is hyper-bent.

Proof. According to assertion (1) of Lemma 11, h_β is hyper-bent if and only if $\sum_{v\in V}\chi(h_0(v)) = -1$. Now, $\sum_{v\in V}\chi(h_0(v)) = \sum_{v\in V}\chi(f_0(v^d)) = \sum_{v\in V}\chi(f_0(v))$ since the mapping $v \mapsto v^d$ is then a permutation of V if $\frac{2^m+1}{3}$ and d are co-prime. The result follows again from assertion (1) of Lemma 11.

The case where $b = 1$. In this subsection, we are interested in characterizing the hyper-bentness of the Boolean function f_1 whose polynomial form is $f_1(x) = \sum_{r \in R} Tr_1^n(a_r x^{r(2^m-1)}) + Tr_1^2(x^{\frac{2^n-1}{3}})$. In this case one can give a characterization of the bentness, analogous to the assertion (2) of Theorem 13.

Theorem 15. *Let $n = 2m$ be an even integer with m odd. Let f_1 be the Boolean function defined on $\mathbb{F}_{2^n}$ by*

$$f_1(x) = \sum_{r \in R} Tr_1^n(a_r x^{r(2^m-1)}) + Tr_1^2(x^{\frac{2^n-1}{3}}).$$

Let g be the related function defined on $\mathbb{F}_{2^m}$ by $g(x) = \sum_{r \in R} Tr_1^m(a_r D_r(x))$, where $D_r(x)$ is the Dickson polynomial of degree r.
Then, f_1 is hyper-bent if and only if,

$$2 \sum_{x \in \mathbb{F}_{2^m}^\star, Tr_1^m(x^{-1})=1} \chi\Big(g(D_3(x))\Big) - 3 \sum_{x \in \mathbb{F}_{2^m}^\star, Tr_1^m(x^{-1})=1} \chi\Big(g(x)\Big) = 2.$$

Proof. Note that

$$2 \sum_{v \in V} \chi(f_0(v)) - \sum_{u \in U} \chi(f_0(u)) = \frac{2}{3} \sum_{u \in U} \chi(f_0(u^3)) - \sum_{u \in U} \chi(f_0(u))$$

$$= -\frac{1}{3} + \frac{4}{3} \sum_{x \in \mathbb{F}_{2^m}^\star, Tr_1^m(x^{-1})=1} \chi\Big(g(D_3(x))\Big) - 2 \sum_{x \in \mathbb{F}_{2^m}^\star, Tr_1^m(x^{-1})=1} \chi\Big(g(x)\Big)$$

according to Lemma 12. One then concludes using Lemma 11 that states that f_1 is hyper-bent if and only if

$$2 \sum_{v \in V} \chi(f_0(v)) - \sum_{u \in U} \chi(f_0(u)) = 1.$$

On can also prove the similar result to Proposition 14.

Proposition 16. *Let $n = 2m$ be an even integer with m odd. Suppose that $m \not\equiv 3 \pmod 6$. Let d be a positive integer such that $\gcd(d, 2^m + 1) = 3$. Let β be a primitive element of $\mathbb{F}_4$. Let f_β be the function defined by (2) and h_1 be the function whose expression is*

$$\sum_{r \in R} Tr_1^n(a_r x^{dr(2^m-1)}) + Tr_1^2(x^{\frac{2^n-1}{3}})$$

If f_β is hyper-bent then, h_1 is hyper-bent.

Proof. Set $h_0(x) := \sum_{r \in R} Tr_1^n(a_r x^{dr(2^m-1)})$. One has (since $\gcd(d, 2^m + 1) = 3$)

$$\sum_{v \in V} \chi(h_0(v)) = \sum_{v \in V} \chi(f_0(v^d)) = \sum_{v \in V} \chi(f_0(v^3)) = \sum_{v \in V} \chi(f_0(v))$$

since the mapping $v \mapsto v^3$ is a permutation when $m \not\equiv 3$ (mod 6). On the other hand, note that (since $\gcd(d, 2^m + 1) = 3$)

$$\sum_{u \in U} \chi(h_0(u)) = \sum_{u \in U} \chi(f_0(u^d)) = \sum_{u \in U} \chi(f_0(u^3)) = 3 \sum_{v \in V} \chi(f_0(v)).$$

Now, $\sum_{v \in V} \chi(f_0(v)) = -1$ according to Lemma 11, since f_β is hyper-bent. Hence, $2 \sum_{v \in V} \chi(h_0(v)) - \sum_{u \in U} \chi(h_0(u)) = -2 - (-3) = 1$, proving that h_1 is hyper-bent (according to Lemma 11).

Examples

Example 1. To illustrate our results, we describe the set of hyper-bent functions of a particular family of Boolean functions belonging to the class (2), that is, the Boolean functions $f_{\beta^i}, i \in \{0, 1, 2\}$ (studied in [15, 16]) defined on $\mathbb{F}_{2^n}$ ($n = 2m$, m odd) as:

$$f_{\beta^i}(x) = Tr_1^n(ax^{2^m - 1}) + Tr_1^2(\beta^i x^{\frac{2^n - 1}{3}}), \quad \forall x \in \mathbb{F}_{2^n}$$

where $a \in \mathbb{F}_{2^m}^\star$ and β is a primitive element of $\mathbb{F}_4$.

In this case, $f_0(x) = Tr_1^n(ax^{2^m - 1})$ is the Dillon function and the related function g is defined by $g(x) = Tr_1^m(ax)$.

According to Lemma 11, f_β is hyper-bent if and only if f_{β^2} is hyper-bent and, according to Theorem 13,

f_β is hyper-bent if and only if $\sum_{x \in \mathbb{F}_{2^m}^\star, Tr_1^m(1/x)=1} \chi(g(D_3(x))) = -2$. But

$$\sum_{x \in \mathbb{F}_{2^m}^\star, Tr_1^m(1/x)=1} \chi(g(D_3(x)))$$

$$= \sum_{x \in \mathbb{F}_{2^m}^\star} \chi(Tr_1^m(a(x^3 + x))) - \sum_{x \in \mathbb{F}_{2^m}^\star, Tr_1^m(1/x)=0} \chi(Tr_1^m(a(x^3 + x)))$$

$$= C_m(a, a) - 1 - \sum_{x \in \mathbb{F}_{2^m}^\star, Tr_1^m(1/x)=0} \chi(Tr_1^m(a(x^3 + x)))$$

$$= C_m(a, a) - 1 - \sum_{x \in \mathbb{F}_{2^m}^\star, Tr_1^m(1/x)=0} \chi(Tr_1^m(ax)).$$

In the last equality, we use that the mapping $x \mapsto D_3(x) := x^3 + x$ is a permutation on the set of $\mathbb{F}_{2^m}^\star$ such that $Tr_1^m(1/x) = 0$ (see for instance [6, Lemma 7]). Now, according to Charpin, Helleseth and Zinoviev [6],

$$\sum_{x \in \mathbb{F}_{2^m}^\star, Tr_1^m(1/x)=0} \chi(Tr_1^m(ax)) = \frac{K_m(a)}{2} - 1.$$

Hence, we get that
$\sum_{x \in \mathbb{F}_{2^m}^\star, Tr_1^m(1/x)=1} \chi(g(D_3(x))) = C_m(a, a) - \frac{K_m(a)}{2}$. Therefore, f_β (resp. f_{β^2})

is hyper-bent if and only if $K_m(a) - 2C_m(a,a) = 4$. The mapping $x \mapsto x^3$ being a permutation on $\mathbb{F}_{2^m}$ for m odd, then every element $a \in \mathbb{F}_{2^m}$ can be (uniquely) written as $a = a'^3$ with $a' \in \mathbb{F}_{2^m}$. One has $C_m(a,a) = \sum_{x \in \mathbb{F}_{2^m}} \chi(Tr_1^m((a'x)^3 + ax) = C_m(1, a^{2/3})$.

Hence, according to Proposition 6 (note that $Tr_1^m(a^{2/3}) = Tr_1^m(a^{1/3})$), the function f_β (resp. f_{β^2}) is hyper-bent if and only if,

$$K_m(a) = \begin{cases} 4 \text{ if } Tr_1^m(a^{1/3}) = 0 \\ 4 \pm \left(\frac{2}{m}\right) 2^{(m+3)/2} \text{ if } Tr_1^m(a^{1/3}) = 1 \end{cases}$$

However, using Proposition 5, the value $4 \pm \left(\frac{2}{m}\right) 2^{(m+3)/2}$ does not belong to $[-2^{(m+2)/2}+1, 2^{(m+2)/2}+1]$ for every $m > 3$. This proves that if $Tr_1^m(a^{1/3}) = 0$ then, the Boolean function f_β (resp. f_{β^2}) is hyper-bent whenever $K_m(a) = 4$ while, when $K_m(a) \neq 4$, f_β (resp. f_{β^2}) is not hyper-bent. Otherwise, if $Tr_1^m(a^{1/3}) = 1$ (which implies that $K_m(a) \neq 4$) then, the function f_β (resp. f_{β^2}) cannot be hyper-bent when $m > 3$.

In the other hand, according to Theorem 15, f_1 is hyper-bent if and only if,

$$2 \sum_{x \in \mathbb{F}_{2^m}, Tr_1^m(x^{-1})=1} \chi\left(g(D_3(x))\right) - 3 \sum_{x \in \mathbb{F}_{2^m}^\star, Tr_1^m(x^{-1})=1} \chi\left(g(x)\right) = 2.$$

We have seen that $\sum_{x \in \mathbb{F}_{2^m}^\star, Tr_1^m(1/x)=1} \chi(g(D_3(x))) = C_m(a,a) - \frac{K_m(a)}{2}$.

Furthermore, according to Charpin, Helleseth and Zinoviev [6],

$$\sum_{x \in \mathbb{F}_{2^m}^\star, Tr_1^m(x^{-1})=1} \chi\left(g(x)\right) = \sum_{x \in \mathbb{F}_{2^m}^\star, Tr_1^m(1/x)=1} \chi(Tr_1^m(ax)) = -\frac{K_m(a)}{2}.$$

Therefore, f_1 is hyper-bent if and only if $K_m(a) + 4C_m(a,a) = 4$.

Now, one has $C_m(a,a) = C_m(1, a^{2/3})$. Hence, according to Proposition 6, f_1 is hyper-bent if and only if,

$$K_m(a) = \begin{cases} 4 \text{ if } Tr_1^m(a^{1/3}) = 0 \\ 4 \pm \left(\frac{2}{m}\right) 2^{(m+5)/2} \text{ if } Tr_1^m(a^{1/3}) = 1 \end{cases}$$

However, using Proposition 5, the value $4 \pm \left(\frac{2}{m}\right) 2^{(m+5)/2}$ does not belong to $[-2^{(m+2)/2}+1, 2^{(m+2)/2}+1]$ for every $m > 3$. This proves that if $Tr_1^m(a^{1/3}) = 0$ then, the Boolean function f_1 is hyper-bent whenever $K_m(a) = 4$ while, when $K_m(a) \neq 4$, f_1 is not hyper-bent. Otherwise, if $Tr_1^m(a^{1/3}) = 1$ (which implies that $K_m(a) \neq 4$) then, the function f_1 cannot be hyper-bent when $m > 3$. One recovers then the results given in [15] (Theorem 12).

Example 2. To illustrate again our results, let consider the set of functions of a particular family of Boolean functions belonging to the subclass (2), that is, the Boolean functions h_β (studied in [17]) defined on $\mathbb{F}_{2^n}$ ($n = 2m$, m odd) as:

$$h_\beta(x) = Tr_1^n(ax^{3(2^m-1)}) + Tr_1^2(\beta x^{\frac{2^n-1}{3}}), \quad \forall x \in \mathbb{F}_{2^n}$$

where $a \in \mathbb{F}_{2^m}^\star$ and β is a primitive element of $\mathbb{F}_4$. Suppose $m \not\equiv 3 \pmod 6$. According to Proposition 14 and the results of Example 1, one can deduce that if $Tr_1^m(a^{1/3}) = 0$ then the Boolean function h_β is hyper-bent whenever $K_m(a) = 4$ and if $Tr_1^m(a^{1/3}) = 1$ then, the Boolean function h_β is not hyper-bent. Thus, one recovers the results given in [17] (Theorem 17).

Table 1. Exponents i and j such that (α^i, α^j) satisfy Conjecture 1 for $n = 10$

i=1	j= 0, 1, 2, 3, 5, 7, 8, 9, 11, 12, 13, 14, 17, 20, 22, 24, 26, 27, 29
i=2	j= 0, 2, 3, 4, 6, 9, 10, 13, 14, 16, 17, 18, 21, 22, 23, 24, 26, 27, 28
i=4	j= 0, 1, 3, 4, 5, 6, 8, 11, 12, 13, 15, 17, 18, 20, 21, 23, 25, 26, 28
i=7	j= 0, 3, 4, 5, 7, 8, 10, 11, 12, 14, 16, 18, 19, 23, 26, 27, 28, 29, 30
i=8	j= 0, 2, 3, 5, 6, 8, 9, 10, 11, 12, 15 16, 19, 21, 22, 24, 25, 26, 30,
i=14	j= 0, 1, 5, 6, 7, 8, 10, 14, 15, 16, 20, 21, 22, 23, 24, 25, 27, 28, 29
i=16	j= 0, 1, 4, 6, 7, 10, 11 12, 13, 16, 17, 18, 19, 20, 21, 22, 24, 29, 30
i=19	j= 0, 4, 5, 6, 7, 8, 9, 13, 14, 15, 17, 18, 19, 21, 25, 27, 29, 2, 30
i=25	j= 0, 1, 2, 3, 4, 7, 9, 15, 18, 19, 20, 22, 23, 24, 25, 26, 28, 29, 30
i=28	j= 0, 1, 2, 9, 10, 11, 12, 13, 14, 15 16, 17, 19, 20, 23, 25, 27, 28, 30

Table 2. Number of pairs (a, a') such that $K_m(a) = 4$ and $\mathcal{S}(a, a') = -1$

n	14	18	22
Number of pairs	882	3978	13948

We now make a conjecture. We need for that to introduce some notations. Let $I := \{x \in \mathbb{F}_{2^m}^\star \mid x = c^3 + c, Tr_1^m(c^{-1}) = 1\}$ and set, for a, $a' \in \mathbb{F}_{2^m}$,

$$\mathcal{S}(a, a') := \sum_{x \in I} \chi(Tr_1^m(a(x + x^3) + a' x^5)).$$

Conjecture 1. For every $a \in \mathbb{F}_{2^m}^\star$ such that $K_m(a) = 4$, the set $\mathfrak{S}_a = \{a' \in \mathbb{F}_{2^m}^\star \mid \mathcal{S}(a, a') = -1\}$ is non empty.

Fact 1. By a computer program, we have checked that the conjecture holds for all $n = 2m$ up to $n = 26$.

Proposition 17. *Let $n = 2m$ with m odd. Suppose that conjecture 1 holds. Let $a \in \mathbb{F}_{2^m}^\star$ such that $K_m(a) = 4$, $a' \in \mathfrak{S}_a$ ($\neq \emptyset$) and β is a primitive element of $\mathbb{F}_4$. Then, the Boolean function f defined on $\mathbb{F}_{2^n}$ whose polynomial form equals*

$$Tr_1^n((a + a')x^{3(2^m - 1)}) + Tr_1^n(a' x^{5(2^m - 1)}) + Tr_1^2(\beta x^{\frac{2^n - 1}{3}})$$

is hyper-bent.

Proof. Let g be the Boolean function defined on $\mathbb{F}_{2^m}$ as

$$g(x) = Tr_1^m((a + a')D_3(x)) + Tr_1^m(a' D_5(x)).$$

According to Theorem 13,

$$f \text{ is hyper-bent if and only if} \sum_{x \in \mathbb{F}_{2^m}^\star, Tr_1^m(x^{-1})=1} \chi(g(D_3(x))) = -2.$$

Now, according to Charpin *et al.* *[6]* *(Lemma 6), the map* $x \mapsto D_3(x)$ *is 3-to-1 from* $\{x \in \mathbb{F}_{2^m} \setminus \mathbb{F}_2 \mid Tr_1^m(x^{-1}) = 1\}$ *to* $I := \{x \in \mathbb{F}_{2^m}^* \mid x = c^3 + c, Tr_1^m(c^{-1}) = 1\}$. *Thus, the former condition can be reworded as* $1 + 3 \sum_{x \in I} \chi(g(x)) = -2$ *that is,* $\sum_{x \in I} \chi(g(x)) = -1$. *Recall now that* $D_3(x) = x + x^3$ *and,* $D_5(x) = x + x^3 + x^5$. *So* $g(x) = Tr_1^m(a(x + x^3) + a'x^5)$.

Remark 3. We have made an exhaustive search by a computer program for $n \in \{10, 14, 18, 22\}$ of all sets $\mathfrak{S}_a$ for each value a such that $K_m(a) = 4$. Let ζ be a primitive element of $\mathbb{F}_{2^{10}}$ (whose minimal polynomial is $x^{10} + x^7 + 1$) and set $\alpha = \zeta^{33}$ (so that α is a primitive element of $\mathbb{F}_{2^5}$).

We list in Table 1 all the pairs of indices (i, j) such that $K_5(\alpha^i) = 4$ and $\alpha^j \in \mathfrak{S}_{\alpha^i}$. We have also found all pairs (i, j) for $n \in \{14, 18, 22\}$. Due to their number, we do not list them like for $n = 10$ but we only give in Table 2 the numbers of pairs that we found (including the case where $K_m(a) = 4$ and $S(a, 0) = -1$).

4   Conclusion

In this paper, we generalize the results of [15–17] to multiple trace terms functions. We provide several characterizations of hyper-bentness by means of exponential sums involving Dickson polynomials. The characterizations introduced in this paper provide new methods for exploring theoritically or by computer search for possible hyper-bent functions of the form (2). In this paper, we have restrict ourselves to the case where the coefficients a_r in (2) are in $\mathbb{F}_{2^m}$. A natural expansion of those characterizations should be to investigate their generalizations to the case where some of the coefficients are in $\mathbb{F}_{2^n}$, but not in $\mathbb{F}_{2^m}$.

References

1. Canteaut, A., Charpin, P., Kyureghyan, G.: A New Class of Monomial Bent Functions. Finite Fields and Their Applications 14(1), 221–241 (2008)
2. Carlet, C.: Boolean Functions for Cryptography and Error Correcting Codes. In: Crama, Y., Hammer, P.L. (eds.) Chapter of the monography, Boolean Models and Methods in Mathematics, Computer Science, and Engineering, pp. 257–397. Cambridge University Press, Cambridge (2010)
3. Carlet, C., Gaborit, P.: Hyperbent functions and cyclic codes. Journal of Combinatorial Theory, Series A 113(3), 466–482 (2006)
4. Carlitz, L.: Explicit evualation of certain exponential sums. Math. Scand. 44, 5–16 (1979)
5. Charpin, P., Gong, G.: Hyperbent functions, Kloosterman sums and Dickson polynomials. IEEE Trans. Inform.Theory 9(54), 4230–4238 (2008)
6. Charpin, P., Helleseth, T., Zinoviev, V.: Divisibility properties of Kloosterman sums over finite fields of characteristic two. In: ISIT 2008, Toronto, Canada, July 6–11, 2008, pp. 2608–2612 (2008)
7. Charpin, P., Kyureghyan, G.: Cubic monomial bent functions: A subclass of $\mathcal{M}$. SIAM, J. Discr. Math. 22(2), 650–665 (2008)
8. Dillon, J.: Elementary Hadamard difference sets. PhD dissertation, University of Maryland (1974)

9. Dillon, J.F., Dobbertin, H.: New cyclic difference sets with Singer parameters. Finite Fields and Their Applications 10(3), 342–389 (2004)
10. Dobbertin, H., Leander, G., Canteaut, A., Carlet, C., Felke, P., Gaborit, P.: Construction of bent functions via Niho Power Functions. Journal of Combinatorial therory, Serie A 113, 779–798 (2006)
11. Gold, R.: Maximal recursive sequences with 3-valued recursive crosscorrelation functions. IEEE Trans. Inform. Theory 14(1), 154–156 (1968)
12. Gologlu, F.: Almost Bent and Almost Perfect Nonlinear Functions, Exponential Sums, Geometries ans Sequences. PhD dissertation, University of Magdeburg (2009)
13. Lachaud, G., Wolfmann, J.: The weights of the orthogonals of the extended quadratic binary Goppa codes. IEEE Trans. Inform. Theory 36(3), 686–692 (1990)
14. Leander, G.: Monomial Bent Functions. IEEE Trans. Inform. Theory 2(52), 738–743 (2006)
15. Mesnager, S.: A new class of bent and hyper-bent boolean functions in polynomial forms. Journal Design, Codes and Cryptography (in press)
16. Mesnager, S.: A new class of bent boolean functions in polynomial forms. In: Proceedings of international Workshop on Coding and Cryptography, WCC 2009, pp. 5–18 (2009)
17. Mesnager, S.: A new family of hyper-bent boolean functions in polynomial form. In: Parker, M.G. (ed.) IMACC 2009. LNCS, vol. 5921, pp. 402–417. Springer, Heidelberg (2009)
18. Rothaus, O.: On "bent" functions. J. Combin. Theory Ser. A 20, 300–305 (1976)
19. Youssef, A.M., Gong, G.: Hyper-bent functions. In: Pfitzmann, B. (ed.) EUROCRYPT 2001. LNCS, vol. 2045, pp. 406–419. Springer, Heidelberg (2001)
20. Yu, N.Y., Gong, G.: Construction of quadratic Bent functions in polynomial forms. IEEE Trans. Inform. Theory 7(52), 3291–3299 (2006)

On the Efficiency and Security of Pairing-Based Protocols in the Type 1 and Type 4 Settings

Sanjit Chatterjee[1], Darrel Hankerson[2], and Alfred Menezes[3]

[1] Department of Combinatorics & Optimization, University of Waterloo
s2chatte@uwaterloo.ca
[2] Department of Mathematics and Statistics, Auburn University
hankedr@auburn.edu
[3] Department of Combinatorics & Optimization, University of Waterloo
ajmeneze@uwaterloo.ca

Abstract. We focus on the implementation and security aspects of cryptographic protocols that use Type 1 and Type 4 pairings. On the implementation front, we report improved timings for Type 1 pairings derived from supersingular elliptic curves in characteristic 2 and 3 and the first timings for supersingular genus-2 curves in characteristic 2 at the 128-bit security level. In the case of Type 4 pairings, our main contribution is a new method for hashing into $\mathbb{G}_2$ which makes the Type 4 setting almost as efficient as Type 3. On the security front, for some well-known protocols we discuss to what extent the security arguments are tenable when one moves to genus-2 curves in the Type 1 case. In Type 4, we observe that the Boneh-Shacham group signature scheme, the very first protocol for which Type 4 setting was introduced in the literature, is trivially insecure, and we describe a small modification that appears to restore its security.

1 Introduction

Bilinear pairings have become an extremely useful instrument in the cryptographer's toolbox. Initial breakthroughs such as the one-round tripartite key agreement protocol of Joux [23] and a practical solution to the problem of identity-based encryption by Boneh and Franklin [5] have led to an almost exponential volume of research to find novel cryptographic applications of pairings.

At an abstract level, for three groups $\mathbb{G}_1$, $\mathbb{G}_2$ and $\mathbb{G}_T$, a pairing is a function $e : \mathbb{G}_1 \times \mathbb{G}_2 \to \mathbb{G}_T$ that is bilinear and non-degenerate. For cryptographic applications we also need the pairing to be efficiently computable. In concrete settings such cryptographically suitable bilinear pairings can be realized over elliptic curves or, more generally, over hyperelliptic curves and abelian varieties. Naturally the groups $\mathbb{G}_1, \mathbb{G}_2$ and $\mathbb{G}_T$ as well as the pairing function are constrained by the underlying mathematical structure over which they are defined.

However, as noted by Galbraith, Paterson and Smart [17], protocol designers sometimes treat the bilinear pairing as a "black box". As a result the designers may gloss over such important structural constraints and the subtleties they introduce in the protocols and their security arguments. This in turn may lead to

M.A. Hasan and T. Helleseth (Eds.): WAIFI 2010, LNCS 6087, pp. 114–134, 2010.

erroneous or misleading claims about the efficiency and security of pairing-based protocols. For example, protocols employing bilinear pairings sometimes assume that the groups $\mathbb{G}_1$ and $\mathbb{G}_2$ also possess some additional properties such as efficient hashing into $\mathbb{G}_2$ or the existence of an efficiently computable isomorphism $\psi : \mathbb{G}_2 \rightarrow \mathbb{G}_1$. Different types of pairings can be realized that possess the properties required of a particular protocol, but not all protocols can be implemented using the same type of pairing.

This motivated a classification of bilinear pairings into different types based on the concrete structures of the underlying groups [17]. The focus of that work was on three types of pairings where the groups $\mathbb{G}_1$, $\mathbb{G}_2$ and $\mathbb{G}_T$ are of the same prime order n. When $\mathbb{G}_1 = \mathbb{G}_2$, the pairing is said to be symmetric (called Type 1 in [17]). The pairing is asymmetric when $\mathbb{G}_1 \neq \mathbb{G}_2$. If there is an efficiently computable isomorphism $\psi : \mathbb{G}_2 \rightarrow \mathbb{G}_1$ then the pairing is said to be of Type 2; if no such isomorphism is known it is called a Type 3 pairing. In either case no efficiently computable isomorphism from $\mathbb{G}_1$ to $\mathbb{G}_2$ is known.

Symmetric pairings (Type 1) are derived from supersingular (hyper)elliptic curves whereas asymmetric pairings are derived from ordinary curves. Known examples of such pairings are the Weil and Tate pairings and their modifications such as the eta pairing [2], the ate pairing [22], and the R-ate pairing [25].

Cryptographic protocols employing pairings are usually described in the symmetric setting, allowing for a relatively simpler description of the protocol and its security argument. However, current research indicates that, at higher security levels, Type 1 pairings are expected to be slower on many platforms. So from the point of view of efficient implementation, Type 2 and Type 3 are considered better choices. And, for a protocol originally proposed in the symmetric setting, it is usually possible to translate the protocol description and the security argument to the asymmetric setting.[1]

In the asymmetric setting, current research suggests that Type 3 is overall a better choice [17]. This is because of the reduced cost of pairing evaluation and also the relatively smaller size of elements of $\mathbb{G}_2$ which in turn reduces the cost of other operations such as group operations in $\mathbb{G}_2$ or testing membership in $\mathbb{G}_2$. The major functional distinctions between the Type 3 and Type 2 settings are that, first of all, in the former it is possible to hash into $\mathbb{G}_2$, which is infeasible in the latter; and, secondly, whereas there is an efficiently computable isomorphism $\psi : \mathbb{G}_2 \rightarrow \mathbb{G}_1$ in Type 2, no such efficiently computable map is known for Type 3. Because in some cases the description of a protocol or its security argument employed the map ψ, it was earlier thought that either such protocols cannot be implemented in Type 3 [17] or a stronger complexity assumption was needed [6,33]. Contrary to this belief it has been recently argued that any protocol or security argument in Type 2 has a natural, efficient, and secure counterpart in Type 3 [11]. Hence, in the asymmetric setting there appears to be no good reason to use Type 2 instead of Type 3.

---

[1] We are not aware of any protocol that has to be necessarily restricted to the symmetric setting.

However, not all pairing-based protocols available in the literature can be implemented in Type 3 (or Type 2). For example, consider the case of the group signature scheme of Boneh and Shacham [7] with verifier-local revocation. In this protocol a random element of $\mathbb{G}_2$ is first obtained through hashing into $\mathbb{G}_2$ and then one applies the map ψ on this element to obtain the corresponding element of $\mathbb{G}_1$. As observed in [33], the protocol cannot be implemented in Type 2 because in that setting we do not have any algorithm to securely hash into $\mathbb{G}_2$, and furthermore cannot be implemented in Type 3 because in that case we do not know how to compute ψ for a random element of $\mathbb{G}_2$.

Perhaps realizing this shortcoming of Type 2 (and Type 3), Shacham in his PhD thesis [32] introduced a new kind of pairing. In this setting, while $\mathbb{G}_1$ and $\mathbb{G}_T$ are cyclic groups of prime order n, $\mathbb{G}_2$ is taken to be a group of exponent n, whose order is some power of n. This was later termed a Type 4 pairing [12,17]. Like Type 2 and Type 3, a Type 4 pairing can be realized over ordinary elliptic or hyperelliptic curves. But unlike Type 2 or Type 3, here one can both hash into $\mathbb{G}_2$ and also have an efficiently computable homomorphism $\psi : \mathbb{G}_2 \to \mathbb{G}_1$. However, the hashing into $\mathbb{G}_2$ is reported to be quite expensive and there is a small probability that the pairing can be degenerate. The Boneh-Shacham group signature scheme of [7] is described in the Type 4 setting in [32] with a standard reductionist security argument. Several other protocols that use a Type 4 pairing have been proposed [29,8] based on the Boneh-Shacham scheme. Thus, protocols that require hashing into $\mathbb{G}_2$ followed by an application of ψ can be implemented in the Type 4 setting although the protocol description may require some special care, and as noted in [12] the security argument can become cumbersome.

Protocols such as the Boneh-Shacham group signature scheme [7] can also be easily implemented in Type 1 because here $\mathbb{G}_1 = \mathbb{G}_2$ and hashing into $\mathbb{G}_1$ is very efficient. Also recall that most pairing-based protocols were originally proposed in this setting.[2] The main drawback of Type 1 is that the bitlengths of the elements of $\mathbb{G}_1$ will be larger (because of the smaller embedding degrees than what is achievable with asymmetric pairings) and, as a result, pairing computation and operations in $\mathbb{G}_1$ can be expected to be slower at high security levels. However, instructions on next-generation processors such as the forthcoming Intel machines may make Type 1 in characteristic 2 (and 3) fields an attractive choice. Some authors [2,31] have also proposed to use genus-2 curves in the symmetric setting and use degenerate divisors to speed the pairing computation.

Our contribution. For efficient and secure implementation of the majority of pairing-based protocols it suffices to work in the Type 3 setting. However, as the preceding discussion suggests, we also need to consider the issues of efficient and secure implementation of protocols in the Type 1 and Type 4 settings. While the question of efficiency does not require any additional justification, we draw attention to the question of security of a cryptographic protocol in these settings for the following reasons. A protocol described in the Type 1 setting

---

[2] We note that not all protocols can be implemented securely in Type 1, e.g., those requiring the extended Diffie-Hellman problem (XDH or SXDH) to be hard [4,9,13].

which is implemented over a genus-2 curve may require a rewriting of the security argument to check whether the original security assurance is indeed maintained in this setting. Similarly, protocols described in the Type 4 setting may require special scrutiny because of the structure of the group $\mathbb{G}_2$, in particular its effect on the way the pairing is actually employed in the protocol. In this work we report on both these aspects of efficiency and security in the Type 1 and Type 4 settings.

Type 1. Following recent work of Beuchat et al. [3] and Aranha et al. [1], we provide improved timings for software implementation of Type 1 pairings over elliptic curves in characteristic 2 and characteristic 3 fields. We also report the first pairing timings for supersingular genus-2 curves at the 128-bit security level. We next take a look at the security arguments of some well-known protocols when implemented with these genus-2 curves and with degenerate divisors. Our analysis shows that for the Boneh-Lynn-Shacham (BLS) signature scheme [6] one needs a new hardness assumption that is trivially equivalent to the security of the scheme. In other words, the reductionist argument does not provide any meaningful assurance about the actual security of the protocol in this setting. A similar analysis is carried out for the Boneh-Franklin IBE scheme [5] and we observe that here also one needs to modify the original security assumption.

Type 4. As already mentioned, the main motivation for working in Type 4 is that it is possible to hash into $\mathbb{G}_2$. However, in terms of efficiency that appears to be a major limitation of the Type 4 setting as hashing into $\mathbb{G}_2$ has been reported to be computationally quite expensive [12,17]. Here we propose a new technique to hash into $\mathbb{G}_2$ which is surprisingly cheap. This method is built upon the shorter representation of elements of $\mathbb{G}_2$ proposed in [10] in the context of the Type 2 setting. We also report the performance benefits that can be obtained for pairing evaluation and other operations involving elements of $\mathbb{G}_2$ in the Type 4 setting. As we have already noted, Type 4 pairings should be carefully used in cryptographic protocols. We show that the Boneh-Shacham group signature scheme as described in Shacham's thesis [32] is trivially insecure. We describe a small modification that appears to restore security. The signature now contains an element of $\mathbb{G}_2$, however with our new representation of elements of $\mathbb{G}_2$ the corresponding increase in the signature size is not very significant.

Organization. The remainder of the paper is organized as follows. In §2 we report the pairing computation times in Type 1 when using degenerate divisors in genus-2 curves over characteristic 2 fields at the 128-bit security level and also discuss the security aspects of the BLS signature and Boneh-Franklin IBE schemes in this setting. In §3 we describe the implementation aspects of Type 4 pairings derived from ordinary elliptic curves having even embedding degree and show how one can efficiently hash into $\mathbb{G}_2$. We then show that the Boneh-Shacham group signature scheme as described [32] is insecure and how a small modification appears to restore security without significant performance penalty.

Notation. In the remainder of this paper, the first component $\mathbb{G}_1$ of the domain of a pairing $e : \mathbb{G}_1 \times \mathbb{G}_2 \to \mathbb{G}_T$ is the order-n subgroup of $E(\mathbb{F}_q)$ or $J_C(\mathbb{F}_q)$, where E is an elliptic curve defined over $\mathbb{F}_q$ and J_C is the divisor class group of a genus-2 hyperelliptic curve C defined over $\mathbb{F}_q$. If e is a Type 1 pairing, then $\mathbb{G}_2 = \mathbb{G}_1$. If e is a Type 2, Type 3 or Type 4 pairing, then $\mathbb{G}_2 = \mathbb{T}$, $\mathbb{T}_0$, $E[n]$, respectively, where $E[n]$ is the n-torsion group of E, $\mathbb{T}_0$ is the Trace-0 subgroup of $E[n]$ and $\mathbb{T}$ is any order-n subgroup of $E[n]$ different from $\mathbb{G}_1$ and $\mathbb{T}_0$; cf. §3 for further details.

2   Type 1 Pairings on Supersingular Genus-2 Curves

In this section, we give the context for performance comparisons at the 128-bit security level for pairings based on supersingular genus-2 curves defined over characteristic 2 finite fields against those built on elliptic curves. Security aspects of the BLS signature scheme and the Boneh-Franklin IBE scheme are discussed in the genus-2 setting.

2.1   Type 1 Pairings

We briefly describe three specific symmetric pairings derived from supersingular elliptic and hyperelliptic curves defined over fields of small characteristic; see [2] for details. The elliptic curves E are defined over $\mathbb{F}_{2^{1223}}$ and $\mathbb{F}_{3^{509}}$, and have embedding degrees 4 and 6, respectively. The genus-2 curve C is defined over $\mathbb{F}_{2^{439}}$ and has embedding degree 12. The pairings are $e : \mathbb{G}_1 \times \mathbb{G}_1 \to \mathbb{G}_T$, where $\mathbb{G}_1$ is the subgroup of prime-order n of $E(\mathbb{F}_{2^{1223}})$, $E(\mathbb{F}_{3^{509}})$ or $J_C(\mathbb{F}_{2^{439}})$, and $\mathbb{G}_T$ is the order-n subgroup of $\mathbb{F}_{2^{4 \cdot 1223}}^*$, $\mathbb{F}_{3^{6 \cdot 509}}^*$ or $\mathbb{F}_{2^{12 \cdot 439}}^*$, respectively. These pairings attain the 128-bit security level because Pollard's rho method for computing discrete logarithms in $E(\mathbb{F}_{2^{1223}})$, $E(\mathbb{F}_{3^{509}})$ and $J_C(\mathbb{F}_{2^{439}})$ has running time at least 2^{128}, as do the index-calculus algorithms for computing discrete logarithms in the extension fields $\mathbb{F}_{2^{4 \cdot 1223}}$, $\mathbb{F}_{3^{6 \cdot 509}}$ and $\mathbb{F}_{2^{12 \cdot 439}}$ [27].

For genus 2, we focus on the most favourable case where the pairing is on degenerate divisors, each of which is essentially a point on the curve. The pairing algorithms given in [2] for the cases under consideration are similar in the sense that there is a "Miller evaluation" loop, followed by an exponentiation in the extension field to select a canonical representative. The final exponentiation is relatively inexpensive, and so the pairing cost can be estimated by counting field multiplications in the main loop.

Elliptic curve over characteristic 2 field. Let $q = 2^{1223}$. We chose the representation $\mathbb{F}_{2^{1223}} = \mathbb{F}_2[z]/(z^{1223} + z^{255} + 1)$. Squaring is inexpensive relative to multiplication, and square roots are likewise inexpensive in this representations since $\sqrt{z} = z^{612} + z^{128}$ and $\sqrt{c} = \sum c_{2i} z^i + \sqrt{z} \sum c_{2i+1} z^i$ for $c = \sum c_i z^i \in \mathbb{F}_{2^{1223}}$. The extension field $\mathbb{F}_{q^4}$ is represented using tower extensions $\mathbb{F}_{q^2} = \mathbb{F}_q[u]/(u^2 + u + 1)$ and $\mathbb{F}_{q^4} = \mathbb{F}_{q^2}[v]/(v^2 + v + u)$.

The supersingular elliptic curve $E_1/\mathbb{F}_{2^{1223}} : y^2 + y = x^3 + x$ has embedding degree 4. We have $\#E_1(\mathbb{F}_{2^{1223}}) = 5n$ where $n = (2^{1223} + 2^{612} + 1)/5$ is a 1221-bit prime. The doubling formula is $(x, y) \mapsto (x^4 + 1, x^4 + y^4 + 1)$, and hence the cost of doubling a point is relatively small.

Barreto, Galbraith, Ó hÉigeartaigh and Scott [2] give an algorithm for computing the η_T pairing, with cost estimated as $612 \times 7 = 4284$ $\mathbb{F}_q$-multiplications. The estimate is based on the number of multiplications in the main loop, and ignores the relatively minor cost of the final exponentiation (1 inversion in $\mathbb{F}_{q^4}$, 3 multiplications in $\mathbb{F}_{q^4}$, and 612 squarings in $\mathbb{F}_q$).

Elliptic curve over characteristic 3 field. Let $q = 3^{509}$. We chose the representation $\mathbb{F}_{3^{509}} = \mathbb{F}_3[z]/(z^{509} - z^{318} - z^{191} + z^{127} + 1)$. Cubing is inexpensive relative to multiplication, and the choice of reduction polynomial enables cube roots to be computed significantly faster than an $\mathbb{F}_q$-multiplication since $z^{1/3} = z^{467} + z^{361} - z^{276} + z^{255} + z^{170} + z^{85}$ and $z^{2/3} = -z^{234} + z^{128} - z^{43}$. The extension field $\mathbb{F}_{q^6}$ is represented using tower extensions $\mathbb{F}_{q^3} = \mathbb{F}_q[u]/(u^3 - u - 1)$ and $\mathbb{F}_{q^6} = \mathbb{F}_{q^3}[v]/(v^2 + 1)$.

The supersingular elliptic curve $E_2/\mathbb{F}_{3^{509}} : y^2 = x^3 - x + 1$ has embedding degree 6. We have $\#E_2(\mathbb{F}_{3^{509}}) = 7n$ where $n = (3^{509} - 3^{255} + 1)/7$ is an 804-bit prime. The tripling formula is $(x, y) \mapsto (x^9 - 1, -y^9)$, and hence the cost of tripling a point is relatively small.

The algorithm of Barreto, Galbraith, Ó hÉigeartaigh and Scott [2] for computing the η_T pairing has a cost estimate of $255 \times 14 = 3570$ $\mathbb{F}_q$-multiplications. As in the characteristic 2 case, the relatively minor cost of the final exponentiation has been ignored.

Genus 2 curve over characteristic 2 field. Let $m = 439$ and $q = 2^m$. We chose the representation $\mathbb{F}_{2^{439}} = \mathbb{F}_2[z]/(z^{439} + z^{49} + 1)$. Squaring and square root are inexpensive relative to multiplication in this representation, with $\sqrt{z} = z^{220} + z^{25}$. The extension field $\mathbb{F}_{q^{12}}$ is represented using tower extensions $\mathbb{F}_{q^6} = \mathbb{F}_q[w]/(w^6 + w^5 + w^3 + w^2 + 1)$ and $\mathbb{F}_{q^{12}} = \mathbb{F}_{q^6}[s]/(s^2 + s + w^5 + w^3)$.

The curve $C/\mathbb{F}_{2^{439}} : y^2 + y = x^5 + x^3$ has embedding degree 12. The divisor class group J_C has $\#J_C(\mathbb{F}_q) = 2^{2m} + 2^{(3m+1)/2} + 2^m + 2^{(m+1)/2} + 1 = 13n$, where n is an 875-bit prime. The pairing is defined for divisors $D = (P_1) + (P_2) - 2(\infty)$ where P_i are points on the curve; however the computation is faster for degenerate divisors where the support consists of a single point [2,26]. If $D = (P) - (\infty)$ is such a degenerate divisor (with $P \in C(\mathbb{F}_q)$), then it is not necessarily the case that jD is degenerate; however, $8D$ is degenerate [2]. Furthermore, this octupling is relatively inexpensive, and is given by $8D = (\phi\pi^6 P) - (\infty)$ where $\pi(x, y) = (x^2, y^2)$ and $\phi(x, y) = (x + 1, y + x^2 + 1)$. Exploiting this octupling, the algorithm in [2] for η_T on degenerate divisors has an approximate cost of $219 \cdot 69 = 15111$ $\mathbb{F}_q$-multiplications (see [30] for additional details).

Comparisons. Barreto et al. [2] give experimental data for pairing times at the "950-bit" and "1230-bit" security levels, where the level is in terms of the bitsize of the extension field $\mathbb{F}_{q^k}$. Times (in milliseconds) for the η_T pairing in the 1230-bit case on a 3 GHz Intel Pentium 4 are given in Table 1.

Their work shows significant incentive to use genus-2 curves in the case that the pairing is on degenerate divisors. However, field multiplication for $\mathbb{F}_{2^{103}}$ exploited 128-bit single-instruction multiple-data (SIMD) registers on the Pentium 4, while the other fields used only 32-bit registers. The rationale for limiting the

Table 1 Times (in milliseconds) from [2] for the η_T pairing at the "1230-bit security level" on a 3 GHz Intel Pentium 4

		$C(\mathbb{F}_{2^{103}})$	
$E(\mathbb{F}_{2^{307}})$	$E(\mathbb{F}_{3^{127}})$	degenerate	general
3.50	5.36	1.87	6.42

wide registers to the genus-2 case was that "Great potential savings can be realized if an element of the base field can be represented in a single machine word, rather than using a multi-precision representation" and a factor 2 acceleration was reported for field multiplication via the wide registers.

This difference in implementations is especially significant since the Pentium 4 is 32-bit. The techniques are perhaps less elegant when applied to larger fields, but similar acceleration can be obtained via wide registers for the fields in the other pairings. For example, [20] examined the acceleration offered by SIMD registers for pairings at a higher security level using the fields $\mathbb{F}_{2^{1223}}$ and $\mathbb{F}_{3^{509}}$, but did not consider an example from genus 2. Beuchat et al. [3] and Aranha et al. [1] subsequently demonstrated significantly faster field arithmetic on platforms considered in [20].

In short, the implementation techniques for the times in [2] favour the genus-2 curve. If the registers used in $\mathbb{F}_{2^{103}}$ were applied to $\mathbb{F}_{2^{307}}$, then we would expect that the pairing times would be significantly closer. On the other hand, we are interested in the 128-bit security level, where the higher embedding degree of the genus-2 curve is an advantage. Our intent here is to give a meaningful comparison at the 128-bit security level among the various pairings on a "reference platform" using whatever methods are believed to be fastest in each scenario. The Pentium 4 is no longer of primary interest, and so we chose the popular 64-bit Intel Core2.

Timings for our implementations appear in Table 2. Pairings for the Barreto-Naehrig (BN) curve (see §3.1) over a prime field are expected to be fastest at this security level, in part because the embedding degree is 12 and the platform possesses a relatively fast integer multiplier on 64-bit operands. Details on this timing using the MIRACL library appear in [20].

Beuchat et al. [3] discuss optimization strategies and set benchmarks for pairing times in the elliptic curve cases over characteristic 2 and 3. As in [1], a focus is on parallelizing the pairing computation, although the times for a single-core computation were also impressive. The times in Table 2 are faster, but are consistent in the sense that characteristic 3 offers an advantage. On the other hand, this advantage is not as large as in [3], mainly due to the difference in characteristic 2 multiplication.

Compared with [2], applying the wide registers across fields has narrowed differences. Genus 2 has lost much of the performance advantage, although it may still be attractive from an implementation and keysize perspective for protocols having pairings on degenerate divisors. The gap between the pairing from the BN curve and those over characteristic 2 and 3 is perhaps narrower than expected.[3]

---

[3] The comparison in [3] is against the slower ate pairing, which gives the timing for the BN curve as 15×10^6 cycles.

Table 2. Timings (in clock cycles) on an Intel Core2. Field operations in characteristic 2 and 3 use 128-bit SIMD registers and exploit shift and shuffle instructions introduced with SSSE3. The timing for the BN curve is from [20].

Field, curve, and pairing	Field mult	Pairing
$E/\mathbb{F}_{p256}$, R-ate	.31	10
$E/\mathbb{F}_{3^{509}}$, η_T	3.86	15.8
$E/\mathbb{F}_{2^{1223}}$, η_T	3.84	19.0
$C/\mathbb{F}_{2^{439}}$, η_T on degenerate divisors	.86	16.4

Units: 10^3 cycles 10^6 cycles

The experimental data in Table 1 gives a factor 3.4 penalty for a pairing on general divisors. In special cases, the cost will be less. Nondegenerate divisors $(P_1)+(P_2)-2(\infty)$ are of two forms, either $P_i \in C(\mathbb{F}_q)$ or P_1 and P_2 are conjugates in $C(\mathbb{F}_{q^2})\backslash C(\mathbb{F}_q)$. A pairing on divisors can be calculated as a product of pairings on points; e.g., in the case where $P_i \in C(\mathbb{F}_q)$, $\eta_T((P_1)+(P_2)-2(\infty),(P)-(\infty)) = \eta_T(P_1,P)\eta_T(P_2,P)$ at twice the cost of a pairing on degenerate divisors. However, this approach may not be the most efficient when points lie in $C(\mathbb{F}_{q^2})\backslash C(\mathbb{F}_q)$ [2]. Lee and Lee [26] give explicit formulas for the pairing on general divisors, with estimated cost (from field multiplications, where an $\mathbb{F}_{q^{12}}$-multiplication is counted as 45 $\mathbb{F}_q$-multiplications) as a factor 4 over the pairing on degenerate divisors.

Implementation notes. Compared with [3] and [20], the characteristic 2 multiplier (described in [1]) uses twice as much data-dependent precomputation but fewer shift operations. Some of the improvement against [20] was achieved by reducing the number of move operations (a weakness underestimated in [20]), although a portion of the acceleration was obtained by exploiting a shift operation introduced with the Supplemental Streaming SIMD Extension 3 (SSSE3).[4] The faster shift is also useful in characteristic 3 – additions are more expensive than in characteristic 2, but field multiplication performs more shifting. Multiplication for $\mathbb{F}_{2^{1223}}$ is via one application of Karatsuba where elements are split at 616 bits (a multiple of 8 that allows fast shifting and eliminates the "fixup" required in combing on n-word input when both inputs have length greater than $nW - w$ for word-length W and comb width w). Multiplication in the 439-bit field is via combing directly on field elements. Combing uses two tables, each of 16 elements.

The strategies for characteristic 2 and 3 are similar, in part because an $\mathbb{F}_3$-element is represented as a pair (a_0, a_1) of bits and addition involves only bitwise operations. Harrison et al. [21] proposed an addition using 7 XOR ($\oplus$) and OR ($\vee$) operations via the sequence: $t \leftarrow (a_0 \vee b_1) \oplus (a_1 \vee b_0)$, $c_0 \leftarrow (a_1 \vee b_1) \oplus t$, $c_1 \leftarrow (a_0 \vee b_0) \oplus t$. The number of operations was reduced to 6 by Kawahara et al. [24] who reported 7–8% improvement in field multiplication on an AMD

---

[4] SSSE3 was also exploited in [1] to obtain very fast squaring and root for a parallel implementation that performed these operations in excess; these accelerations give only minor reduction in pairing times here (e.g., 6% for the pairing over $\mathbb{F}_{2^{1223}}$).

Opteron (a processor similar to the Intel Core2) using "non-standard" encodings of $\mathbb{F}_3$-elements. They also gave a 6-operation addition in the encoding suggested by [21] using XOR and ANDN ($x \wedge \bar{y}$).

The SIMD instruction set on the Core2 includes ANDN, although Beuchat et al. [3] reported that the 7-op addition "consistently yields a shorter computation time" than the 6-op variant with ANDN, and speculated that ANDN on the Core2 "is implemented less efficiently" than XOR and OR. However, these instructions have the same timings [14], and our experimental data is that field multiplication is faster with the 6-op variant. We suspect that the discrepancy is due to register allocation strategy in the accumulation portion of the multiplication method. A variation on the formulation in [21] is proposed in [3]: $t \leftarrow (a_0 \vee a_1) \wedge (b_0 \vee b_1)$, $c_0 \leftarrow (a_0 \vee b_0) \oplus t$, $c_1 \leftarrow (a_1 \vee b_1) \oplus t$. Specifics are not given on why this resulted in faster code, but we note that it permits simpler register tracking in the accumulation portion of field multiplication and has one operation on accumulator registers only. In this sense, the formulation in [21] and the 6-op variant for $c \leftarrow c + b$ are less pleasant. Compilers can be quite sensitive to the precise form of the code; however, the 6-op variant can be coded without increasing dependency chains, and we expect this formulation to be fastest provided that unnecessary moves are avoided. Experimentally, we observed roughly 10% faster times for field multiplication.

As in [3], we use the loop-unrolling technique of [19] along with the $\mathbb{F}_{3^{6m}}$ multiplication of [18] (requiring 15 multiplications and 67 additions in $\mathbb{F}_{3^m}$) to accelerate the pairing computation in characteristic 3. This reduces the cost from 14 to an effective 12.5 $\mathbb{F}_{3^m}$ multiplications in each iteration of the Miller loop. Some incremental accelerations noted in [3] were not implemented; for example, a few tables of precomputation in the evaluation loop of the pairing computation can be reused (a width-4 comb requires 81 elements of precomputation, although half are obtained by simple negation).

Timings were done on a 2.4 GHz Intel Core2-quad running Sun Solaris, using the GNU C 4.1 compiler with some fragments written in assembly. Most of the SIMD operations are via intrinsics, with SSSE3 instructions accessed via assembly.

2.2   Security of Protocols Using Degenerate Divisors

The principal motivation for considering hyperelliptic curves and degenerate divisors for pairing-based protocols is to speed the pairing computation at higher security levels. Naturally we need the assurance that the protocol, when implemented in this setting, maintains its original security guarantee. The question of security received some attention in [15,2], however the main emphasis in those works is on efficient pairing computation. Here we take a closer look at the security argument of two well-known protocols when implemented in the setting of §2.1.

BLS signature scheme. We first describe the BLS signature scheme [6] using symmetric pairings (Type 1). We then present two variants of the scheme depending upon which particular elements are chosen to be degenerate and examine the resulting effect on the security argument.

Let $e : \mathbb{G}_1 \times \mathbb{G}_1 \to \mathbb{G}_T$ be a Type 1 pairing on a genus-2 curve, and let $H : \{0,1\}^* \to \mathbb{G}_1$ be a hash function. Let P_1 be a known generator of $\mathbb{G}_1$. The public parameters of the system are $\langle \mathbb{G}_1, \mathbb{G}_T, H, P_1 \rangle$.

Alice's private key is an integer $x \in_R [0, n-1]$ and her corresponding public key is $X = xP_1$. To sign a message M, Alice computes $Q = H(M)$ and then $\sigma = xQ$ as her signature on M. To verify, Bob computes $Q = H(M)$ and accepts σ as a valid signature on M if and only if

$$e(\sigma, P_1) = e(Q, X). \tag{1}$$

Correctness of the verification algorithm follows because of the bilinearity property of e, i.e.,

$$e(\sigma, P_1) = e(xQ, P_1) = e(Q, xP_1) = e(Q, X).$$

Security of the scheme is based on the hardness of the computational Diffie-Hellman problem (DHP) in $\mathbb{G}_1$ assuming H to be a random oracle. Recall that the DHP in $\mathbb{G}_1 = \langle P_1 \rangle$ is the following: given X (where $X = xP_1$ for some $x \in_R [0, n-1]$) and $Q \in_R \mathbb{G}_1$, compute xQ. The essential ideas behind the reductionist security argument are as follows. Given a DHP instance (X, Q), the simulator sets the challenge public key as X and runs the BLS adversary $\mathcal{A}$. The simulator responds to all hash queries $H(M)$ made by $\mathcal{A}$, except for a randomly chosen distinguished query, by selecting $a \in_R [0, n-1]$ and setting $H(M) = aP_1$; the response to the distinguished hash query $H(M^*)$ is $H(M^*) = Q$. The simulator responds to signing queries $M \neq M^*$ by setting $\sigma = aX$. If $\mathcal{A}$ eventually produces a forged signature σ^* on M^*, then the simulator has been successful in obtaining the solution σ^* to the DHP instance (X, Q).

Recall from §2.1 that C is a supersingular genus-2 curve over $\mathbb{F}_q$, $q = 2^m$, and $\mathbb{G}_1$ is the set of n-torsion points in $J_C(\mathbb{F}_q)$ where $n \approx q^2$ is prime. Let $\mathcal{D}$ be the set of degenerate divisors in $\mathbb{G}_1$; then $\#\mathcal{D} \approx q$ since $\#C(\mathbb{F}_q) \approx q$. For efficient implementation we would like to have some (if possible all) of the elements of $\mathbb{G}_1$ used in the protocol to lie in $\mathcal{D}$. However, the choice is constrained by how these elements are actually generated in the protocol, and whether the protocol environment can be properly simulated in the security argument. We further elaborate on these issues based on the following two versions of BLS. We use calligraphic fonts for degenerate divisors to distinguish them from general divisors.

BLS-1a: The key generation algorithm chooses a random element $\mathcal{P}_1$ of $\mathcal{D}$ as the system parameter and a hash function $H : \{0,1\}^* \to \mathcal{D}$. Both these tasks can be accomplished without any security penalty and is as efficient as working in the elliptic curve setting.[5] Then, with overwhelming probability, Alice's public

---

[5] In [16, §7] the concern was raised that hashing to the set of degenerate divisors in $J_C(\mathbb{F}_q)$ instead of to the set of general divisors can lead to a loss security. This is because hash collisions in the former case can be found in $O(q^{1/2})$ time using generic algorithms, whereas collision finding in the latter case takes $O(q)$ time. However, the concern is not an issue in our setting with $q = 2^{439}$ because then $\sqrt{q} \approx 2^{219}$ which is significantly greater than the target security level of 2^{128}.

key $x\mathcal{F}_1$ will *not* be a degenerate divisor. Since the range of H is $\mathcal{D}$, $Q = H(M)$ is a degenerate divisor. But $\sigma = xQ$ will likely be non-degenerate (with overwhelming probability). As a result, one of the arguments in each of the two pairing computations in the verification equation (1) is a degenerate divisor while the other is non-degenerate. This still makes the pairing computation faster compared to the case when both arguments are general divisors.

Next we investigate to what extent the original security argument of BLS is applicable in the case of BLS-1a. The DHP (with respect to the generator $\mathcal{P}_1$ of $\mathbb{G}_1$) is the problem of determining xQ, given $X = x\mathcal{P}_1 \in \mathbb{G}_1$ for some unknown $x \in_R [0, n-1]$ and $Q \in_R \mathbb{G}_1$. The natural choice would be to argue security of BLS-1a based on the hardness of the following variant of DHP. Given $X = x\mathcal{P}_1 \in \mathbb{G}_1$ for some unknown $x \in_R [0, n-1]$ and $Q \in_R \mathcal{D}$, compute xQ — we call this problem DHP*. The following shows that DHP and DHP* are computationally equivalent.

Lemma 1. *The DHP and DHP* problems are computationally equivalent.*

Proof. It is clear that DHP* reduces to DHP. To prove the converse, suppose that we are given a DHP instance (X, Q) and an oracle for solving DHP*. If $Q \in \mathcal{D}$ then the DHP*-oracle can be used to compute xQ. If $Q \notin \mathcal{D}$, say $Q = (P_1) + (P_2) - 2(\infty)$, there are two cases to consider. Let us say that Q is of type A if $P_1, P_2 \in C(\mathbb{F}_q)$ and of type B if $P_1, P_2 \in C(\mathbb{F}_{q^2}) \setminus C(\mathbb{F}_q)$.

Suppose first that Q is of type A. This case can be recognized because the Mumford representation (see [28]) of Q will take the form (a, b), where $a, b \in \mathbb{F}_q[x]$ with $\deg(a) = 2$, $\deg(b) \leq 1$, and where the roots of a belong to $\mathbb{F}_q$. More explicitly, if $a(x) = (x - u_1)(x - u_2)$ with $u_1, u_2 \in \mathbb{F}_q$, then $P_1 = (u_1, v_1)$ and $P_2 = (u_2, v_2)$ where $v_1 = b(u_1)$ and $v_2 = b(u_2)$. Thus, we can efficiently write $Q = Q_1 + Q_2$, where $Q_1 = (P_1) - (\infty)$ and $Q_2 = (P_2) - (\infty)$ are degenerate divisors. The DHP*-oracle can then be used to compute xQ_1 and xQ_2, from which $xQ = xQ_1 + xQ_2$ is immediately obtained.

Suppose now that Q is of type B. In this case, we can multiply Q by randomly-selected integers $\ell \in [1, n-1]$ until the resulting divisor Q' is of type A. The expected number of trials is 2, since the number of type A divisors in $J_C(\mathbb{F}_q)$ is approximately $q^2/2$, as is the number of type B divisors in $J_C(\mathbb{F}_q)$ (this follows because $\#C(\mathbb{F}_q) \approx q$ and $\#C(\mathbb{F}_{q^2}) \approx q^2$). As above, one can then compute xQ' and hence $xQ = \ell^{-1}(xQ')$. □

Now, given a DHP* instance (X, Q), the simulator sets the challenge public key as X and interacts with the BLS-1a adversary $\mathcal{A}$. To properly answer $\mathcal{A}$'s signing query on a message M, the simulator has to "program" the random oracle in such a way that it outputs some $\mathcal{H} \in_R \mathcal{D}$ for which the simulator knows the discrete log with respect to $\mathcal{P}_1$. Recall that this was trivially accomplished for the original protocol — the simulator first chose $a \in_R [0, n-1]$ and then returned $a\mathcal{P}_1$. However, to apply this strategy in the simulation of BLS-1a, the simulator must satisfy the additional constraint that $a\mathcal{P}_1$ lies in $\mathcal{D}$.

The simulator could easily satisfy this condition if given some fixed $\mathcal{P} \in \mathcal{D}$ she has some mechanism to choose a random a such that $a\mathcal{P}$ also belongs to $\mathcal{D}$.

The only known way to guarantee this in our genus-2 setting is to choose a to be a power of 8, i.e., if $\mathcal{P} \in \mathcal{D}$ then $8^i\mathcal{P}$ is also a degenerate divisor for any integer i. However, as the following lemma indicates, the hash output will then be confined to an extremely small subset of $\mathcal{D}$ (and thus the simulation will fail).

Lemma 2. *Let $\mathcal{P}$ be a degenerate divisor of order n in $J_C(\mathbb{F}_q)$, where C is the supersingular genus-2 curve over $\mathbb{F}_q$ (with $q = 2^m$) defined in §2.1. Then there are exactly $4m$ degenerate divisors of the form $8^i\mathcal{P}$.*

Proof. We have $q^{12} \equiv 1 \pmod{n}$, and so $8^{4m} \equiv 1 \pmod{n}$. Hence, the order of 8 modulo n is in $\{1, 2, 4, m, 2m, 4m\}$. Since $n > 8^4$, $8^m \not\equiv 1 \pmod{n}$ and $8^{2m} \not\equiv 1 \pmod{n}$, the order of 8 modulo n must be $4m$. $\square$

One way to circumvent this problem is to define a new problem which we call DHP* with oracle access and denote by DHP$^*_\mathcal{O}$. In addition to the DHP* instance $X = x\mathcal{P}_1$ and $\mathcal{Q}$, the solver (i.e., the BLS-1a simulator in the present context) is given access to an oracle $\mathcal{O}$. Each time it is invoked, the oracle $\mathcal{O}$ returns a random $\mathcal{P} \in \mathcal{D}$ along with $x\mathcal{P}$. It is easy to argue that the security of BLS-1a is equivalent to the hardness of DHP$^*_\mathcal{O}$. For example, when reducing DHP$^*_\mathcal{O}$ to the problem of breaking BLS-1a, the simulator returns $\mathcal{P}$ when $\mathcal{A}$ queries the random oracle on some message M, and subsequently $x\mathcal{P}$ in response to a signature query on M (where the simulator obtains $(\mathcal{P}, x\mathcal{P})$ from its oracle $\mathcal{O}$). At some point, $\mathcal{A}$ returns a valid forgery on some message M^* whose hash value has been set to $\mathcal{Q}$. The simulator returns this signature as the solution to the given DHP$^*_\mathcal{O}$ instance.

However, there is a circularity in the whole argument — the assumption that it is hard to solve DHP$^*_\mathcal{O}$ is nothing but a rephrasing of the assertion that it is hard to forge a BLS-1a signature. Currently we do not know any way out of this circularity based on the known security argument for BLS. Neither is there any evidence to suggest that BLS-1a is insecure.

BLS-1b: Alternatively, we can keep the range of the hash function H to be $\mathbb{G}_1$ and choose only the fixed system parameter $\mathcal{P}_1$ to be a degenerate divisor. With this modification, we still make some efficiency gains in the verification algorithm namely in the evaluation of $e(\sigma, \mathcal{P}_1)$. The known security argument for BLS with respect to DHP can now be easily adapted for BLS-1b.

Boneh-Franklin identity-based encryption scheme. The situation is similar for the BF-IBE scheme [5]. We assume the reader is familiar with the basic idea of the protocol. Suppose that the public parameters are $\langle \mathbb{G}_1, \mathbb{G}_T, H, P_1 \rangle$, the Key Generation Centre's public key is $D_{\mathsf{pub}} \in \mathbb{G}_1$, and the public key corresponding to an arbitrary identity ID is obtained as $Q_{\mathsf{ID}} = H(\mathsf{ID})$. Encryption involves the computation of a pairing value $e(D_{\mathsf{pub}}, Q_{\mathsf{ID}})$. The hardness of BF-IBE is based on the so-called bilinear Diffie-Hellman (BDH) problem — given aP_1, bP_1, cP_1 for $a, b, c \in_R [0, n-1]$, compute $e(P_1, P_1)^{abc}$.

In [2], Barreto et al. suggest that without loss of security it is possible to choose both D_{pub} and Q_{ID} to be degenerate divisors so that encryption involves pairing of two degenerate divisors. We note that the known security argument of BF-IBE suffers from the same problem that we encountered in BLS-1a, namely it is not possible to simulate the random oracle H with range $\mathcal{D}$. The security argument does however go through if one chooses only D_{pub} to be degenerate. Now in the encryption algorithm only one of the arguments to the pairing function is a degenerate divisor while the other is a general divisor. In this case, the corresponding instance of the BDH problem contains one degenerate divisor and two general divisors. As was done in Lemma 1, one can prove that this variant of BDH is equivalent to the original BDH problem.

Remark 1. We have not found any pairing-based protocols in the literature that can be implemented so that both arguments to one or more of the pairing functions are degenerate divisors, and where the original security argument in the elliptic curve setting can be carried over to the genus-2 setting. Thus, the speed benefits of using pairings in the genus-2 setting where both arguments are degenerate divisors do not seem to be directly applicable to known protocols.

3   Type 4 Pairings

Let E be an ordinary elliptic curve defined over the finite field $\mathbb{F}_q$. Let n be a prime divisor of $\#E(\mathbb{F}_q)$ satisfying $\gcd(n, q) = 1$, and let k (the embedding degree) be the smallest positive integer such that $n \mid q^k - 1$. We will assume that k is even. Since $k > 1$, we have $E[n] \subseteq E(\mathbb{F}_{q^k})$. We will further assume that $n^3 \nmid \#E(\mathbb{F}_{q^k})$. Let $\mathbb{G}_T$ be the order-n subgroup of $\mathbb{F}_{q^k}^*$. The (full) Tate pairing is a non-degenerate bilinear function $\hat{e} : E[n] \times E[n] \to \mathbb{G}_T$ and can be defined as follows:

$$\hat{e}(P, Q) = \left(\frac{f_{n,P}(Q + R)}{f_{n,P}(R)} \right)^{(q^k - 1)/n}, \tag{2}$$

where $R \in E(\mathbb{F}_{q^k})$ with $R \notin \{\infty, P, -Q, P - Q\}$, and where the *Miller function* $f_{n,P}$ is a function whose only zeros and poles in E are a zero of order n at P and a pole of order n at ∞.

Let $\mathbb{G}_1 = E(\mathbb{F}_q)[n]$. If the first component of the domain of $\hat{e}$ is restricted to $\mathbb{G}_1$, then the definition of $\hat{e} : \mathbb{G}_1 \times E[n] \to \mathbb{G}_T$ simplifies to $\hat{e}(P, Q) = (f_{n,P}(Q))^{(q^k - 1)/n}$. Such a mapping $\hat{e}$ is called a *Type 4 pairing* [32] because the second component of the domain of $\hat{e}$ is the full n-torsion group $E[n]$. The Trace function Tr defined by $\text{Tr}(P) = \sum_{i=0}^{k-1} \pi^i(P)$, where π denotes the q-th power Frobenius, is an efficiently-computable homomorphism from $E[n]$ to $\mathbb{G}_1$. The kernel of Tr, called the *Trace-0 group*, is an order-n subgroup of $E[n]$. Hashing onto $\mathbb{G}_1$ can be efficiently computed by first hashing to an x-coordinate of $E(\mathbb{F}_q)$, then solving a quadratic equation over $\mathbb{F}_q$ to find the corresponding y-coordinate, and finally multiplying the resulting point by the cofactor $h_1 = \#E(\mathbb{F}_q)/n$ to obtain an n-torsion point. Hashing onto $E[n]$ can be accomplished in a similar fashion, by first hashing onto a random point in $E(\mathbb{F}_{q^k})$ and then multiplying by the cofactor $h_k = \#E(\mathbb{F}_{q^k})/n^2$. However, hashing onto $E[n]$ is considerably more expensive than hashing onto $\mathbb{G}_1$ since computations now take place in the larger field $\mathbb{F}_{q^k}$ instead of in $\mathbb{F}_q$, and moreover the cofactor $h_k \approx q^{k-2}$ can be quite large. For the case of BN curves, Chen, Cheng and Smart [12] estimated that the cost of hashing onto $E[n]$ is about 540 times that of performing a point multiplication in $\mathbb{G}_1$ (and estimated the cost of hashing onto $\mathbb{G}_1$ as "free"). This expensive hashing is a major drawback of Type 4 pairings. In the next section, we show that the representation for $E[n]$ introduced in [10] can be used to speed hashing into $E[n]$. In particular, for the case of BN curves, we estimate that our new method for hashing into $E[n]$ is less than 3 times as costly as a point multiplication in $\mathbb{G}_1$.

3.1   On Efficient Implementation

Following [17], we denote by D the CM discriminant of E and set

$$e = \begin{cases} \gcd(k,6), & \text{if } D = -3, \\ \gcd(k,4), & \text{if } D = -4, \\ 2, & \text{if } D < -4, \end{cases} \tag{3}$$

and $d = k/e$. Then E has a unique degree-e twist $\tilde{E}$ defined over $\mathbb{F}_{q^d}$ such that $n \mid \#\tilde{E}(\mathbb{F}_{q^d})$ [22]. Let $\tilde{P}_2 \in \tilde{E}(\mathbb{F}_{q^d})$ be a point of order n, and let $\tilde{\mathbb{T}}_0 = \langle \tilde{P}_2 \rangle$. Then there is a monomorphism $\phi : \tilde{\mathbb{T}}_0 \to E(\mathbb{F}_{q^k})$ such that $P_2 = \phi(\tilde{P}_2) \notin \mathbb{G}_1$. The group $\mathbb{T}_0 = \langle P_2 \rangle$ is the Trace-0 subgroup of $E[n]$. The monomorphism ϕ can be defined so that $\phi : \tilde{\mathbb{T}}_0 \to \mathbb{T}_0$ can be efficiently computed in both directions; therefore we can identify $\tilde{\mathbb{T}}_0$ and $\mathbb{T}_0$, and consequently $\mathbb{T}_0$ can be viewed as having coordinates in $\mathbb{F}_{q^d}$ (instead of in the larger field $\mathbb{F}_{q^k}$).

We have $E[n] \cong \mathbb{G}_1 \times \mathbb{T}_0$. Define the homomorphism $\psi : E[n] \to \mathbb{G}_1$ by $\psi(Q) = \frac{1}{k}\mathrm{Tr}(Q)$. Then it is easy to verify that $Q - \psi(Q) \in \mathbb{T}_0$ for all $Q \in E[n]$ and consequently the map $\rho : Q \mapsto Q - \psi(Q)$ is a homomorphism from $E[n]$ onto $\mathbb{T}_0$. Thus, the map $\phi : E[n] \to \mathbb{G}_1 \times \mathbb{T}_0$ defined by $\phi(Q) = (\psi(Q), \rho(Q))$ is an efficiently-computable isomorphism, whose inverse, given by $(Q_1, Q_2) \mapsto Q_1 + Q_2$, is also efficiently computable. Hence, without loss of generality, the elements of $E[n]$ can be represented as pairs of points (Q_1, Q_2), where $Q_1 \in \mathbb{G}_1$ and $Q_2 \in \mathbb{T}_0$.

With this representation for $E[n]$, hashing onto $E[n]$ can be defined as $H(m) = (H_1(m), H_2(m))$, where H_1 and H_2 are hash functions with ranges $\mathbb{G}_1$ and $\mathbb{T}_0$, respectively. This is expected to be faster than the conventional hashing method outlined in the beginning of this section because hashing onto $\mathbb{G}_1$ and $\mathbb{T}_0$ requires arithmetic in $\mathbb{F}_q$ and $\mathbb{F}_{q^d}$, respectively, rather than in $\mathbb{F}_{q^k}$. Observe that if H_1 and H_2 are modeled as random oracles, then H is also a random oracle.

The ate [22] and R-ate [25] pairings are fast Type 3 pairings from $\mathbb{G}_1 \times \mathbb{T}_0$ to $\mathbb{G}_T$ defined by $e_3(P, Q) = \hat{e}(Q, P)^N$ for some fixed integer N. Now, define the Type 4 pairing $e_4 : \mathbb{G}_1 \times E[n] \to \mathbb{G}_T$ by $e_4(P, Q) = e_3(P, \hat{Q})$, where $\hat{Q} = Q - \pi^{k/2}(Q)$. Note that if $Q = (Q_1, Q_2)$, then $\hat{Q} = (\infty, 2Q_2)$. Thus, e_4 is a bilinear pairing and can be computed in essentially the same time as the Type 3 pairing e_3. The pairing e_4 is non-degenerate in the sense that (i) for each $P \in \mathbb{G}_1 \setminus \{\infty\}$, there exists $Q \in E[n]$ such that $e_4(P, Q) \neq 1$; and (ii) for each $Q \in E[n] \setminus \mathbb{G}_1$, there exists $P \in \mathbb{G}_1$ such that $e_4(P, Q) \neq 1$.

Table 3. Bitlengths of elements in $\mathbb{G}_1$, $\mathbb{T}_0$, $E[n]$ and $\mathbb{G}_T$, and estimated costs (in terms of $\mathbb{F}_p$ multiplications) of basic operations for Type 3 and Type 4 pairings derived from a particular BN elliptic curve

	Type 3	Type 4
Bitlength of elements in $\mathbb{G}_1$	257	257
Bitlength of elements in $\mathbb{T}_0/E[n]$	513	770
Bitlength of elements in $\mathbb{G}_T$	1,024	1,024
Compressing elements in $\mathbb{G}_1$	free	free
Compressing elements in $\mathbb{T}_0/E[n]$	free	free
Decompressing elements in $\mathbb{G}_1$	$315m$	$315m$
Decompressing elements in $\mathbb{T}_0/E[n]$	$674m$	$989m$
Addition in $\mathbb{G}_1$	$11m$	$11m$
Doubling in $\mathbb{G}_1$	$7m$	$7m$
Addition in $\mathbb{T}_0/E[n]$	$30m$	$41m$
Doubling in $\mathbb{T}_0/E[n]$	$17m$	$24m$
Exponentiation in $\mathbb{G}_1$	$1{,}533m$	$1{,}533m$
Exponentiation in $\mathbb{T}_0/E[n]$	$3{,}052m$	$4{,}585m$
Fixed-base exponentiation in $\mathbb{T}_0$	$718m$	$718m$
Fixed-base exponentiation in $\mathbb{T}_0/E[n]$	$1{,}906m$	$2{,}624m$
Hashing into $\mathbb{G}_1$	$315m$	$315m$
Hashing into $\mathbb{T}_0/E[n]$	$3{,}726m$	$4{,}041m$
e_n/R_n Pairing	$15{,}175m$	$15{,}175m$
Testing membership in $\mathbb{G}_1$	free	free
Testing membership in $\mathbb{T}_0/E[n]$	$3{,}052m$	$3{,}052m$

Remark 2. In cryptographic applications of Type 4 pairings, one can ensure that hash values $H(m) = (H_1(m), H_2(m))$ do not lie in $\mathbb{G}_1$ or $\mathbb{T}_0$ by defining H_1 and H_2 to have ranges $\mathbb{G}_1 \setminus \{\infty\}$ and $\mathbb{T}_0 \setminus \{\infty\}$, respectively. This ensures that $\psi(H(m)) \neq \infty$ and $e_4(P, H(m)) \neq 1$ for $P \neq \infty$.

For concreteness, we consider the BN curve $E/\mathbb{F}_p : Y^2 = X^2 + 3$ with BN parameters $z = 600000000001F2D$ that was studied in [10]. This curve has the property that $n = \#E(\mathbb{F}_p)$ is a 256-bit prime, and the embedding degree is $k = 12$. For this particular curve, Table 3 lists the bitlengths of elements in $\mathbb{G}_1$, $\mathbb{T}_0$, $E[n]$ and $\mathbb{G}_T$, and the estimated costs of performing essential operations in these groups; for detailed explanations see [10]. Table 3 demonstrates that Type 4 pairings have very similar performance attributes as Type 3 pairings.

3.2   On Secure Implementation

As we have observed earlier, the only motivation to consider the Type 4 setting for implementation of a protocol is when the protocol requires hashing into the second component $\mathbb{G}_2$ of the pairing's domain followed by an application of ψ on the hash output. However, for Type 4 pairings, $\mathbb{G}_2 = E[n]$ has order n^2, which can be a fundamental distinction affecting the functionality and security of a protocol in the Type 4 setting. We demonstrate this with the very first protocol for which the Type 4 setting was introduced in the literature.

Boneh-Shacham group signature scheme. In a group signature scheme every member of the group has a secret key but there is a single public key for the entire group. The signer-anonymity property of such a scheme finds application, for example, in privacy preserving attestation [7]. Revocation of a user may be critical for such an application, e.g., when the user's secret key is compromised.

Boneh and Shacham proposed a short group signature scheme [7] with an interesting property that given a list of revoked users a verifier can locally check whether the signature has been generated by one of them. This is called a verifier-local revocation (VLR) group signature. They defined a security model for VLR group signature, proposed a construction based on asymmetric pairings, provided a security proof, and discussed the efficiency of the scheme in the elliptic curve setting.

The original description [7] of the Boneh-Shacham group signature scheme (BS-VLR) includes a hash function whose range is $\mathbb{G}_2 \times \mathbb{G}_2$ and also employs the map ψ on the components of the outputs of this hash function. As noted elsewhere [33,11], this protocol cannot be implemented in either the Type 2 or the Type 3 setting. Later in his Ph.D. thesis [32], Shacham introduced Type 4 pairings and reproduced the BS-VLR scheme in that setting without further modification.

Here we take another look at the BS-VLR scheme from [32] and demonstrate that the protocol as described does not achieve its desired functionality and in fact is *not* secure. The protocol description is quite involved and so is its security proof. We recall only those parts of the protocol that are relevant to our discussion. Interested readers are referred to §7.4 of Shacham's thesis [32] as well as the original paper of Boneh and Shacham [7] for the elaborate details.

BS-VLR group signature scheme: The protocol employs a Type 4 pairing $e : \mathbb{G}_1 \times E[n] \rightarrow \mathbb{G}_T$. In order to maintain consistency with [7,32], we use multiplicative notation for $\mathbb{G}_1$ and $E[n]$. The group public key is $gpk = (g_1, g_2, w)$, where $g_2 \in_R E[n]$, $g_1 = \psi(g_2)$, and $w = g_2^\gamma$ for some $\gamma \in_R [1, n-1]$. Suppose that the group consists of N members. The private key of the ith member is $gsk[i] = (A_i, x_i)$ where $x_i \in_R [1, n-1]$ and $A_i = g_1^{1/(\gamma + x_i)}$. The revocation token corresponding to this private key is A_i which is made public in a revocation list (RL) when membership of i is revoked from the group.

The signer i computes, among other items, two elements $T_1, T_2 \in \mathbb{G}_1$ in the following way:

1. $(\hat{u}, \hat{v}) \leftarrow H_0(gpk, M, r)$ where M is the message to be signed, $r \in_R [1, n-1]$ is a nonce, and H_0 is a hash function with range $E[n] \times E[n]$ (treated as a random oracle in the security proof); cf. Remark 2.
2. $u \leftarrow \psi(\hat{u})$ and $v \leftarrow \psi(\hat{v})$.
3. $T_1 \leftarrow u^\alpha$ and $T_2 \leftarrow A_i v^\alpha$, where $\alpha \in_R [1, n-1]$.

T_1, T_2 and r are then sent as part of the group signature σ on M. Note that given r, the verifier can easily obtain $\hat{u}$ and $\hat{v}$.

Verification is a two-step procedure — signature check and revocation check. The signature is accepted as valid only if both these checks are successful. We do not describe the first step where the verifier performs the standard check for validity of the signature σ on M under the group public key gpk. The insecurity of the protocol lies in the revocation check step, which we reproduce verbatim from [32].

Revocation check. For each element $A \in \mathsf{RL}$, check whether A is encoded in (T_1, T_2) by checking if

$$e(T_2/A, \hat{u}) \stackrel{?}{=} e(T_1, \hat{v}). \tag{4}$$

If no element of RL is encoded in (T_1, T_2), the signer of σ has not been revoked.

In other words, suppose the group member i who generated the signature has already been revoked, i.e., $A_i \in \mathsf{RL}$. Then for $A = A_i$, the left side of (4) becomes

$$e(T_2/A_i, \hat{u}) = e(v^\alpha, \hat{u}) = e(v, \hat{u})^\alpha,$$

while the right side becomes

$$e(T_1, \hat{v}) = e(u^\alpha, \hat{v}) = e(u, \hat{v})^\alpha.$$

It is assumed that $e(v, \hat{u})^\alpha$ and $e(u, \hat{v})^\alpha$ are equal, in which case equation (4) holds. As a result the verifier can link the signature to the revoked member i and hence reject it.

In fact, for a signature generated by a revoked user, equation (4) trivially holds if we are in the Type 2 or Type 3 settings where $\mathbb{G}_1$ and $\mathbb{G}_2$ are cyclic groups of the same prime order n. (Simply write $\hat{u} = \hat{v}^x$ for some $x \in [0, n-1]$, and note that $u = \psi(\hat{u}) = \psi(\hat{v}^x) = v^x$.) But recall that the protocol is now described in the Type 4 setting where $\mathbb{G}_2 = E[n]$ is a group of order n^2. Notice that $E[n]$ has $n + 1$ different subgroups of order n, two of which are $\mathbb{G}_1$ and $\mathbb{T}_0$. Suppose that $\mathbb{T}$ is any order-n subgroup of $E[n]$ other than $\mathbb{G}_1$ and $\mathbb{T}_0$. In the Type 4 setting, if $\hat{u}$ and $\hat{v}$ are in the same subgroup $\mathbb{T}$, then equation (4) holds. Conversely, if $\hat{u}$ and $\hat{v}$ are in different subgroups $\mathbb{T}$, then equation (4) only holds with negligible probability. However, $\hat{u}$ and $\hat{v}$ are obtained through hashing to random points in $E[n]$, and so the probability that they both belong to the same subgroup $\mathbb{T}$ is negligible. In fact, the inability to deterministically hash to a particular subgroup $\mathbb{T}$ is the sole reason to describe the protocol in the Type 4 setting instead of Type 2. So with an overwhelming probability equation (4) will not be satisfied and a signature generated by the revoked member i will pass the revocation check.

The security definition of the BS-VLR signature scheme requires that the protocol must satisfy the correctness, traceability, and selfless-anonymity properties, of which the first two are relevant to our discussion. Informally speaking, correctness means that every properly generated signature is accepted as valid if and only if the corresponding signer is not revoked, whereas traceability means that an adversary should not be able to forge a signature that cannot be traced to a revoked user. Neither of these is satisfied for the BS-VLR signature scheme as we have already explained. The wrong assertion in Theorem 7.4.4 of [32] regarding the correctness of the scheme renders the proof of Theorem 7.4.8 regarding traceability meaningless.

Modified BS-VLR signature scheme in Type 4: We now describe a small modification to the protocol that appears to restore security. One apparent drawback is that the signature in the modified protocol contains an element of $E[n]$ and may no longer be considered as "short", which was one of the original motivations of the construction. Fortunately, our new representation of $E[n]$ as discussed in §3.1 (cf. Table 3) helps to maintain the relatively small signature length.

To begin with, we note that the problem with the original protocol [32] does not stem from any intrinsic structural weakness. Rather, it is because of a technical issue related to the structure of $\mathbb{G}_2$. For example, it is possible to securely implement the protocol in the Type 1 setting (where $\mathbb{G}_2 = \mathbb{G}_1$) though the signature length increases. To keep this length short one has to work in the asymmetric setting; and since the protocol requires both hashing into $\mathbb{G}_2$ and the map ψ, the only known option is Type 4. However, that means $\mathbb{G}_2$ is no longer a cyclic group of prime order n, but a group of order n^2. Hence, we cannot expect that two (or more) randomly generated elements will lie in the same order-n subgroup $\mathbb{T}$ of $\mathbb{G}_2$.

Keeping this in mind, the problem of the BS-VLR signature scheme in Type 4 can be easily fixed with a simple modification. The essential idea is the following. For $\hat{u}, \hat{v} \in_R E[n]$, even though in general one cannot expect that $e(\psi(\hat{v}), \hat{u})$ will be equal to $e(\psi(\hat{u}), \hat{v})$, bilinearity of e ensures that

$$e(\psi(\hat{v})^\alpha, \hat{u}) = e(\psi(\hat{v}), \hat{u}^\alpha) \tag{5}$$

for all $\alpha \in [0, n-1]$. So we make the following changes to the protocol.

1. The key generation algorithm remains unchanged. But note that g_2 is a random order-n element of $E[n]$ which can be obtained by hashing into $E[n]$ as discussed in §3.1.
2. The hash function H_0 has range $E[n] \times \mathbb{G}_1$ (instead of $E[n] \times E[n]$).
3. In the signing algorithm, compute $(\hat{u}, v) = H_0(gpk, M, r)$ and $\hat{T}_1 = \hat{u}^\alpha$. Then use $\hat{T}_1$ and $\hat{u}$ to compute the helper value $R_3 = \hat{T}_1^{r_x} \cdot \hat{u}^{-r_\delta} \in E[n]$, and use $\hat{T}_1$ to compute the challenge value c. Send $\hat{T}_1$ (*not* T_1) as part of the signature.
4. In the verification algorithm, use $\hat{T}_1$ and $\hat{u}$ (instead of T_1 and u) to rederive $R_3 \in E[n]$, use $T_1 = \psi(\hat{T}_1)$ to rederive R_1 in the signature check process, and use $\hat{T}_1$ to rederive the challenge value c. Use $\hat{T}_1$ (*not* T_1) in the revocation check step, i.e., for each $A \in \mathsf{RL}$, determine whether A is encoded in $(\hat{T}_1, T_2)$ by checking if

$$e(T_2/A, \hat{u}) = e(v, \hat{T}_1). \tag{6}$$

The only noticeable differences with the original scheme is that the signature now contains $\hat{T}_1 \in E[n]$ instead of $T_1 \in \mathbb{G}_1$ and the revocation check is performed based on $\hat{T}_1$. We briefly analyze the resulting effect on the security.

It is easy to check that the modified scheme satisfies the correctness property. For a signature generated by an honest user, the original argument of Theorem 7.4.4 in [32] applies when $\hat{T}_1$ (and not T_1) is sent as part of the signature. In particular, for a signature generated by a revoked member, equation (6) is exactly in the form of (5) and hence that signature will not pass the revocation check and will be rejected.

The selfless-anonymity and traceability properties of the original scheme are established through involved reductionist security arguments in Lemma 7.4.7 and Theorem 7.4.8 of [32]. Recall that the traceability property is violated because the correctness property does not hold for the original scheme. In fact we do not find any flaw per se in the proofs of these two theorems if we assume that $\mathbb{G}_2$ is a group of prime order n.

However, in the Type 4 setting $\mathbb{G}_2$ is the set of all n-torsion points $E[n]$, which is a group of order n^2. Still it is possible to carry over the original security arguments with some modifications. We do not reproduce the complete arguments here but only emphasize that in the selfless-anonymity game (Lemma 7.4.7) the following variant of the Decision Linear problem should be used: Given a 6-tuple $(u_0, u_1, h_0 = u_0^a, h_1 = u_1^b, v, Z)$, where $u_0, u_1 \in_R E[n]$, $a, b \in_R [1, n-1]$, $v \in_R \mathbb{G}_1$, and either $Z = v^{a+b}$ or $Z \in_R \mathbb{G}_1$, decide whether $Z = v^{a+b}$. Furthermore, the elements u_0 and u_1 have to be appropriately randomized when answering signature queries on behalf of users i_0 and i_1. This randomization is possible because elements of $E[n]$ can be represented as described in §3.1; for further details see the full version of this paper.

We have identified two other protocols in the literature that extend or apply the idea of the BS-VLR signature scheme. These are the VLR signature with backward unlinkability due to Nakanishi and Funabiki [29] and the remote biometric authentication protocol due to Bringer et al. [8]. All our observations regarding the BS-VLR scheme apply to these protocols as well.

Protocols that employ asymmetric pairings and which utilize hashing into $\mathbb{G}_2$ followed by an application of the map ψ can only be instantiated in the Type 4 setting. In fact, to the best of our understanding, one should only resort to Type 4 for these kinds of protocols, since any other protocol employing an asymmetric pairing can be more efficiently instantiated in the Type 3 setting. However, when describing a protocol in the Type 4 setting or arguing its security, protocol designers should be cautious of the fact that $\mathbb{G}_2$ is no longer a prime-order group like $\mathbb{G}_1$ or $\mathbb{G}_T$. Not doing so may critically affect the functionality and security of the protocol as illustrated by the examination of BS-VLR.

4   Concluding Remarks

We presented the first timings for Type 1 pairings derived from supersingular genus-2 curves in characteristic 2 at the 128-bit level, and showed that hashing to the group $\mathbb{G}_2$ in Type 4 pairings is not nearly as costly as previously believed. Furthermore, we demonstrated some pitfalls that can arise when designing protocols and formulating reductionist security arguments in the Type 1 and Type 4 settings.

References

1. Aranha, D., López, J., Hankerson, D.: High-speed parallel software implementation of the η_T pairing. In: Pieprzyk, J. (ed.) CT-RSA 2010. LNCS, vol. 5985, pp. 89–105. Springer, Heidelberg (2010)
2. Barreto, P., Galbraith, S., hÉigeartaigh, C.Ó., Scott, M.: Efficient pairing computation on supersingular abelian varieties. Designs, Codes and Cryptography 42, 239–271 (2007)
3. Beuchat, J.-L., López-Trejo, E., Martínez-Ramos, L., Mitsunari, S., Rodríguez-Henríquez, F.: Multi-core implementation of the Tate pairing over supersingular elliptic curves. In: Miyaji, A., Echizen, I., Okamoto, T. (eds.) CANS 2009. LNCS, vol. 5888, pp. 413–432. Springer, Heidelberg (2009), http://eprint.iacr.org/2009/276
4. Boneh, D., Boyen, X., Shacham, H.: Short group signatures. In: Franklin, M. (ed.) CRYPTO 2004. LNCS, vol. 3152, pp. 41–55. Springer, Heidelberg (2004)
5. Boneh, D., Franklin, M.: Identity-based encryption from the Weil pairing. SIAM Journal on Computing 32, 586–615 (2003)
6. Boneh, D., Lynn, B., Shacham, H.: Short signatures from the Weil pairing. Journal of Cryptology 17, 297–319 (2004)
7. Boneh, D., Shacham, H.: Group signatures with verifier-local revocation. In: 11th ACM Conference on Computer and Communications Security – CCS 2004, pp. 168–177 (2004)
8. Bringer, J., Chabanne, H., Pointcheval, D., Zimmer, S.: An application of the Boneh and Shacham group signature scheme to biometric authentication. In: Matsuura, K., Fujisaki, E. (eds.) IWSEC 2008. LNCS, vol. 5312, pp. 219–230. Springer, Heidelberg (2008)
9. Camenisch, J., Hohenberger, S., Lysyanskaya, A.: Compact E-cash. In: Cramer, R. (ed.) EUROCRYPT 2005. LNCS, vol. 3494, pp. 302–321. Springer, Heidelberg (2005)
10. Chatterjee, S., Hankerson, D., Knapp, E., Menezes, A.: Comparing two pairing-based aggregate signature schemes. Designs, Codes and Cryptography 55, 141–167 (2010)
11. Chatterjee, S., Menezes, A.: On cryptographic protocols employing asymmetric pairings – the role of ψ revisited. Cryptology ePrint Archive, Report 2009/480 (2009)
12. Chen, L., Cheng, Z., Smart, N.: Identity-based key agreement protocols from pairings. International Journal of Information Security 6, 213–241 (2007)
13. Delerablée, C., Pointcheval, D.: Dynamic fully anonymous short group signatures. In: Nguyên, P.Q. (ed.) VIETCRYPT 2006. LNCS, vol. 4341, pp. 193–210. Springer, Heidelberg (2006)
14. Fog, A.: Instruction Tables: Lists of instruction latencies, throughputs and micro-operation breakdowns for Intel, AMD and VIA CPUs (2009), http://www.agner.org/optimize/
15. Frey, G., Lange, T.: Fast bilinear maps from the Tate-Lichtenbaum pairing on hyperelliptic curves. In: Hess, F., Pauli, S., Pohst, M. (eds.) ANTS 2006. LNCS, vol. 4076, pp. 466–479. Springer, Heidelberg (2006)
16. Galbraith, S., Hess, F., Vercauteren, F.: Hyperelliptic pairings. In: Takagi, T., Okamoto, T., Okamoto, E., Okamoto, T. (eds.) Pairing 2007. LNCS, vol. 4575, pp. 108–131. Springer, Heidelberg (2007)

17. Galbraith, S., Paterson, K., Smart, N.: Pairings for cryptographers. Discrete Applied Mathematics 156, 3113–3121 (2008)
18. Gorla, E., Puttmann, C., Shokrollahi, J.: Explicit formulas for efficient multiplication in $\mathbb{F}_{3^{6m}}$. In: Adams, C., Miri, A., Wiener, M. (eds.) SAC 2007. LNCS, vol. 4876, pp. 173–183. Springer, Heidelberg (2007)
19. Granger, R., Page, D., Stam, M.: On small characteristic algebraic tori in pairing-based cryptography. LMS Journal of Computation and Mathematics 9, 64–85 (2006)
20. Hankerson, D., Menezes, A., Scott, M.: Software implementation of pairings. In: Joye, M., Neven, G. (eds.) Identity-Based Cryptography. IOS Press, Amsterdam (2008)
21. Harrison, K., Page, D., Smart, N.P.: Software implementation of finite fields of characteristic three, for use in pairing-based cryptosystems. LMS Journal of Computation and Mathematics 5, 181–193 (2000)
22. Hess, F., Smart, N., Vercauteren, F.: The eta pairing revisited. IEEE Trans. Information Theory 52, 4595–4602 (2006)
23. Joux, A.: A one round protocol for tripartite Diffie-Hellman. Journal of Cryptology 17, 263–276 (2004)
24. Kawahara, Y., Aoki, K., Takagi, T.: Faster implementation of η_T pairing over $GF(3^m)$ using minimum number of logical instructions for $GF(3)$-addition. In: Galbraith, S.D., Paterson, K.G. (eds.) Pairing 2008. LNCS, vol. 5209, pp. 282–296. Springer, Heidelberg (2008)
25. Lee, E., Lee, H., Park, C.: Efficient and generalized pairing computation on abelian varieties. IEEE Trans. Information Theory 55, 1793–1803 (2009)
26. Lee, E., Lee, Y.: Tate pairing computation on the divisors of hyperelliptic curves of genus 2. Journal of the Korean Mathematical Society 45, 1057–1073 (2008)
27. Lenstra, A.: Unbelievable security: Matching AES security using public key systems. In: Boyd, C. (ed.) ASIACRYPT 2001. LNCS, vol. 2248, pp. 67–86. Springer, Heidelberg (2001)
28. Menezes, A., Wu, Y., Zuccherato, R.: An elementary introduction to hyperelliptic curves. Appendix in Algebraic Aspects of Cryptography. Springer, Heidelberg (1998)
29. Nakanishi, T., Funabiki, N.: A short verifier-local revocation group signature scheme with backward unlinkability. In: Yoshiura, H., Sakurai, K., Rannenberg, K., Murayama, Y., Kawamura, S.-i. (eds.) IWSEC 2006. LNCS, vol. 4266, pp. 17–32. Springer, Heidelberg (2006)
30. Ó hÉigeartaigh, C.: Pairing computation on hyperelliptic curves of genus 2. PhD thesis, Dublin City University (2006)
31. Ó hÉigeartaigh, C., Scott, M.: Pairing calculation on supersingular genus 2 curves. In: Biham, E., Youssef, A.M. (eds.) SAC 2006. LNCS, vol. 4356, pp. 302–316. Springer, Heidelberg (2007)
32. Shacham, H.: New paradigms in signature schemes, PhD thesis, Stanford University (2005)
33. Smart, N., Vercauteren, F.: On computable isomorphisms in efficient pairing-based systems. Discrete Applied Mathematics 155, 538–547 (2007)

Switching Construction of Planar Functions on Finite Fields

Alexander Pott and Yue Zhou

Department of Mathematics, Otto-von-Guericke-University Magdeburg,
39106 Magdeburg, Germany
alexander.pott@ovgu.de, yue.zhou@st.ovgu.de

Abstract. A function $f : \mathbb{F}_{p^n} \to \mathbb{F}_{p^n}$ is planar, if $f(x+a) - f(x) = b$ has precisely one solution for all $a, b \in \mathbb{F}_{p^n}$, $a \neq 0$. In this paper, we discuss possible extensions of the switching idea developed in [1] to the case of planar functions. We show that some of the known planar functions can be constructed from each other by switching.

Keywords: planar function, equivalence of functions, semifield, projective plane.

1 Introduction and Preliminaries

The construction of almost perfect nonlinear functions $\mathbb{F}_{2^n} \to \mathbb{F}_{2^n}$ and planar functions $\mathbb{F}_{p^n} \to \mathbb{F}_{p^n}$ (p odd) is of interest due to the connection with geometry, and application in cryptography, but also in themselves as remarkable classes of mappings on finite fields.

In this paper, we discuss a possible extension of the so called switching construction in [1] to the non-binary case. We consider the known planar functions and their projections on fields of order 3^n with $n \leqslant 6$.

Definition 1. *A semifield F is a set with two binary operations, addition $+$ and multiplication $*$, which satisfy the following axioms:*

- *$(F, +)$ is a group, with identity element 0;*
- *$(F, *)$ is a quasigroup;*
- *$0 * a = a * 0 = 0$ for all a;*
- *The left and right distributive laws hold;*
- *There is an element $e \in F$ such that $e * x = x * e = x$ for all $x \in F$.*

Furthermore, if F satisfies all the axioms above, except possibly the last one, then F is called a presemifield.

In the earlier literature (predating 1965), semifields were also called *division algebras* or *distributive quasifields*. The study of semifields was initiated by Dickson [2], shortly after the classification of finite fields. Until now, semifields have become an attracting topic in many different areas of mathematics, such as difference sets,

M.A. Hasan and T. Helleseth (Eds.): WAIFI 2010, LNCS 6087, pp. 135–150, 2010.

coding theory, group theory and finite geometry. Although the definition extends to infinite sets, this article is only concerned with the finite case.

A finite field is a trivial example of a semifield. The first non-trivial examples were constructed by Dickson [2]. In [3], Knuth showed that the additive group of a presemifield F is an elementary abelian group, and the additive order of the elements in F is called the characteristic of F. Hence, any finite presemifield can be represented by $(\mathbb{F}_{p^n}, +, *)$. Here $(\mathbb{F}_{p^n}, +)$ is the additive group of $\mathbb{F}_{p^n}$ and $x * y = \varphi(x, y)$, where φ is a mapping from $\mathbb{F}_{p^n} \times \mathbb{F}_{p^n}$ to $\mathbb{F}_{p^n}$.

On the other hand, there is a well-known correspondence, via coordinatisation, between commutative presemifields and translation planes of Lenz-Barlotti type V.1 and above, see [4]. In [5], Albert showed that two presemifields coordinatise isomorphic planes if and only if they are isotopic:

Definition 2. *Let $F_1 = (\mathbb{F}_{p^n}, +, *)$ and $F_2 = (\mathbb{F}_{p^n}, +, \star)$ be two presemifields. If there exist three linearized permutation polynomials $L, M, N \in \mathbb{F}_{p^n}[x]$ such that*

$$M(x) \star N(y) = L(x * y)$$

for any $x, y \in \mathbb{F}_{p^n}$, then F_1 and F_2 are called isotopic, *and the triple (M, N, L) is an* isotopism *between F_1 and F_2. Furthermore, if there exists an isotopism of the form (N, N, L) between F_1 and F_2, then F_1 and F_2 are* strongly isotopic.

We refer the reader to [6] for more background on finite fields, in particular about linearized polynomials. Let $F = (\mathbb{F}_{p^n}, +, *)$ be a presemifield, and $a \in F$. If we define a new multiplication $\star$ by the rule

$$(x * a) \star (a * y) := x * y$$

we obtain a semifield $(F, +, \star)$ with unit $a * a$. There are many semifields associated with a presemifield, but they are all isotopic.

Next, we give the definition of planar functions, which was introduced by Dembowski and Ostrom in [7] to describe affine planes possessing a collineation group with specific properties.

Definition 3. *Let p be an odd prime. A function $f : \mathbb{F}_{p^n} \to \mathbb{F}_{p^n}$ is called a* planar function, *or* perfect nonlinear *(PN), if for each $a \in \mathbb{F}_{p^n}^*$, $\Delta_f(x, a) := f(x + a) - f(x)$ is a bijection on $\mathbb{F}_{p^n}$.*

Planar functions can also be defined using *differential uniformity*: A function $f : \mathbb{F}_{p^n} \to \mathbb{F}_{p^n}$ is called *differentially δ-uniform*, if for every $a \neq 0$ and every $b \in \mathbb{F}_{p^n}$

$$\#\{x | f(x + a) - f(x) = b\} \leq \delta \ .$$

It is obvious that differentially 1-uniform functions are PN or planar functions. Furthermore, it is easy to see that planar functions can only exist for p odd. For p even, differentially 2-uniform functions are called *almost perfect nonlinear* (*APN*). We refer the readers to [8,9,10] for more background on planar and APN functions.

A *Dembowski-Ostrom* (DO) polynomial $D \in \mathbb{F}_{p^n}[x]$ is a polynomial

$$D(x) = \sum_{i,j} a_{ij} x^{p^i + p^j} \ .$$

Obviously, $\Delta_D(x, a) = D(x + a) - D(x) - D(a)$ is a linearized polynomial for any nonzero a. If we replace $D(x)$ by $D(x) + c$, then $\Delta_D(x, a)$ should be defined as a linear mapping $D(x + a) - D(x) - D(a) + D(0)$, and all proof in this paper go through with this small modification. In [11], Coulter and Henderson proved that planar DO polynomials are equivalent to commutative semifields with odd characteristic. This relation between the two concepts is identical to the equivalence between bilinear forms and quadratic forms in finite fields with odd characteristic. If $*$ is the presemifield product, then the corresponding planar function is $f(x) = x * x$. When the planar DO polynomial f is given, then the corresponding presemifield product is

$$x * y = (1/2)(f(x + y) - f(x) - f(y)) \ .$$

Now, we introduce group ring notation. This notion is also quite useful to describe the equivalence and the "switching construction" of planar functions.

Let $\mathbb{K}$ be an arbitrary field, and let $(G, +)$ be an additively written abelian group of finite order. The group algebra $\mathbb{K}[G]$ consists of all "formal" sums

$$\sum_{g \in G} a_g g, \quad a_g \in \mathbb{K} \ .$$

We define componentwise addition

$$\sum_{g \in G} a_g g + \sum_{g \in G} b_g g = \sum_{g \in G} (a_g + b_g) g$$

and multiplication by

$$\sum_{g \in G} a_g g \cdot \sum_{g \in G} b_g g = \sum_{g \in G} (\sum_{h \in G} a_h \cdot b_{g-h}) g \ .$$

These two operations together with the scalar multiplication $\lambda \sum_{g \in G} a_g g = \sum_{g \in G} (\lambda a_g) g$, the set $\mathbb{K}[G]$ becomes an algebra, named *group algebra*. The dimension of this algebra as a $\mathbb{K}-$vectorspace is $|G|$. Given a function $f : \mathbb{F}_p^n \to \mathbb{F}_p^m$, we associate a group algebra element G_f in $\mathbb{K}[\mathbb{F}_p^n \times \mathbb{F}_p^m]$ with it:

$$G_f = \sum_{v \in \mathbb{F}_p^n} (v, f(v)) \ .$$

The coefficients of the group elements in G_f are just 0 or 1. More generally, any subset T of a group G can be identified with the element $\sum_{g \in T} g$, where the coefficients of all elements in T are 1, and the coefficients of elements not in T are 0. Hence, G_f is the group algebra element corresponding to the "graph" of the function f, which consists of all pairs $(v, f(v))$, $v \in \mathbb{F}_p^n$.

There are three important equivalence relations of functions over finite fields, for which differential uniformity is invariant.

Definition 4. *Two functions* $f, g : \mathbb{F}_p^n \to \mathbb{F}_p^m$ *are called* CCZ-equivalent *if there is an automorphism* Ψ *of the group* $\mathbb{F}_p^n \times \mathbb{F}_p^m$ *and an element* $(u, v) \in \mathbb{F}_p^n \times \mathbb{F}_p^m$ *such that*

$$\Psi(G_f) = G_g \cdot (u, v) \ .$$

Furthermore, if Ψ *fixes the subgroup* $\{(0, y) | y \in \mathbb{F}_p^m\}$ *setwise, then* f *and* g *are called* extended affine(EA)-equivalent. *Additionally, if* Ψ *also fixes the set* $\{(x, 0) | x \in \mathbb{F}_p^n\}$, *then* f *and* g *are called* affine equivalent.

Generally speaking, EA-equivalence implies CCZ-equivalence, but not vice versa, see [12]. However, if planar functions f and g are CCZ-equivalent, then they are also EA-equivalent [13]. Since the correspondence between commutative pre-semifields with odd characteristic and planar functions as we mentioned above, the strong isotopism of two commutative presemifields is equivalent to the affine equivalence of the corresponding planar DO functions. Furthermore, in [11], Coulter and Henderson derived strong conditions for when two commutative presemifields are isotopic, for example, they showed that any two commutative presemifields of order p^n with n odd are isotopic if and only in they are strongly isotopic. Moreover, it is still an open problem to find two commutative presemifields which are not strongly isotopic but isotopic.

The known families of planar functions, and corresponding commutative semifields which are defined in arbitrary odd characteristic are the following:

1.
$$x^2$$

over $\mathbb{F}_{p^n}$, which corresponds to the finite field $\mathbb{F}_{p^n}$;

2.
$$x^{p^k+1}$$

over $\mathbb{F}_{p^n}$, with $\frac{n}{\gcd(k,n)}$ odd, which corresponds to Albert's commutative twisted fields [14];

3. the functions over $\mathbb{F}_{p^{2k}}$, corresponding to Dickson semifields [2];

4.
$$x^{1+q'} - vx^{q^2+q'q}$$

over $\mathbb{F}_{q^3}$, where p is an odd prime, $q = p^s$, $q' = p^t$, $s' = s/\gcd(s,t)$, $t' = t/\gcd(s,t)$, s' is odd, $\mathrm{ord}(v) = q^2 + q + 1$, and at least one of the following conditions hold:

$$s' + t' \equiv 0 \mod 3,$$

$$q \equiv q' \equiv 1 \mod 3 \ .$$

This planar function family corresponds to Zha-Kyureghyan-Wang (ZKW) semifields [15,16];

5.
$$x^{1+q'} - vx^{q^3+q'q}$$

over $\mathbb{F}_{q^4}$, where p is an odd prime, $q = p^s$, $q' = p^t$ such that $2s/\gcd(2s,t)$ is odd, $q \equiv q' \equiv 1 \mod 4$, and $\mathrm{ord}(v) = q^3 + q^2 + q + 1$. It corresponds to Bierbrauer semifields [17];

6.

$$\mathrm{Tr}(x^{q+1}) + \mathrm{Tr}(\beta x^{p^s+1})\omega$$

over $\mathbb{F}_{q^2}$, where p is an odd prime, $q = p^m$, $\mathrm{Tr}(\cdot)$ is the trace function from $\mathbb{F}_{q^2}$ to $\mathbb{F}_q$, $\omega, \beta \in \mathbb{F}_{q^2}$, $\mathrm{Tr}(\omega) = 0$ and s is a positive integer such that the followings hold:

(a) β^{q-1} is not contained in the subgroup of order $(q+1)/\gcd(q+1, p^s+1)$ in $(\mathbb{F}_{q^2}, *)$;

(b) There is no $0 \neq a \in \mathbb{F}_{q^2}$, such that $\mathrm{Tr}(a) = 0$ and $a^{p^s} = -a$.

This planar function family was firstly discovered by Budaghyan and Helleseth in [18], as two independent families, which belongs to the one discovered by Bierbrauer in [19]. However, it is still not known whether Bierbrauer's family properly contains them, so we call the corresponding semifields Budaghyan-Helleseth-Bierbrauer(BHB) semifields;

7.

$$\mathrm{Tr}(x^2) + G(x^{q^2+1})$$

over $\mathbb{F}_{q^{2m}}$, where q is a power of an odd prime p, $m = 2k + 1$, $\mathrm{Tr}(\cdot)$ is the trace function from $\mathbb{F}_{q^{2m}}$ to $\mathbb{F}_{q^m}$, and $G(x) = h(x - x^{q^m})$, where $h \in \mathbb{F}_{q^{2m}}[x]$ is defined as

$$h(x) = \sum_{i=0}^{k}(-1)^i x^{q^{2i}} + \sum_{j=0}^{k-1}(-1)^{k+j} x^{q^{2j+1}} \quad .$$

This planar function family corresponds to Bierbrauer's generalization of the semifield discovered by Lunardon, Marino, Polverino and Trombetti over q^6, see [19,20]. As in case (6), they should be called Lunardon-Marino-Polverino-Trombetti-Bierbrauer (LMPTB) semifields [19].

For $p = 3$, there are some more families as follows:

8.

$$x^{10} \pm x^6 - x^2$$

over $\mathbb{F}_{3^n}$, with n odd, corresponding to the Coulter-Matthews and Ding-Yuan (CMDY) semifields [21,22][1];

9. the functions over $\mathbb{F}_{3^{2k}}$, corresponding to the Cohen-Ganley (CG) semifields [23];

10. the functions over $\mathbb{F}_{3^{2k}}$, with k odd, corresponding to the Ganley semifields [24];

11.

$$x^2 + x^{90}$$

over $\mathbb{F}_{3^5}$, corresponding to the At-Cohen-Weng (ACW) semifield [25,26];

---

[1] It has been pointed out by a reviewer that this family is also described in the master thesis by Bodil Stakkestad Kristensen, Department of Informatics, University of Bergen, 1997.

12. the function over $\mathbb{F}_{3^8}$, corresponding to the Coulter-Henderson-Kosick semi-field [27];
13. the function over $\mathbb{F}_{3^{10}}$, corresponding to the Penttila-Williams semifield [28];

In [29], the polynomial representations of those planar functions corresponding to the Dickson, Cohen-Ganley, Ganley and Penttila-Williams semifields can be found. We refer to a survey contained in the very interesting paper [30] for more information about those commutative semifields. The isotopy relations involving BHB and LMPTB semifields are still not known, however, except for this pair, every two of the semifield families mentioned above are generally not isotopic with each other.

The only known planar functions which are not DO polynomials are:

14.
$$x^{\frac{3^k+1}{2}}$$

over $\mathbb{F}_{3^n}$, where k is odd and $\gcd(k,n) = 1$. We call it Coulter-Matthews (CM) planar functions [21], and they do not correspond to any semifield.

2 Switching Construction

Now we consider certain projection homomorphisms on the group algebra $\mathbb{K}[G]$, which is defined in Section 1. Let U be a subgroup of G. Then the canonical homomorphism $\varphi_U : G \to G/U$ defined by $\varphi_U(g) := g + U$ (denoted by $\bar{g}$) can be extended to a homomorphism $\varphi_U : \mathbb{K}[G] \to \mathbb{K}[G/U]$. To be precise, let $D = \sum a_g g \in \mathbb{K}[G]$, then $\varphi(D) = \sum_{\bar{g} \in G/U} (\sum_{h \in g+U} a_h) \cdot \bar{g}$. Furthermore, if D corresponds to a set in G, i.e. D has only coefficients 0 and 1, then the coefficient of $\bar{g}$ is $|D \cap (g + U)|$.

Definition 5. *Let* $f, g : \mathbb{F}_p^n \to \mathbb{F}_p^n$ *be two functions, and let* U *be a subgroup of* $\{0\} \times \mathbb{F}_p^n$. *We call* f *and* g switching neighbors *with respect to* U *if* $\varphi_U(G_f) = \varphi_U(G_g)$ *and* $1 \leqslant \dim(U) < n$. *Furthermore,* f, g *are called* switching neighbors in the narrow sense *if* $\dim(U) = 1$.

In this definition, it is required that $\dim(U) < n$. Otherwise, if $\dim(U) = n$ is allowed, then every planar functions are switching neighbors. If f, g are switching neighbors with respect to U, then g can be obtained from f by first projecting G_f onto $\varphi_U(G_f)$, and then "lifting" this element to G_g. In fact, this project and lift method turns out to be very powerful for the construction of new APN functions. In [1], it is shown that many APN functions can be constructed by switching, and even a new nonquadratic APN function over $\mathbb{F}_{2^6}$ has been found. So, it is natural to consider these questions:

Question 1. Is it possible that two planar functions are switching neighbors?

Question 2. Can we use the switching idea for the construction of new planar functions?

One of the difficulties to generalize the idea in [1] to odd characteristic finite field is that the linear restrictions in the even characteristic case for the switching become nonlinear conditions in the odd characteristic case. Due to the limitation of computer capacity, we only consider the switching for the case $p = 3$.

Theorem 1. *Assume that $f : \mathbb{F}_3^n \to \mathbb{F}_3^n$ is a planar function. Let $u \in \mathbb{F}_3^n \setminus \{0\}$, and let $\delta : \mathbb{F}_3^n \to \mathbb{F}_3$. Then $f(x) + \delta(x) \cdot u$ is a planar function if and only if*

$$\sum_{i=0}^{2} \Delta_\delta(x_i, a) = 0, \quad \text{and} \quad \Delta_\delta(x_1, a) - \Delta_\delta(x_2, a) \neq 1 , \tag{1}$$

for all $0 \neq a, x_i \in \mathbb{F}_3^n$ with

$$\Delta_f(x_i, a) = b + i \cdot u , \tag{2}$$

for $i = 0, 1, 2$ and $b \in \mathbb{F}_3^n$.

Proof. Since f is a planar function, the three equations

$$\Delta_f(x, a) = b + i \cdot u , \quad i = 0, 1, 2$$

have one solution for each i, denoted by x_i, $i = 0, 1, 2$. Now, we consider the value of $\Delta_\delta(x_i, a)$. Notice that $f + \delta \cdot u$ is planar, if and only if

$$\{\Delta_f(x_i, a) + \Delta_\delta(x_i, a) \cdot u | i = 0, 1, 2\} = \{b, b + u, b + 2u\}$$

That means the vector $(\Delta_\delta(x_0, a), \Delta_\delta(x_1, a), \Delta_\delta(x_2, a))$ belongs to

$$\{(i, i, i) | i = 0, 1, 2\} \cup \{(i, i + 1, i + 2) | i = 0, 1, 2\} ,$$

which is equivalent to (1).   □

Theorem 1 suggests a strategy to find the $p-$ary function δ such that $f + \delta \cdot u$ is a planar function: Determine all the x_i and $x_i + a$ such that (2) holds. Then they give rise to linear constraints $\sum_{i=0}^{2} \Delta_\delta(x_i, a) = 0$, and nonlinear constraints $\Delta_\delta(x_1, a) - \Delta_\delta(x_2, a) \neq 1$. Finally, find out whether these planar functions obtained from the switching construction above are new.

For the $p > 3$ case, the nonlinear conditions can not be written as some linear inequalities as in Theorem 1, and it seems quite difficult to make an efficient MAGMA [31] program to do the switching construction.

3   EA-equivalence of Planar Functions

In Section 1, we mentioned the EA-equivalence of planar functions. Similarly to APN functions, we can also build links between linear codes and planar functions to investigate the equivalence. The two definitions of equivalence of linear codes, which are used in this paper, are as follows,

Definition 6. *Let* $\mathbb{F}$ *be an arbitrary field. Two linear codes in* $\mathbb{F}^n$ *are monomially equivalent if each can be obtained from the other by permuting the coordinate positions in* $\mathbb{F}^n$ *and multiplying each coordinate by a non-zero field element. The codes will be said to be* permutation equivalent *if a permutation of the coordinate positions suffices to take one to the other.*

By coding theory, the two equivalences mentioned above can be represented by monomial matrices and permutation matrices, respectively, which we multiply from the right side of the generator matrix of the code. Furthermore, the set of monomial matrices that map the linear code C to itself forms the group MAut(C) called the *monomial automorphism group of* C. Similarly, the set of permutation matrices that map C to itself forms another group PAut(C) named the *permutation automorphism group of* C.

Let $f : \mathbb{F}_p^n \to \mathbb{F}_p^m$ be any function. Define a matrix $M_f \in \mathbb{F}_p^{(m+n+1,p^n)}$ as follows

$$M_f = \begin{pmatrix} \cdots & 1 & \cdots \\ \cdots & x & \cdots \\ \cdots & f(x) & \cdots \end{pmatrix}_{x \in \mathbb{F}_p^n} \tag{3}$$

Then we can construct a code C_f over $\mathbb{F}_p$ by the generator matrix M_f. Different from the codes corresponding to APN functions over $\mathbb{F}_{2^n}$, the weight distributions of the codes from all the known planar functions are the same (In [32], this is obtained only for CM planar functions and some DO planar functions, but this result can be generalized to all DO planar functions). Moreover, we can not tell whether f is a planar function by the minimum distance of $C_f^\perp$ as in the APN cases. However, it can still be used to test the equivalence of two functions, due to the following proposition:

Proposition 1. *Let* p *be a prime,* m *and* n *be integers. Two functions* $f, g :$ $\mathbb{F}_p^n \to \mathbb{F}_p^m$ *are CCZ-equivalent, if and only if the corresponding codes* C_f *and* C_g *are permutation equivalent.*

Proof. Assume that C_f and C_g are permutation equivalent, then we have a permutation matrix P and an $(n+m+1) \times (n+m+1)$ matrix L with full rank, such that

$$L \cdot M_f \cdot P = M_g .$$

That means there are $u \in \mathbb{F}_p^n$, $v \in \mathbb{F}_p^m$ and a matrix $\tilde{L}$ with full rank, such that

$$\tilde{L} \cdot \begin{pmatrix} \cdots & x & \cdots \\ \cdots & f(x) & \cdots \end{pmatrix} \cdot P = \begin{pmatrix} \cdots & x & \cdots \\ \cdots & g(x) & \cdots \end{pmatrix} + \begin{pmatrix} u \\ v \end{pmatrix} .$$

Therefore, by the definition of CCZ-equivalence, f and g are CCZ-equivalent. The proof of the converse is the same. □

MAGMA only provide us the command to tell whether two codes are monomially equivalent, however for the codes C_f and C_g from the planar functions f, g, monomial and permutation equivalence are identical, because there are only $p-1$

code words $(i, i, \cdots, i)$ with weight p^n $(0 < i < p)$, see [32]. Meanwhile, for a linear code C, we can also calculate $\mathrm{PAut}(C)$ and $\mathrm{MAut}(C)$, which are important invariances and useful to investigate some properties of the code. Next, we will give some properties of $\mathrm{PAut}(C_f)$ and $\mathrm{MAut}(C_f)$, where C_f is defined by the generator matrix M_f in (3).

For any linear mappings $l_1, l_2 : \mathbb{F}_p^n \to \mathbb{F}_p$, since

$$l_1(x^p) + l_2(f(x)^p) = l_1(x) + l_2(f(x)) \ ,$$

we have $\mathrm{Gal}(\mathbb{F}_{p^n}) \subseteq \mathrm{PAut}(C_f)$, where $\mathrm{Gal}(\mathbb{F}_{p^n})$ denote the Galois group of the field extension $\mathbb{F}_{p^n}$ of $\mathbb{F}_p$.

Moreover, let C be a cyclic code of length $p^n - 1$ over $\mathbb{F}_p$. Then we call the set $D_C = \{i | \alpha^i$ is a zero of $C, 1 \leqslant i \leqslant p^n - 1\}$ the defining set of C, where $\alpha \in \mathbb{F}_{p^n}$ is a primitive element of $\mathbb{F}_{p^n}$. If C^{ext} is the extended code of C (this extension is also referred to as *adding an overall parity check*), i.e. C^{ext} is an extended cyclic code, then we say that C^{ext} has defining set $D_{C^{\mathrm{ext}}} := D_C \cup \{0\}$. We refer to [33,34] for more background about cyclic and extended cyclic codes.

Now, assume that $f = x^d \in \mathbb{F}_{p^n}[x]$, which is a monomial. Then $C_f^{\perp}$ is an extended cyclic code, and the defining set of $C_f^{\perp}$ is $\hat{T} = \{0\} \cup \{p^i | i = 0, 1, \ldots, n - 1\} \cup \{d \cdot p^i | i = 0, 1, \ldots, n - 1\}$. According to the definition of a cyclic code, notice that $\mathrm{PAut}(C) = \mathrm{PAut}(C^{\perp})$, the following lemma can be easily verified.

Lemma 1. *Let $f = x^d \in \mathbb{F}_{p^n}[x]$, and linear code C_f are defined by the generator matrix M_f in (3). Then $(\mathbb{F}_{p^n}^*, *) \subseteq \mathrm{PAut}(C_f)$.*

Furthermore, let $\mathrm{GA}_m(q) = \mathbb{F}_q^m \rtimes \mathrm{GL}_m(\mathbb{F}_q)$ be the *general affine group* over the vector space $\mathbb{F}_q^m$. For any $\sigma_{a,b} \in \mathrm{GA}_1(q) = \{\sigma_{u,v} | u \in \mathbb{F}_q^*, v \in \mathbb{F}_q\}$, we have $\sigma_{a,b}(g) = a \cdot g + b$, where $g \in \mathbb{F}_q$. Moreover, if C is an extended cyclic code of length p^n over $\mathbb{F}_p$, with $\mathrm{GA}_1(p^n) \subseteq \mathrm{PAut}(C)$, then C is called *affine-invariant*.

We can define a partial ordering $\preceq$ on $S := \{0, 1, \ldots, p^n - 1\}$. If $s, t \in S$ have p-adic expansions

$$s = \sum_{i=0}^{n-1} s_i p^i, \quad t = \sum_{i=0}^{n-1} t_i p^i \ .$$

with $0 \leq s_i, t_i < p$ for $0 \leq i < n$, then we say $s \preceq t$ provided $s_i \leq t_i$ for $0 \leq i < n$. To prove our result about $\mathrm{PAut}(C_f)$, we need the following two propositions,

Proposition 2. *Let p be a prime, and C be an extended cyclic $p-$ary code over $\mathbb{F}_p$ of length p^n with defining set $D_C \subset S = \{0, 1, \ldots, p^n - 1\}$. Then,*

1. *(Kasami, Lin and Peterson [35])*
   C is affine-invariant if and only if for all $s \in D_C$, we have $t \preceq s \Rightarrow t \in D_C$;
2. *(Berger and Charpin [36])*
   Let a be a divisor of n. Assume $n > 1$ and C is affine-invariant, then $\mathrm{GA}_{n/a}(p^a) \subset \mathrm{PAut}(C)$ if and only if the following condition holds:

   for all $s \in D_C$ and all $j \in S$ with $j \preceq s$, we have $s + j(p^a - 1) \in D_C$

   where $s + j(p^a - 1)$ is computed modulo $p^n - 1$.

Proposition 3. *[33, Theorem 3.1 in Chapter 17] Let G be a subgroup of the symmetric group S_I on I, where $I = \mathbb{F}_{p^n}$ with p a prime and $n > 1$. If G contains $\mathrm{GA}_1(p^n)$, then one of the following holds:*

1. $G = S_I$,
2. $p = 2$ and $G = A_I$, the alternating group on I, or
3. *there exists a divisor a of n such that $\mathrm{GA}_{n/a}(p^a) \subseteq G \subseteq \Gamma\mathrm{A}_{n/a}(p^a)$, where $\Gamma\mathrm{A}_{n/a}(p^a) = \mathrm{GA}_{n/a}(p^a) \rtimes \mathrm{Gal}(\mathbb{F}_{p^a})$ is the semi-affine group of $\mathbb{F}_{p^a}$.*

We can prove:

Theorem 2. *Let $f = x^d \in \mathbb{F}_{p^n}[x]$ and $d = p^i + p^j$ for some $0 \leq i, j < n - 1$. Then the linear code C_f defined by the generator matrix M_f has the following permutation automorphism group:*

$$\mathrm{PAut}(C_f) = \Gamma\mathrm{A}_1(p^n) = \mathrm{GA}_1(p^n) \rtimes \mathrm{Gal}(\mathbb{F}_{p^n}) \ .$$

Proof. By Proposition 2, it can be verified that $\mathrm{GA}_{n/a}(p^a) \subseteq \mathrm{PAut}(C_f^\perp)$ if and only if $a = n$. Furthermore, since $\mathrm{Gal}(\mathbb{F}_{p^n}) \subseteq \mathrm{PAut}(C_f^\perp)$, by Proposition 3, it can be shown that $\mathrm{PAut}(C_f^\perp) = \Gamma\mathrm{A}_1(p^n)$. Since $\mathrm{PAut}(C_f) = \mathrm{PAut}(C_f^\perp)$, we prove the claim in the theorem. $\square$

On the other hand, when f is a DO polynomial, we can prove:

Theorem 3. *Let $f \in \mathbb{F}_{p^n}[x]$ be a DO polynomial, l be a linear mapping from $\mathbb{F}_{p^n}$ to $\mathbb{F}_{p^m}$. The linear code $C_{l \circ f}$ is defined by the generator matrix $M_{l \circ f}$. Then the elementary abelian group $(\mathbb{F}_{p^n}, +)$ and $\mathrm{Gal}(\mathbb{F}_{p^n})$ are subgroups of $\mathrm{PAut}(C_{l \circ f})$.*

Proof. It is trivial to prove that $\mathrm{Gal}(\mathbb{F}_{p^n}) \subseteq \mathrm{PAut}(C_{l \circ f})$. We only prove $(\mathbb{F}_{p^n}, +) \subseteq \mathrm{PAut}(C_{l \circ f})$ below. For any $a \in \mathbb{F}_{p^n}$, define the matrix $M_{l \circ f}^{(a)}$ as follows

$$M_{l \circ f}^{(a)} = \begin{pmatrix} \cdots & 1 & \cdots \\ \cdots & x + a & \cdots \\ \cdots & l \circ f(x + a) & \cdots \end{pmatrix}_{x \in \mathbb{F}_p^n} .$$

It is obvious that $M_{l \circ f}^{(a)}$ can be obtained by permutating the columns of $M_{l \circ f}$. Let $C_{l \circ f}^{(a)}$ be the code generated by $M_{l \circ f}^{(a)}$. If we show that $C_{l \circ f}^{(a)} = C_{l \circ f}$, then we prove the claim.

Let $l_1 : \mathbb{F}_{p^n} \to \mathbb{F}_p$ and $l_2 : \mathbb{F}_{p^m} \to \mathbb{F}_p$ be two linear mappings, and $c \in \mathbb{F}_p$, then any codeword in C_f can be written as

$$(l_1(x) + l_2 \circ l \circ f(x) + c)_{x \in \mathbb{F}_p^n} \ .$$

Since f is a DO-polynomial, $u(x) = l_2 \circ l(f(x + a) - f(x) - f(a))$ is a linear mapping. Define a linear mapping $l_3 : \mathbb{F}_{p^n} \to \mathbb{F}_p$ by

$$l_3(x) = l_1(x) - u(x) \ .$$

Then it can be proved that

$$l_1(x) + l_2 \circ l \circ f(x) = l_3(x + a) + l_2 \circ l \circ f(x + a) + d \ ,$$

where $d = l_2 \circ l \circ f(a) - l_3(a)$, for any $x, a \in \mathbb{F}_{p^n}$. Therefore, $C_{l \circ f} = C_{l \circ f}^{(a)}$. $\square$

4   Computational Results

Now, let us describe the switching construction of planar functions on $\mathbb{F}_{3^n}$ in the following steps. First, we "project" all known functions to $n-1$-dimensional subspaces, calculate how many inequivalent classes there are. Then, we do the "lift" for all the inequivalent projections, construct $f'(x) = f(x) + \delta(x) \cdot u$. Finally, we test whether f' is inequivalent to known planar functions.

Due to the nonlinear conditions in Theorem 1, we can only do the exhaustive search for the switching of all known planar functions on $\mathbb{F}_{3^n}$ with $n \leq 6$. However, all the planar functions obtained this way are equivalent to known ones.

Next, we investigate the number of equivalent m-dimensional projections of two known inequivalent planar functions, for every $0 < m < n$. If l is a projection from $\mathbb{F}_p^n$ to $\mathbb{F}_p^m$, then l can be expressed as an $m \times n$ matrix, and there are $\frac{\prod_{i=n-m+1}^{n}(p^i-1)}{\prod_{i=1}^{m}(p^i-1)}$ such projections. When $m = 1$, the function $l \circ f(x)$ can always be expressed by $\mathrm{tr}(a \cdot f(x))$, where $a \in \mathbb{F}_{p^n}^*$ and $\mathrm{tr}(\cdot)$ is the trace function from $\mathbb{F}_{p^n}$ to $\mathbb{F}_p$. If f is a planar DO-polynomial, then $\mathrm{tr}(a \cdot f(x))$ is a nondegenerate quadratic form on $\mathbb{F}_p$. Furthermore, it can be proved that, under the affine transformations of $p-$ary functions, there are two quadratic forms for even n, and only one quadratic form for odd n. Moreover, it is obvious that $\mathrm{tr}(a \cdot g(x))$ is not quadratic, when g is a non-DO monomial function, for example CM functions $x^{\frac{3^k+1}{2}}$ with odd $k > 1$ and $\gcd(n,k) = 1$. Therefore we have:

Proposition 4. *Let g be a non-DO monomial planar function on $\mathbb{F}_{p^n}$, then g is not a switching neighbor of any DO planar functions on $\mathbb{F}_{p^n}$ with respect to any $m-$dimensional subspace $U \in \mathbb{F}_p^n$ with $0 < m < n$.*

Note 1. For APN functions, the claim in Proposition 4 does not hold. That means, even if f and g are CCZ-equivalent, then it is still possible that there is an integer m, such that $l_1 \circ f$ is not CCZ-equivalent with $l_2 \circ g$ for projections $l_1, l_2 : \mathbb{F}_p^n \to \mathbb{F}_p^m$. It happens because CCZ-equivalence is strictly more general than EA-equivalence for APN functions. For example, it can be shown that x^{2^3+1}, x^{2^3-1} are CCZ-equivalent but not EA-equivalent on $\mathbb{F}_{2^5}$ (actually, they are inverse for each other). However, it can be verified by MAGMA that, for any $0 < m < 5$ and any projection $l_1, l_2 : \mathbb{F}_p^n \to \mathbb{F}_p^m$, $l_1 \circ f$ and $l_2 \circ g$ are never CCZ-equivalent with each other.

In the following, we only do the calculation for the planar functions on $\mathbb{F}_{3^n}$ with $n \geqslant 3$, since x^2 is the only planar function on $\mathbb{F}_3$ and on $\mathbb{F}_{3^2}$.

4.1   The $\mathbb{F}_{3^3}$ Case

In [11], Coulter and Henderson showed that there are only two inequivalent planar DO functions x^2 and x^{p+1} over $\mathbb{F}_{p^3}$. Obviously, the CM planar function

Table 1. Switching Neighbors with respect to all $l : \mathbb{F}_3^3 \to \mathbb{F}_3^2, \mathbb{F}_3$

	x^2	x^4
x^2	1	1
x^4	1	1

(a) $\mathbb{F}_3^3 \to \mathbb{F}_3^2$,

	x^2	x^4
x^2	1	1
x^4	1	1

(b) $\mathbb{F}_3^3 \to \mathbb{F}_3$

family does not provide any other functions here, so we have only 2 known inequivalent planar functions over $\mathbb{F}_{3^3}$, and by Theorem 3, the permutation automorphism groups of the corresponding codes are both $\Gamma A_1(p^n)$. The numbers in Table 1 denote, for two planar functions $f, g : \mathbb{F}_3^3 \to \mathbb{F}_3^3$, how many inequivalent projections $l, k : \mathbb{F}_3^3 \to \mathbb{F}_3^m$ there are, such that $l \circ f$ is equivalent to $k \circ g$.

4.2 The $\mathbb{F}_{3^4}$ Case

By the lists in Section 1, all the known planar functions on $\mathbb{F}_{3^4}$ are CM functions and those from the following commutative semifields of order 3^4: finite field, Dickson semifields, BHB semifields and CG semifields. However, by MAGMA program, we know that all the planar functions from Dickson, BHB and CG semifields are EA-equivalent. Hence, we list the only 3 known inequivalent planar functions on $\mathbb{F}_{3^4}$ in Table 2, with the order of the permutation automorphism groups of corresponding codes. Furthermore, as in Subsection 4.1, we have Table 3. It is worth noting that two inequivalent examples (x^2, BHB) are switching neighbors.

4.3 The $\mathbb{F}_{3^5}$ Case

From the lists in Section 1, Table 4 contains all the known inequivalent planar functions on $\mathbb{F}_{3^5}$. No two functions are switching neighbors with respect to 1-dimension linear space. Therefore, we do not write down the entire matrix of the number of switching neighbor since all off-diagonal elements are 0 (Table 5).

Table 2. All known inequivalent planar functions over $\mathbb{F}_{3^4}$

No.	Families	Functions	$|\text{PAut}(C_f)|/|\text{Gal}(\mathbb{F}_{3^4})|$
1	Finite Field	x^2	$|GA_1(3^4)|$
2	Dickson, BHB, CG	$x^4 + x^{10} - x^{36}$	$16 \cdot |\mathbb{F}_{3^4}|$
3	CM	x^{14}	$|\mathbb{F}_{3^4}^*|$

Table 3. Switching Neighbors with respect to all $l : \mathbb{F}_3^4 \to \mathbb{F}_3^3, \mathbb{F}_3^2, \mathbb{F}_3$

	x^2	BHB	x^{14}
x^2	2	1	0
BHB	1	4	0
x^{14}	0	0	2

(a) $\mathbb{F}_3^4 \to \mathbb{F}_3^3$

	x^2	BHB	x^{14}
x^2	5	5	0
BHB	5	7	0
x^{14}	0	0	6

(b) $\mathbb{F}_3^4 \to \mathbb{F}_3^2$

	x^2	BHB	x^{14}
x^2	2	2	0
BHB	2	2	0
x^{14}	0	0	2

(c) $\mathbb{F}_3^4 \to \mathbb{F}_3$

Table 4. All known inequivalent planar functions over $\mathbb{F}_{3^5}$

No.	Families	Functions	$|\mathrm{PAut}(C_f)|/|\mathrm{Gal}(\mathbb{F}_{3^5})|$
1	Finite Field	x^2	$|\mathrm{GA}_1(3^5)|$
2	Albert	x^4	$|\mathrm{GA}_1(3^5)|$
3	Albert	x^{10}	$|\mathrm{GA}_1(3^5)|$
4	CMDY[1]	$x^{10} + x^6 - x^2$	$2 \cdot |\mathbb{F}_{3^5}|$
5	CMDY[2]	$x^{10} - x^6 - x^2$	$2 \cdot |\mathbb{F}_{3^5}|$
6	ACW	$x^2 + x^{90}$	$22 \cdot |\mathbb{F}_{3^5}|$
7	CM	x^{14}	$|\mathbb{F}_{3^5}|$

Table 5. Number of inequivalent $l \circ f$ for all $l : \mathbb{F}_3^5 \to \mathbb{F}_3^4$

f	x^2	x^4	x^{10}	CMDY[1]	CMDY[2]	$x^2 + x^{90}$	x^{14}
Number	1	1	1	25	25	3	1

Table 6. Switching Neighbors with respect to all $l : \mathbb{F}_3^5 \to \mathbb{F}_3^3$

	x^2	x^4	x^{10}	CMDY[1]	CMDY[2]	$x^2 + x^{90}$	x^{14}
x^2	2	0	0	0	0	0	0
x^4	0	2	0	0	2	0	0
x^{10}	0	0	2	0	0	0	0
CMDY[1]	0	0	0	239	14	3	0
CMDY[2]	0	2	0	14	230	1	0
$x^2 + x^{90}$	0	0	0	3	1	22	0
x^{14}	0	0	0	0	0	0	2

Table 7. Switching Neighbors with respect to all $l : \mathbb{F}_3^5 \to \mathbb{F}_3^2$

	x^2	x^4	x^{10}	CMDY[1]	CMDY[2]	$x^2 + x^{90}$	x^{14}
x^2	2	2	2	2	2	2	0
x^4	2	2	2	2	2	2	0
x^{10}	2	2	2	2	2	2	0
CMDY[1]	2	2	2	4	4	4	0
CMDY[2]	2	2	2	4	4	4	0
$x^2 + x^{90}$	2	2	2	4	4	4	0
x^{14}	0	0	0	0	0	0	2

Furthermore, Table 6 and 7 show their switching neighbors with respect to all the projections from $\mathbb{F}_3^5$ to $\mathbb{F}_3^3$ and $\mathbb{F}_3^2$ respectively. As we mentioned at the beginning of this section, with respect to all the projections from $\mathbb{F}_3^5$ to $\mathbb{F}_3$, all the DO planar functions share only one quadratic form on $\mathbb{F}_p$, except for the CM function, which is a nonquadratic function on $\mathbb{F}_p$ under these projection. Hence, we do not give another table to describe the $\mathbb{F}_3^5 \to \mathbb{F}_3$ case.

Table 8. All known inequivalent planar functions over $\mathbb{F}_{3^6}$

No.	Families	Functions	$\frac{|\mathrm{PAut}(C_f)|}{|\mathrm{Gal}(\mathbb{F}_{3^6})|}$
1	Finite Field	x^2	$|\mathrm{GA}_1(3^6)|$
2	Albert	x^{10}	$|\mathrm{GA}_1(3^6)|$
3	Dickson	$x^{162} + x^{84} + \alpha^{58}x^{54} + \alpha^{58}x^{28} + x^6 + \alpha^{531}x^2$	$26 \cdot |\mathbb{F}_{3^6}|$
4	BHB	$\alpha^{75}x^{2214} + x^{756} + \alpha^{205}x^{82} + x^{28}$	$52 \cdot |\mathbb{F}_{3^5}|$
5	LMPTB	$2x^{270} + x^{246} + 2x^{90} + x^{82} + x^{54} + 2x^{30} + x^{10} + x^2$	$52 \cdot |\mathbb{F}_{3^5}|$
6	Ganley	$x^{270} + 2x^{244} + \alpha^{449}x^{162} + \alpha^{449}x^{84} + \alpha^{534}x^{54}$ $+2x^{36} + \alpha^{534}x^{28} + x^{10} + \alpha^{449}x^6 + \alpha^{279}x^2$	$13 \cdot |\mathbb{F}_{3^5}|$
7	CG	$x^{486} + x^{252} + \alpha^{561}x^{162} + \alpha^{561}x^{84} + \alpha^{183}x^{54}$ $+\alpha^{183}x^{28} + x^{18} + \alpha^{561}x^6 + \alpha^{209}x^2$	$4 \cdot |\mathbb{F}_{3^5}|$
8	CM	x^{122}	$|\mathbb{F}_{3^5}|$

Table 9. Number of inequivalent $l \circ f$ for all $l : \mathbb{F}_3^6 \to \mathbb{F}_3^5$

f	x^2	x^{10}	BHB	LMPTB	Dickson	Ganley	CG	x^{122}
Number	2	2	7	7	7	12	43	2

4.4   The $\mathbb{F}_{3^6}$ Case

Table 8 contains all the known inequivalent planar functions on $\mathbb{F}_{3^6}$, where α is a primitive element of $\mathbb{F}_{3^6}$ and a root of $x^6 - x^4 + x^2 - x - 1$. It should be noted that, although the codes corresponding to BHB and LMPTB semifields have automorphism groups with the same size, it can be calculated by MAGMA that they are inequivalent with each other.

For all the projections $l : \mathbb{F}_3^6 \to \mathbb{F}_3^5$, the numbers of inequivalent $l \circ f$ are listed in Table 9. We have shown that there is only one equivalent pair which comes from the projections of Dickson and CG planar function respectively, i.e. there is again one case where two inequivalent functions are switching neighbors in the narrow sense. Since there are 11011 projections from $\mathbb{F}_3^6$ to $\mathbb{F}_3^4$ or $\mathbb{F}_3^2$, it is beyond our computation capacity to compute all projection up to equivalence. Hence, we can not give the classification of projections with other dimension here.

5   Conclusion

This paper shows that some, but not many of the known planar functions are switching neighbors in the narrow sense. Therefore, we can be mildly optimistic that a switching construction may provide new examples for some $\mathbb{F}_{p^n}$.

The partially ordered set of codes that can be obtained via projection seems to be nontrivial. It may be interesting to investigate these POSETs in more detail.

The function $l \circ f$ considered in Section 3 and 4 correspond to so called relative difference set [10]. These relative difference sets describe certain designs. It may be interesting (though apparently much more involved) to investigate the non-isomorphic designs that can be obtained via projection $l \circ f$.

References

1. Edel, Y., Pott, A.: A new almost perfect nonlinear function which is not quadratic. Advances in Mathematics of Communications 3(1), 59–81 (2009)
2. Dickson, L.: On commutative linear algebras in which division is always uniquely possible. Transaction of the American Mathematical Society 7, 514–522 (1906)
3. Knuth, D.: Finite semifields and projective planes. PhD thesis, California Institute of Technology, Pasadena, California (1963)
4. Hughes, D., Piper, F. (eds.): Projective Planes. Springer, Berlin (1973)
5. Albert, A.: Finite division algebras and finite planes. In: Combinatorial Analysis: Proceedings of the 10th Symposium in Appled Mathematics Symposia in Appl. Math., vol. 10, pp. 53–70. American Mathematical Society, Providence (1960)
6. Lidl, R., Niederreiter, H.: Finite fields, 2nd edn. Cambridge University Press, Cambridge (1997)
7. Dembowski, P., Ostrom, T.: Planes of order n with collineation groups of order n^2. Mathematische Zeitschrift 103(3), 239–258 (1968)
8. Carlet, C.: Vectorial boolean functions for cryptography. In: Crama, Y., Hammer, P. (eds.) Boolean Methods and Models., vol. 2. Cambridge University Press, Cambridge (in press), http://www-roc.inria.fr/secret/Claude.Carlet/chap-vectorial-fcts.pdf
9. Carlet, C., Ding, C.: Highly nonlinear mappings. J. Complexity 20(2-3), 205–244 (2004)
10. Pott, A.: Nonlinear functions in abelian groups and relative difference sets. Discrete Applied Mathematics 138(1-2), 177–193 (2004)
11. Coulter, R.S., Henderson, M.: Commutative presemifields and semifields. Advances in Mathematics 217(1), 282–304 (2008)
12. Budaghyan, L., Carlet, C., Pott, A.: New classes of almost bent and almost perfect nonlinear polynomials. IEEE Transactions on Information Theory 52(3), 1141–1152 (2006)
13. Kyureghyan, G.M., Pott, A.: Some theorems on planar mappings. In: von zur Gathen, J., Imaña, J.L., Koç, Ç.K. (eds.) WAIFI 2008. LNCS, vol. 5130, pp. 117–122. Springer, Heidelberg (2008)
14. Albert, A.: On nonassociative division algebras. Transaction of the American Mathematical Society 72, 292–309 (1952)
15. Zha, Z., Kyureghyan, G.M., Wang, X.: Perfect nonlinear binomials and their semifields. Finite Fields and Their Applications 15(2), 125–133 (2009)
16. Bierbrauer, J.: New commutative semifields and their nuclei. In: Bras-Amorós, M., Høholdt, T. (eds.) AAECC-18. LNCS, vol. 5527, pp. 179–185. Springer, Heidelberg (2009)
17. Bierbrauer, J.: New semifields, PN and APN functions. Des. Codes Cryptography 54(3), 189–200 (2010)
18. Budaghyan, L., Helleseth, T.: New perfect nonlinear multinomials over $\mathbb{F}_{p^{2k}}$ for any odd prime p. In: Golomb, S.W., Parker, M.G., Pott, A., Winterhof, A. (eds.) SETA 2008. LNCS, vol. 5203, pp. 403–414. Springer, Heidelberg (2008)
19. Bierbrauer, J.: New commutative semifields from projection mappings. Submitted, also presented at the Colloquium on Combinatorics 2009, Magdeburg, Germany (2009)
20. Lunardon, G., Marino, G., Polverino, O., Trombetti, R.: Symplectic spreads and quadric veroneseans. Manuscript, also presented at Finite Fields 2009, Dublin, Ireland (2009)

21. Coulter, R.S., Matthews, R.W.: Planar functions and planes of Lenz-Barlotti class II. Des. Codes Cryptography 10(2), 167–184 (1997)
22. Ding, C., Yuan, J.: A family of skew Hadamard difference sets. J. Comb. Theory Ser. A 113(7), 1526–1535 (2006)
23. Cohen, S., Ganley, M.: Commutative semifields, two-dimensional over their middle nuclei. Journal of Algebra 75, 373–385 (1982)
24. Ganley, M.: Central weak nucleus semifields. European Journal of Combinatorics 2, 339–347 (1981)
25. At, N., Cohen, S.D.: A new tool for assurance of perfect nonlinearity. In: Golomb, S.W., Parker, M.G., Pott, A., Winterhof, A. (eds.) SETA 2008. LNCS, vol. 5203, pp. 415–419. Springer, Heidelberg (2008)
26. Weng, G.: A new planar function. Private communication (2007)
27. Coulter, R.S., Henderson, M., Kosick, P.: Planar polynomials for commutative semifields with specified nuclei. Des. Codes Cryptography 44(1-3), 275–286 (2007)
28. Penttila, T., Williams, B.: Ovoids of parabolic spaces. Geometriae Dedicata 82(1-3), 1–19 (2004)
29. Nakagawa, N.: On functions of finite fields (2006), http://www.math.is.tohoku.ac.jp/~taya/sendaiNC/2006/report/nakagawa.pdf
30. Kantor, W.M.: Commutative semifields and symplectic spreads. Journal of Algebra 270(1), 96–114 (2003)
31. Bosma, W., Cannon, J., Playoust, C.: The MAGMA algebra system I: the user language. J. Symb. Comput. 24(3-4), 235–265 (1997)
32. Li, C., Qu, L., Ling, S.: On the covering structures of two classes of linear codes from perfect nonlinear functions. IEEE Transactions on Information Theory 55(1), 70–82 (2009)
33. Pless, V.S., Huffman, W.C., Brualdi, R.A.: Handbook of Coding Theory, Vol. I, II. Elsevier Science Inc., New York (1998)
34. Huffman, W.C., Pless, V.: Fundamentals of Error-Correcting Codes. Cambridge University Press, Cambridge (2003)
35. Kasami, T., Lin, S., Peterson., W.W.: New generalizations of the Reed-Muller codes part I: primitive codes. IEEE Transactions on Information Theory 14(2), 189–199 (1968)
36. Berger, T., Charpin, P.: The permutation group of affine-invariant extended cyclic codes. IEEE Transactions on Information Theory 42(6), 2194–2209 (1996)

Solving Equation Systems by Agreeing and Learning

Thorsten Ernst Schilling and Håvard Raddum

Selmer Center, University of Bergen
{thorsten.schilling,havard.raddum}@ii.uib.no

Abstract. We study sparse non-linear equation systems defined over a finite field. Representing the equations as symbols and using the Agreeing algorithm we show how to learn and store new knowledge about the system when a guess-and-verify technique is used for solving. Experiments are then presented, showing that our solving algorithm compares favorably to MiniSAT in many instances.

Keywords: agreeing, multivariate equation system, SAT-solving, dynamic learning.

1 Introduction

In this paper we present a dynamic learning strategy to solve systems of equations defined over some finite field where the number of variables occuring in each equation is bounded by some constant l. The algorithm is based on the group of Gluing-Agreeing algorithms by Håvard Raddum and Igor Semaev[1,2]. Solving non-linear systems of equations is a well known NP-complete problem already when all equations are of degree 2; this is known as the MQ-problem [3]. Finding a method to solve such systems efficiently is crucial to algebraic cryptanalysis and could break certain ciphers that can be expressed by a set of algebraic equations, such as AES [4], HFE [5], etc.

Several approaches have been proposed to solve such systems, among them SAT-solving [6], Gröbner-basis algorithms [7] and linearization [4]. Since our algorithm falls into the category of the *guess and verify methods*, we compared our solving technique to a state-of-the-art SAT-solving implementation, namely MiniSAT [8].

We adapt the two past major improvements to the DPLL [9] algorithms, which are *watching* and *dynamic learning* [10]. During the search for a solution the method obtains new information from wrong guesses and requires for many instances much less or almost no guessing to obtain a solution to the equation system. The method we present learns new constraints on vectors over some finite field $\mathbb{F}_q$ and can therefore be seen as a generalization of the most common learning method SAT-solvers use, which operates on single variables over $\mathbb{F}_2$. Like in SAT-solving the learning routine of our algorithm runs in polynomial time. Furthermore we show by experimental results that our approach outperforms MiniSAT for a certain class of equation systems, while there still is space for improvement of the method.

M.A. Hasan and T. Helleseth (Eds.): WAIFI 2010, LNCS 6087, pp. 151–165, 2010.

The paper is organized as follows. In Section 2 we explain the symbol representation of equations and the basic idea for agreeing. Section 3 introduces the concept of pockets, and how pockets efficiently integrate with guessing and agreeing. Section 4 shows how the solving technique can gather new (valuable) information from wrong guesses, and Section 5 compares our proposed method to MiniSAT. Section 6 conculdes the paper.

2　Preliminaries

Let
$$f_0(X_0) = 0, f_1(X_1) = 0, \ldots, f_{m-1}(X_{m-1}) = 0 \tag{1}$$
be an equation system in m equations and $n = |X| = |X_0 \cup X_1 \cup \ldots X_{m-1}|$ variables over some finite field $\mathbb{F}_q$. Equations f_i are often given in their ANF-form using the variables in X_i, but here we will use symbol representation.

Definition 1 (Symbol). *Let* $f_i(X_i) = 0$ *be an equation over some finite field* $\mathbb{F}_q$. *We say that* $S_i = (X_i, V_i)$ *is its corresponding symbol where* X_i *is the set of variables in which the equation* f_i *is defined and* V_i *is the set of vectors over* $\mathbb{F}_q$ *in variables* X_i *for which* $f_i(X_i) = 0$ *is satisfied.*

Following this definition the system (1) can be expressed by a set of symbols $\{S_0, S_1, \ldots, S_{m-1}\}$. The cost of transforming (1) to a set of symbols is clearly dominated by the number of equations and the variables involved per equation. Let $l = max\{|X_i| \mid 0 \le i < m\}$. Transforming the system (1) to a set of symbols can be done in time $O(mq^l)$ and we say that (1) is l-sparse. The examples in this paper will only consider $q = 2$, which is the case for most equation systems arising in practice.

Example 1 (Symbol). Let the equation
$$f_0(X_0 = \{x_0, x_1, x_2\}) = x_0 \oplus x_1 x_2 = 0$$
be given over $\mathbb{F}_2$. In order to construct $S_0 = (X_0, V_0)$ we need to know V_0. Every vector $v_i \in V_0$ represents by definition a solution to $f_0(X_0) = 0$ and by searching over all 2^3 vectors in 3 variables and evaluating them we can compute V_0. Therefore the corresponding symbol is
$$S_0 = (X_0 = \{x_0, x_1, x_2\}, V_0 = \{(0,0,0), (0,0,1), (0,1,0), (1,1,1)\}).$$

Throughout the paper a symbol S_0 is represented in table-form for better readability. For this example S_0 it is

S_0	0 1 2
a_0	0 0 0
a_1	0 0 1
a_2	0 1 0
a_3	1 1 1

where the integers $0, 1, 2$ in the first row indicate the variables x_0, x_1, x_2 and $a_0, \ldots, a_3$ are identifiers of the vectors in V_0.

2.1   Agreeing

In order to find a solution to (1) the Agreeing algorithm attempts to delete vectors from symbols S_i which cannot be part of a common solution. In the following, the projection of a vector v_k on variables A is denoted by $v_k[A]$ and $V[A]$ denotes the set of projections of all vectors $v_k \in V$ on variables A.

Given two symbols $S_i = (X_i, V_i)$ and $S_j = (X_j, V_j)$ with $i \neq j$ we say that S_i and S_j are in a non-agreeing state if there exists at least one vector $a_p \in V_i$ such that $a_p[X_i \cap X_j] \notin V_j[X_i \cap X_j]$. If there exists a solution to the system, each symbol will contain one vector that matches the global solution. The vector a_p cannot be combined with any of the possible assignments in symbol S_j, hence it cannot be part of a solution to the whole system and can be deleted. The deletion of all vectors $a_p \in V_i$ and $b_q \in V_j$ which are incompatible with all vectors in V_j and V_i, respectively, is called agreeing. If by agreeing the set of vectors of a symbol gets empty, there exists no solution to the equation system. The agreeing of all pairs of symbols in a set of symbols $\{S_0, \ldots, S_{m-1}\}$ until no further deletion of vectors can be done is called the Agreeing algorithm.

Example 2 (Agreeing). The following pair of symbols is in a non-agreeing state:

$$
\begin{array}{c|ccc}
S_0 & 0 & 1 & 2 \\
\hline
a_0 & 0 & 0 & 0 \\
a_1 & 0 & 0 & 1 \\
a_2 & 0 & 1 & 0 \\
a_3 & 1 & 1 & 1
\end{array}
\qquad
\begin{array}{c|ccc}
S_1 & 0 & 1 & 3 \\
\hline
b_0 & 0 & 0 & 0 \\
b_1 & 1 & 0 & 1
\end{array}.
$$

The vectors a_2, a_3 differ from each b_j in their projection on common variables x_0, x_1 and can be deleted. Likewise, b_1 cannot be combined with any of the a_i and can also be deleted. After agreeing the symbols become:

$$
\begin{array}{c|ccc}
S_0 & 0 & 1 & 2 \\
\hline
a_0 & 0 & 0 & 0 \\
a_1 & 0 & 0 & 1
\end{array}
\qquad
\begin{array}{c|ccc}
S_1 & 0 & 1 & 3 \\
\hline
b_0 & 0 & 0 & 0
\end{array}.
$$

2.2   Guessing

In the example above a further simplification of the equation system by agreeing is not possible. One has to introduce a *guess* to the system. With Example 2, that can be the deletion of vector a_0. The system is in an agreeing state and there exists only a single vector in V_0 and V_1 which gives us a local solution to the equation system, namely the combination of a_1 and b_0, that is $x_0 = 0, x_1 = 0, x_2 = 1, x_3 = 0$.

Since practical examples of equation systems are fully or almost fully pairwise agreeing, a single run of the Agreeing-algorithm obtains no or little extra information about the solution to the system. Thus guessing a vector $g \in V_i$ and deleting all other $v \in V_i, v \neq g$ of a symbol and verifying the partial solution by agreeing is a way to find a solution. If the guess was wrong the changes to the equation system are undone and another guess is introduced.

3   Pocket-Agreeing

We introduce an improvement of the Agreeing algorithm based on the tuple propagation by I. Semaev [11]. The Pocket Agreeing is closer to a potential software implementation and offers some speed advantages and a simple learning process.

The goal is to implement a software method to verify a guess fast. Another aspect is fast backtracking. That means that when a guess is confirmed as incorrect the guess should be undone fast to avoid unnecessary overhead during the computation.

Definition 2 (Pocket). *Let $S_i = (X_i, V_i)$ and $S_j = (X_j, V_j)$ be two pair-wise agreeing symbols with $X_i \cap X_j = X_{i,j}$ and $|X_{i,j}| > 0$. For every projection $\rho \in V_i[X_{i,j}]$ one creates a pair of pockets*

$$p_\alpha = (\{a \mid a \in V_i \text{ and } a[X_{i,j}] = \rho\}, \beta), p_\beta = (\{b \mid b \in V_j \text{ and } b[X_{i,j}] = \rho\}, \alpha))$$

with α and β as unique identifiers or $\emptyset$. For the pocket $p_\alpha = (A, \beta)$, we use the notation $V(p_\alpha) = A$ and $I(p_\alpha) = \beta$.

The purpose of pockets is to have a system that easily identifies vectors that cannot be part of a global solution. Assume that all the vectors in a pocket p are identified as incompatible with a global solution for the system in its current state, and get deleted. Then we can immediately delete all vectors in pocket $I(p)$ since these have the same assignment of variables also found in p, and so must be inconsistent with a global solution too. Also note that one particular vector from a symbol will in general appear in several different pockets. When a vector is deleted from one pocket it is also simultaneously deleted from all the other pockets where it appears.

Example 3 (Pocket). Given the symbols S_0, S_1 from Example 2 after they are pairwise agreeing, $X_0 \cap X_1 = X_{0,1} = \{0, 1\}$. There exists only one projection $V_0[X_{0,1}] = \{(0, 0)\}$, thus there is only one pair of pockets to create, namely

$$p_0 = (\{a_0, a_1\}, 1)$$
$$p_1 = (\{b_0\}, 0).$$

3.1   Propagation

Given a set of pockets generated from symbols $S_0, \ldots, S_{m-1}$ one can run agreeing through pockets. In order to do so efficiently one assigns a *flag* to each vector in the problem instance instead of actually deleting them. The flag of a vector a_i can have three values: *undefined, marked,* and *selected,* where the flags of all vectors are initially undefined. If a vector a_i is marked, denoted by $\overline{a_i}$, it is not suitable for extending the current partial solution, i.e. it is considered to be deleted. If an a_i is selected, denoted by a_i^+, it is considered to be part of the current partial solution, and cannot be deleted. In other words a_i^+ is *guessed.*

The main rule of propagating information in Pocket-Agreeing for the set of pockets is: "While there is a pocket $p_q = (A, b)$ where all $a_i \in A$ are marked, mark all vectors in the pocket p_b, if $b \neq \emptyset$."

This method is analogous to agreeing, where vectors whose projection is not found in another symbol are deleted. In Pocket-Agreeing equal projections are calculated beforehand, stored as pockets, and instead of being deleted as soon as they are not suitable for extending a partial solution, the vectors are flagged as marked.

3.2   Watching

One technical improvement of the Pocket-Agreeing is the possibility to introduce watches as done in SAT-solving. If one wants to implement the Pocket-Agreeing one has to check constantly if all $a_i \in A$ are marked in a pocket (A, b). Experiments show that this consumes a lot of time during the propagation.

In order to avoid this, one assigns in every pocket p a *watch* $w \in V(p)$. Only if the w gets marked it is checked if all the other $a_i \in V(p)$ are marked too. If w gets marked there are two possible cases to distinguish:

1. All $a_i \in V(p)$ are marked, and by the propagation rule all vectors in the pocket $I(p)$ have to be marked, too.
2. There exists at least one $a_i \in V(p)$ which is not marked. This is then the new watch.

This technique reduces the time used in the propagation phase. Also backtracking, i.e., *undoing* a guess in case it was wrong, is sped up. If at some point in the program the conclusion is reached that the guess was wrong, one wants to undo the changes - namely markings - caused by the last guess, in order to try another guess.

To do so one just undoes the marking of vectors from the last guess, since pockets were not changed. The watches can stay the same, since they were by construction the last vectors which got marked in the pocket, or they are not marked at all.

3.3   Guessing

The process of guessing starts with selecting one symbol S_i where all but one vector from V_i are marked. The remaining vector v^+ gets flagged as selected in order to remember that it is guessed to be part of a correct solution. Then Pocket-Agreeing based on the latest markings is started.

Two possible outcomes of the agreeing are possible:

1. Only non-selected vectors get marked. The system is in an agreeing state.
2. At some point the algorithm marks some g^+, which is by the description above the last vector remaining for some symbol. This is called a *conflict*.

If the system ends in an agreeing state, we pick another symbol, select one of its vectors, mark the others and continue the propagation. In the case of a conflict the extension of the partial solution with the previous guess(es) was not possible, and we must backtrack.

4   Learning

During the computation of a solution to the input equation system, it is natural that wrong guesses occur. It is now interesting *why* these occur, since a wrong guess implies that a wrong branch of the search tree was visited. Usually, the implications that show a guess must be wrong only involve a subset of the introduced markings. The purpose of this section is to identify exactly which markings yield a proof of inconsistency for the system. By storing this information the solver *learns* new facts about the system, and the overall number of guesses needed to find the solution is reduced.

Definition 3 (Implication Graph). *An implication graph G is a directed graph. Its vertices are vectors which are marked.*

For a marked vector a_i the pocket $P(a_i)$ is the pocket where all vectors became marked, and by propagation caused the marking of a_i. If the marking of a_i is due to an introduced guess then $P(a_i) = \emptyset$. The set of directed edges E consists of all markings due to propagation, i.e.:

$$(a_i, a_j) \in E \text{ if } a_i \in V(P(a_j)).$$

Edges (a_i, a_j) are labeled by $P(a_j)$.

Example 4 (Implication Graph). Let the following pockets be given.

$$p_0 = (\{a_0\}, 1)$$
$$p_1 = (\{c_0, c_1\}, 0)$$
$$p_2 = (\{b_0, b_1\}, \emptyset)$$
$$p_3 = (\{c_0\}, 2)$$

Introducing the marking $\overline{a_0}$ would yield the following implication graph.

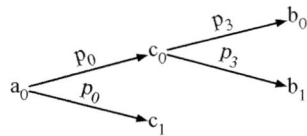

Fig. 1. Example Implication Graph

The implication graph is not unique and depends on the order in which empty pockets are processed.

4.1   Conflict Analysis

Let g^+ be the vector which yielded the conflict, that is it was flagged as selected and by agreeing became marked. The immediate source of the conflict is the marking of all $h_j \in V(P(g^+))$. But for further analysis we are more interested in vectors which caused the conflict by introducing a guess. These are h_j's connected to g^+ in the implication graph, where $P(h_j) = \emptyset$. By analyzing the graph we can find the h_j's recursively:

$$R(g) = \{h_j | h_j \in V(P(g)) \text{ and } P(h_j) = \emptyset\} \cup \bigcup_{\substack{h_j \in V(P(g)) \\ P(h_j) \neq \emptyset}} R(h_j). \qquad (2)$$

$R(g)$ will then be the set of marked vectors due to *guesses*, that caused g to be marked. In other words, $R(g)$ tells us exactly *which* of the introduced guesses that are incompatible with g being part of the solution. This information can be stored as a new pocket, as shown in the following.

4.2   Conflict Construction and Reduction

Assume the marking of g^+ yields a conflict and we have found that $R(g) = \{h_0, h_1, \ldots, h_r\}$ are the marked vectors that imply the marking of g^+. We can now create a new pair of pockets with the implication

$$R(g) \Rightarrow g,$$

i.e., if all vectors in $R(g)$ are marked, then g must be marked. The pockets expressing this are

$$\begin{aligned} p_{s^*} &= (\{h_1, \ldots, h_r\}, t) \\ p_t &= (g, \emptyset). \end{aligned} \qquad (3)$$

However, storing (3) for further computation does not give us any new information, since it is a direct consequence of agreeing. We are more interested in a reduced condition under which we can mark g and exclude it from a common solution during the search process. The following lemma shows how to find a reduced condition for when g can be marked.

Lemma 1. *Let the pockets $p = (\{h_1, \ldots, h_r\}, q)$ and $p_q = (\{g_1, \ldots, g_s\}, \emptyset)$ be given. For any h_j and g_i, let $X_{g_i h_j}$ be the set of variables that are common to both h_j and g_i. Let H be the set of vectors $h_j \in V(p)$ such that $h_j[X_{g_i h_j}] \neq g_i[X_{g_i h_j}]$ for all i. Then marking all vectors in $V(p) \setminus H$ implies marking all vectors in $V(p_q)$.*

Proof: Mark all vectors in $V(p) \setminus H$ and assume that some g_i is part of the solution to the system and should not be marked. Since any vector $h_j \in H$ is different in its projection on $X_{g_i h_j}$ from $g_i[X_{g_i h_j}]$, no vectors in H can be combined with g_i in a global solution, so all vectors in H must be marked. Then the pocket p yields that g_i must also be marked. This conflict shows that g_i cannot be part of the solution to the system after all, so all vectors in $V(p_q)$ should be marked once the vectors in $V(p) \setminus H$ are marked.   □

Using this lemma, we delete from the vectors in $R(g)$ all h_j for which is true that
$$h_j[X_{gh_j}] \neq g[X_{gh_j}],$$
and save the implication in a pair of pockets:
$$p_s = (\{h_j | h_j \in R(g) \text{ and } h_j[X_{gh_j}] = g[X_{gh_j}]\}, t)$$
$$p_t = (g, \emptyset).$$

These two pockets are then added to the list of pockets the system already knows.

From the conflict described above we can also derive further new knowledge. Up until now we have our reduced implication $p_s \Rightarrow p_t$, i.e. if all vectors in p_s are marked, mark the vector $g \in V(p_t)$. Also, it holds for any vector g that

$$g^+ \equiv \overline{c_1}, \overline{g_2}, \ldots, \overline{g_r} \text{ with } g_i \neq g \text{ and } g, g_1, \ldots, g_r \text{ are all vectors in a symbol} \quad (4)$$

Thus g can become an *implicit guess* by marking all other g_i's in the same symbol. From the pair of pockets p_s, p_t we can now further derive that if g is guessed, at least one of the vectors in p_s has to be selected. Otherwise all h_j in p_s would be marked, and the pockets p_s, p_t would yield a conflict. We express this with the following lemma.

Lemma 2. *Let the pockets $p_s = (h_1, \ldots, h_r, t)$ and $p_t = (g, \emptyset)$ be given. For any symbol $S_\gamma = (X_\gamma, V_\gamma)$ such that $V_\gamma \cap V(p_s) \neq \emptyset$ the implication of the following pockets must hold:*

$$p_{s_\gamma} = (\{g_1, \ldots, g_r | g_i \neq g\} \cup (V(p_s) \setminus V_\gamma), t_\gamma)$$
$$p_{t_\gamma} = (V_\gamma \setminus V(p_s), \emptyset)$$

Proof: Let $p_s = (\{a_u, \ldots, a_v, b_x, \ldots, b_y\}, t)$ where $\{b_x, \ldots, b_y\} = V(p_s) \cap V_\gamma$. Then the condition $g^+, \overline{a_u}, \ldots, \overline{a_v}$ implies that one of $b_x, \ldots, b_y$ has to be selected (guessed). Otherwise, if none of $b_x, \ldots, b_y$ are selected all vectors in $V(p_s)$ are marked and g has to be marked too (by $p_s \Rightarrow p_t$). This would be a conflict since g^+ is implicitly selected. Guessing one of $b_x, \ldots, b_y$ implies the marking of all vectors in $V_\gamma \setminus \{b_x, \ldots, b_y\}$, which is exactly the set of vectors in p_{t_γ}.   □

By using Lemma 1, we should also reduce the condition for when the vectors in $V(p_{t_\gamma})$ can be deleted by excluding vectors in $V(p_{s_\gamma})$ that differ in projection on common variables to all vectors in $V(p_{t_\gamma})$.

Remark 1 (Cycle-rule). Lemma 1 is an extension to the cycle-rule by Igor Semaev [12]. The cycle-rule states that through (4) it is possible to delete from an implication $\overline{a_0}, \ldots, \overline{a_r} \Rightarrow \overline{h_0}, \ldots, \overline{h_s}$ those a_i which belong to the same symbol as $h_0, \ldots, h_s$. However, the cycle-rule is extended by removing vectors from $a_0, \ldots, a_r$ which do not belong to the same symbol, but only differ in their projection from the vectors $h_0, \ldots, h_s$. Note that if two vectors belong to the same symbol, they always differ in their projection on common variables.

4.3   Non-chronological Backtracking

After the learning is completed the last guess should be undone and based on the extended pocket database Agreeing should run again. If the system is now in a non-agreeing state it can only be due to newly learnt pockets p_s. Thus any change to the system that does not involve vectors in $V(p_s)$ will necessarily result in a conflict again. Therefore we can jump back to the tree-level at which the last change in an p_s occurred, depending on which pocket yielded the conflict. This way we cut futile branches of the search tree and economize the search in the number of guesses.

Example 5. Let the following equation system be given:

S_0	1 2 3	S_1	2 4 5 6 12	S_2	4 7 8	S_3	1 9 10	S_4	10 11 12	S_5	9 11 12
a_0	0 0 0	b_0	0 1 0 0 0	c_0	1 0 0	d_0	0 0 1	e_0	0 0 1	f_0	0 0 1
a_1	0 1 1	b_1	0 1 0 1 0	c_1	1 0 1	d_1	0 1 0	e_1	0 1 0	f_1	0 1 0
a_2	1 1 0	b_2	0 1 1 0 1	c_2	0 1 0	d_2	1 0 0	e_2	1 0 0	f_2	1 0 0
a_3	1 1 1	b_3	1 0 1 1 1	c_3	0 1 1	d_3	1 1 1	e_3	1 1 1	f_3	1 1 1

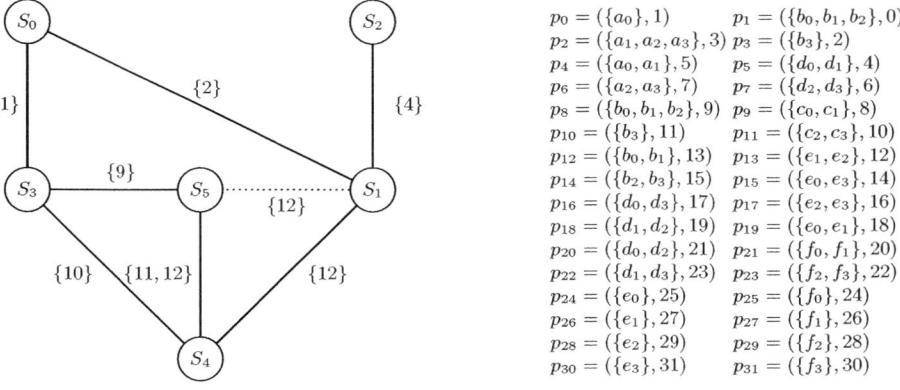

$$
\begin{aligned}
&p_0 = (\{a_0\}, 1) && p_1 = (\{b_0, b_1, b_2\}, 0)\\
&p_2 = (\{a_1, a_2, a_3\}, 3) && p_3 = (\{b_3\}, 2)\\
&p_4 = (\{a_0, a_1\}, 5) && p_5 = (\{d_0, d_1\}, 4)\\
&p_6 = (\{a_2, a_3\}, 7) && p_7 = (\{d_2, d_3\}, 6)\\
&p_8 = (\{b_0, b_1, b_2\}, 9) && p_9 = (\{c_0, c_1\}, 8)\\
&p_{10} = (\{b_3\}, 11) && p_{11} = (\{c_2, c_3\}, 10)\\
&p_{12} = (\{b_0, b_1\}, 13) && p_{13} = (\{e_1, e_2\}, 12)\\
&p_{14} = (\{b_2, b_3\}, 15) && p_{15} = (\{e_0, e_3\}, 14)\\
&p_{16} = (\{d_0, d_3\}, 17) && p_{17} = (\{e_2, e_3\}, 16)\\
&p_{18} = (\{d_1, d_2\}, 19) && p_{19} = (\{e_0, e_1\}, 18)\\
&p_{20} = (\{d_0, d_2\}, 21) && p_{21} = (\{f_0, f_1\}, 20)\\
&p_{22} = (\{d_1, d_3\}, 23) && p_{23} = (\{f_2, f_3\}, 22)\\
&p_{24} = (\{e_0\}, 25) && p_{25} = (\{f_0\}, 24)\\
&p_{26} = (\{e_1\}, 27) && p_{27} = (\{f_1\}, 26)\\
&p_{28} = (\{e_2\}, 29) && p_{29} = (\{f_2\}, 28)\\
&p_{30} = (\{e_3\}, 31) && p_{31} = (\{f_3\}, 30)
\end{aligned}
$$

Fig. 2. The intersection graph and the resulting pockets. Dotted edges in the intersection graph are ignored.

The intersection graph in Figure 2 indicates pairs of symbols from which pockets are generated. The labled edges between symbols show intersections in the sets of variables. No pockets are generated from the pair S_1, S_5 since changes of variable x_{12} will propagate through the path S_1, S_4, S_5 while agreeing.

Assume that by some heuristic the order of symbols to be guessed is $S_0, S_1, S_2,$ S_3, S_4, S_5. The partial solutions a_0^+, b_0^+, c_0^+ are selected in that order. This results in the following equation system after agreeing:

$$
\frac{S_0\,|\,1\ 2\ 3}{a_0\,|\,0\ 0\ 0}, \quad
\frac{S_1\,|\,2\ 4\ 5\ 6\ 12}{b_0\,|\,0\ 1\ 0\ 0\ 0}, \quad
\frac{S_2\,|\,4\ 7\ 8}{c_0\,|\,1\ 0\ 0}, \quad
\begin{array}{c|ccc}
S_3 & 1 & 9 & 10 \\
\hline
d_0 & 0 & 0 & 1 \\
d_1 & 0 & 1 & 0
\end{array}, \quad
\begin{array}{c|ccc}
S_4 & 10 & 11 & 12 \\
\hline
e_1 & 0 & 1 & 0 \\
e_2 & 1 & 0 & 0
\end{array}, \quad
\begin{array}{c|ccc}
S_5 & 9 & 11 & 12 \\
\hline
f_1 & 0 & 1 & 0 \\
f_2 & 1 & 0 & 0
\end{array}.
$$

For a further extension of the partial guess one tries to extend the partial solution by d_0. The resulting implication graph after marking d_1 is shown below. Marking d_1 causes e_1 to be marked by pocket p_{18}, which again causes f_1 and d_0 to be marked by pockets p_{26} and p_{21}. This is clearly a conflict, since d_0 was

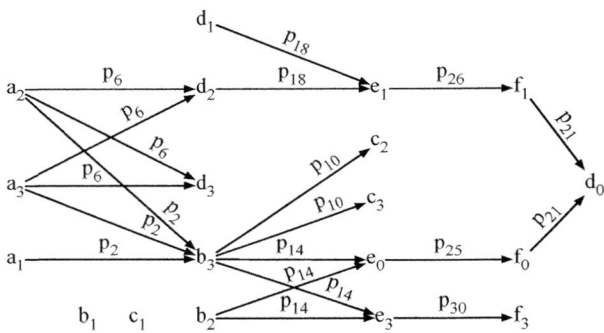

Fig. 3. Implication Graph of guess a_0, b_0, c_0, d_0

previously selected but should be marked now. Now we analyze the source of the conflict in order to learn from it.

$$R(d_0) = \{a_1, a_2, a_3, b_2, d_1\}$$

To create the reduced p_s we compare projections of a_1, a_2, a_3, b_2, d_1 in common variables to projections of d_0. We see that a_2 and a_3 have a different projection than d_0 on their common variable x_1, so these vectors can be excluded from p_s by Lemma 1. d_1 can obviously also be excluded since it belongs to the same symbol as d_0. After this reduction we get:

$$p_{32} = (\{a_1, b_2\}, 33)$$
$$p_{33} = (\{d_0\}, \emptyset).$$

Using Lemma 2 we also derive:

$$p_{34} = (\{d_1, d_2, d_3, a_1\}, 35)$$
$$p_{35} = (\{b_0, b_1, b_3\}, \emptyset)$$
$$p_{36} = (\{d_1, d_2, d_3, b_2\}, 37)$$
$$p_{37} = (\{a_0, a_2, a_3\}, \emptyset)$$

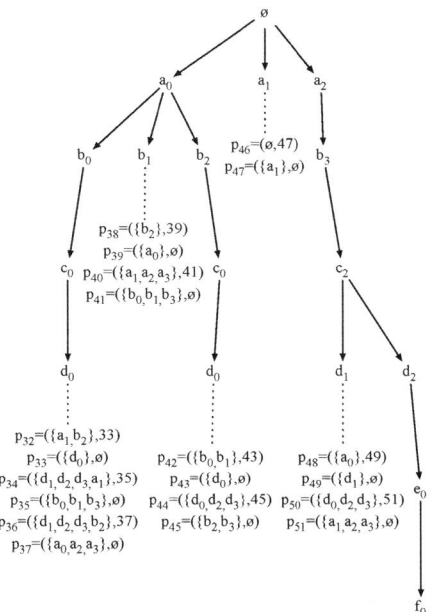

Fig. 4. Search tree with learning

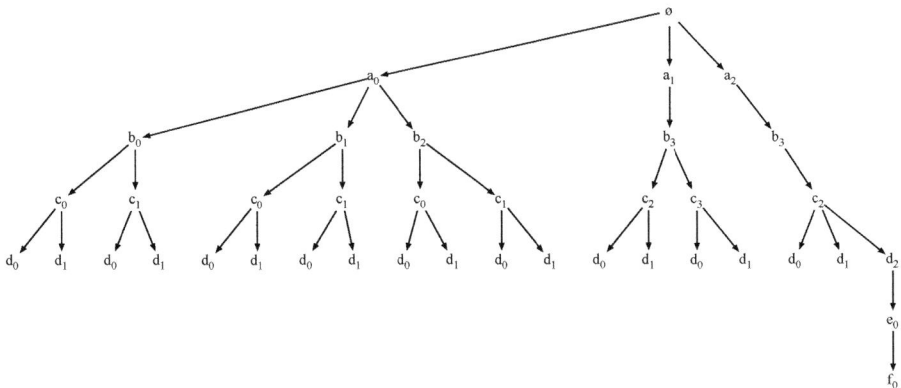

Fig. 5. Naive search tree

After this learning process we agree the system again, with our newly obtained knowledge. The pockets p_{32} and p_{33} cause d_0 to be marked. This implicitly selects d_1^+, which immediately yields another conflict, without introducing any new guess. Thus the guesses a_0^+, b_0^+, c_0^+ cannot all be right. We can immediately read from p_{32} where to backtrack. We see from p_{32} that the guessing of c_0^+ was not a cause for the conflict, otherwise there would be some c_i-vectors in p_{32}. This tells us that if we now backtrack and select, say c_1^+, we will end up in the very same conflict again. Hence we can go back to the point where b_0 got guessed (and b_2 marked) and try selecting another b_j-vector. Bypassing the guesses on all c_i-vectors that would be due in a naive search algorithm saves a lot of time.

Figure 4 shows the decision tree until the first solution is found. Branches not incorporating vectors from all symbols indicate conflicts. Connected to the dotted lines are the newly learned pockets. In comparison the naive search tree, without learning, is depicted in Figure 5.

4.4   Variable-Based Guessing

In the algorithm we have explained, we guess on which of the possible assignments in a symbol that is the correct one. It may look more natural to guess on the value of single variables as is done in SAT-solving. Given an instance $S_0, S_1, \ldots, S_m$ in variables X there exists a simple way to realize variable-based guessing. Instead of establishing a separate mechanism of introducing the guess on a single variable one inserts new symbols of the form $S_{x_i} = (\{x_i\}, \{v_0 = 0, v_1 = 1\})$ for every $x_i \in X$ before the pocket generation. These symbols contain no information but can easily be integrated into the system. Assume one wants to guess that $x_i = 0$. From the newly inserted symbol one just marks v_1 and propagates the guess by agreeing instead of keeping a separate table of all vectors in which x_i occurs as 1 and marking them. Another advantage is that this way of introducing variable guessing integrates with the learning without problems.

Of course this approach works for other fields than $\mathbb{F}_2$, too. Assume an equation system over $\mathbb{F}_q$ then one inserts for every $x_i \in X$ a symbol $S_{x_i} = (\{x_i\}, V_{x_i} = \{v_j | v_j \in \mathbb{F}_q\})$.

5   Experiments

5.1   Results

In order to evaluate the strength of the proposed solving algorithm, several experiments were made with random equation systems over $\mathbb{F}_2$. A software, called *Gluten*, that implements the algorithm was developed. To get a comparison with another solving technique we took a SAT-solver, namely MiniSAT since the guess/verify technique to obtain a solution is similar. Furthermore SAT-solving is a well researched field and MiniSAT among the fastest programs in this field.

Rather than comparing pure solving time we compare the number of variable guesses needed until a solution to the system is obtained. During all the experiments it holds $m = n$, i.e. the number of equations is equal to the number of variables. We make sure the systems have at least one solution. The sparsity l is also fixed to $l = 5$. The ANF degree for the equations we generate will be randomly distributed, but will of course be upper bounded by the sparsity. Furthermore every $m = n$ was tested with 100 randomly generated instances and the arithmetic mean calculated afterwards.

Figures which display both very large and very small values are log-scaled for better readability.

5.2   Random Instances

In this experiment the expected number of roots for every equation is $E(|V_i|) = 2^4$ and binomially distributed, as would be the case when the symbols are obtained from random ANF's.

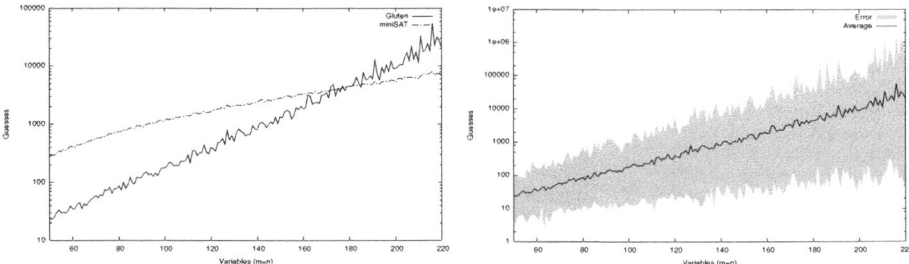

Fig. 6. Gluten vs. MiniSAT (log-scale)   **Fig. 7.** Gluten average and error (log-scale)

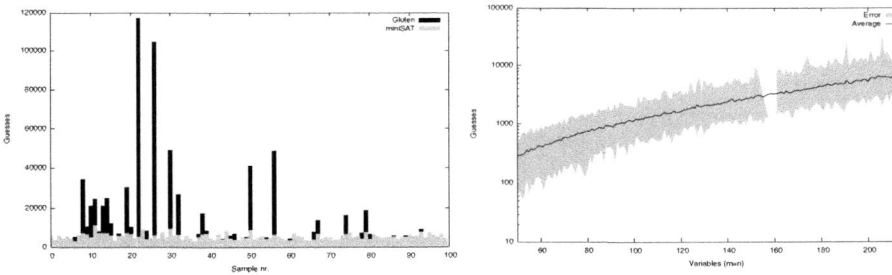

Fig. 8. $n = m = 200$ **Fig. 9.** MiniSAT average and error (log-scale)

In Figure 6 one can see that Gluten performs clearly better up til around $m = n = 170$. Afterwards the average values for MiniSAT stay low while the number of guesses for Gluten rise fast. Figure 7 shows that the error margin is very high to the average in comparison to the error margin of MiniSAT, shown in Figure 9. In other words, Gluten runs into a few cases where it makes an extremely high number of guesses whereas MiniSAT is able to keep its number of guesses not too far from the average.

To get a better comparison of both methods in Figure 8, the case of $n = m = 200$ along with the sample number is given. For every of the 100 samples the black bar indicates the number of variable guesses Gluten took to obtain a solution and the grey bar shows the number of guesses MiniSAT took to find a solution. In approximately $1/3$ of all samples Gluten performs worse, in the rest approximately equally or better.

5.3   Uniformly Distributed Number of Roots

The case when the number of roots in the equations are distributed uniformly at random was also investigated. That means that the size of V_i is taken uniformly at random from $[1, 2^l - 1]$ for each symbol.

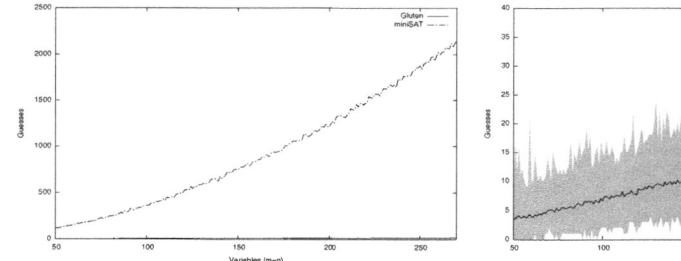

Fig. 10. Gluten vs. MiniSAT   **Fig. 11.** Gluten average and error

In this scenario Gluten performs much better on the whole spectrum of the experimental data. As Figure 10 and 11 shows the number of guesses for Gluten rise linearly while the curve giving the number of MiniSAT's guessings seems to be quadratic (the polynomial $0.0232n^2 + 1.6464n - 15.4$ fits the dashed curve very well). The Gluten values are less than 50; note the different scalings in Figure 10 and 11. It is also interesting to notice that Gluten only needs to make very few guesses, even for systems with over 250 variables.

6   Conclusion and Further Work

We have shown how new knowledge about the equation system can be obtained in polynomial time when guessing partial solutions and running the Agreeing algorithm. New constraints on vectors defining partial solutions can be added and using this, futile search-regions can be pruned. Our experiments show our proposed algorithm performs better than SAT-solving in a large number of instances. In particular, the experimental data shows that it is only necessary to make a small number of guesses to solve systems where the number of roots are uniformly distributed.

Several mechanisms are not yet introduced to our algorithm. Among them are random restarts during the search process or random guesses. It is obvious that a good guessing heuristic is crucial for the success of a solver of this kind. While SAT-solving is well studied and a lot of different search-heuristics are available, this is still an open field and topic for future research for the algorithm proposed in this paper.

References

1. Raddum, H.: MRHS Equation Systems. In: Adams, C., Miri, A., Wiener, M. (eds.) SAC 2007. LNCS, vol. 4876, pp. 232–245. Springer, Heidelberg (2007)
2. Raddum, H., Semaev, I.: Solving Multiple Right Hand Sides linear equations. Designs, Codes and Cryptography 49(1), 147–160 (2008)
3. Courtois, N., Patarin, J.: About the XL Algorithm over GF(2). In: Joye, M. (ed.) CT-RSA 2003. LNCS, vol. 2612, pp. 141–157. Springer, Heidelberg (2003)
4. Courtois, N., Pieprzyk, J.: Cryptanalysis of Block Ciphers with Overdefined Systems of Equations. In: Zheng, Y. (ed.) ASIACRYPT 2002. LNCS, vol. 2501, pp. 267–287. Springer, Heidelberg (2002)
5. Kipnis, A., Shamir, A.: Cryptanalysis of the HFE Public Key Cryptosystem by Relinearization. In: Wiener, M. (ed.) CRYPTO 1999. LNCS, vol. 1666, pp. 19–30. Springer, Heidelberg (1999)
6. Massacci, F., Marraro, L.: Logical cryptanalysis as a SAT problem. Journal of Automated Reasoning 24(1), 165–203 (2000)
7. Faugère, J.: A new efficient algorithm for computing Gröbner bases (F4). Journal of Pure and Applied Algebra 139(1-3), 61–88 (1999)
8. Een, N., Sörensson, N.: Minisat v2.0 (beta). Solver description, SAT Race (2006), http://fmv.jku.at/sat-race-2006/
9. Davis, M., Logemann, G., Loveland, D.: A machine program for theorem-proving. Commun. ACM 5(7), 394–397 (1962)
10. Moskewicz, M., Madigan, C., Zhao, Y., Zhang, L., Malik, S.: Chaff: Engineering an efficient SAT solver. In: Proceedings of the 38th Conference on Design Automation, pp. 530–535. ACM, New York (2001)
11. Semaev, I.: Sparse algebraic equations over finite fields. SIAM Journal on Computing 39(2), 388–409 (2009)
12. Semaev, I., Schilling, T.: Personal correspondence (2009)

Speeding Up Bipartite Modular Multiplication

Miroslav Knežević, Frederik Vercauteren, and Ingrid Verbauwhede

Katholieke Universiteit Leuven
Department of Electrical Engineering - ESAT/SCD-COSIC and IBBT
Kasteelpark Arenberg 10, B-3001 Leuven-Heverlee, Belgium
{mknezevi,fvercaut,iverbauw}@esat.kuleuven.be

Abstract. A large set of moduli, for which the speed of bipartite modular multiplication considerably increases, is proposed in this work. By considering state of the art attacks on public-key cryptosystems, we show that the proposed set is safe to use in practice for both elliptic curve cryptography and RSA cryptosystems. We propose a hardware architecture for the modular multiplier that is based on our method. The results show that, concerning the speed, our proposed architecture outperforms the modular multiplier based on standard bipartite modular multiplication. Additionally, our design consumes less area compared to the standard solutions.

Keywords: Bipartite modular multiplication (BMM), Barrett reduction, Montgomery reduction, Public-key cryptography (PKC).

1 Introduction

Public-key cryptography (PKC), a concept introduced by Diffie and Hellman [9] in the mid 70's, has gained its popularity together with the rapid evolution of today's digital communication systems. The best-known public-key cryptosystems are based on factoring *i.e.* RSA [20] and on the discrete logarithm problem in a large prime field (Diffie-Hellman, ElGamal, Schnorr, DSA) [14] or on an elliptic curve (ECC/HECC) [2]. Based on the hardness of the underlaying mathematical problem, PKC usually deals with large numbers ranging from a few hundreds to a few thousands of bits in size. Consequently, the efficient implementations of the PKC primitives has always been a challenge.

An efficient implementation of the mentioned cryptosystems highly depends on the efficient implementation of modular arithmetic. Namely, modular multiplication forms the basis of modular exponentiation which is the core operation of the RSA cryptosystem. It is also present in many other cryptographic algorithms including those based on ECC and HECC. In particular, if one uses projective coordinates for ECC/HECC, modular multiplication remains the most time consuming operation for ECC. Hence, an efficient implementation of PKC relies on efficient modular multiplication.

Two algorithms for modular multiplication, namely Barrett [3] and Montgomery [15] algorithms are widely used today. Both algorithms avoid multiple-precision divisions, the operation that is considered to be expensive, especially in hardware. The classical modular multiplication algorithm, based on Barrett's reduction, uses single-precision

M.A. Hasan and T. Helleseth (Eds.): WAIFI 2010, LNCS 6087, pp. 166–179, 2010.

multiplications with a precomputed modulus reciprocal instead of expensive divisions [8]. An algorithm that efficiently combines classical and Montgomery multiplications, both in finite fields of characteristic 2, was first proposed by Potgieter [17] in 2002. Published in 2005, a bipartite modular multiplication (BMM) by Kaihara and Takagi [11] extended this approach to the ring of integers.

In this work, we propose a large set of moduli, for which the intermediate quotient evaluation in both Barrett and Montgomery algorithms basically comes for free. Therefore a speed of bipartite modular multiplication, where the Barrett and Montgomery algorithms are the main ingredients, significantly increases. By considering state of the art attacks on public-key cryptosystems, we show that the proposed set is safe to use in practice for both ECC/HECC and RSA cryptosystems. We propose a hardware architecture for the modular multiplier that outperforms the multiplier based on standard BMM method.

The remainder of the paper is structured as follows. Section 2 introduces preliminaries. Section 3 describes the proposed method. In Sect. 4 we give the results of hardware implementations and Sect. 5 discusses the security issues. Section 6 concludes.

2 Preliminaries

In the paper we use the following notations. A multiple-precision n-bit integer A is represented in radix r representation as $A = (A_{n_w-1} \ldots A_0)_r$ where $r = 2^w$; n_w represents the number of digits and is equal to $\lceil n/w \rceil$ where w is a *digit-size*; A_i is called a *digit* and $A_i \in [0, r-1]$.

2.1 Classical and Montgomery Modular Multiplication Methods

Given a modulus M and two elements $X, Y \in \mathbb{Z}_M$ where $\mathbb{Z}_M$ is the ring of integers modulo M, the ordinary modular multiplication is defined as:

$$X \times Y \triangleq XY \bmod M \ .$$

Let the modulus M be an n_w-digit integer, where the radix of each digit is $r = 2^w$. The classical modular multiplication algorithm computes $XY \bmod M$ by interleaving the multiplication and modular reduction phases as it is shown in Algorithm 1. The calculation of the intermediate quotient q_C at step 4 of the algorithm is done by utilizing integer division which is considered as an expensive operation, especially in hardware. The idea of using the precomputed reciprocal of the modulus M and simple shift and multiplication operations instead of division originally comes from Barrett [3]. To explain the basic idea, we rewrite the intermediate quotient q_C as:

$$q_C = \left\lfloor \frac{Z}{M} \right\rfloor = \left\lfloor \frac{\frac{Z}{2^{n+\beta}} \frac{2^{n+\alpha}}{M}}{2^{\alpha-\beta}} \right\rfloor \geq \left\lfloor \frac{\left\lfloor \frac{Z}{2^{n+\beta}} \right\rfloor \left\lfloor \frac{2^{n+\alpha}}{M} \right\rfloor}{2^{\alpha-\beta}} \right\rfloor = \hat{q} \ .$$

The value $\hat{q}$ represents an estimation of the intermediate quotient q_C. In most of the cryptographic applications, the modulus M is fixed during the many modular multiplications and hence the value $\lfloor 2^{n+\alpha}/M \rfloor$ can be precomputed and reused multiple times.

Algorithm 1. Classical modular multiplication algorithm.

Input: $X = (X_{n_w-1} \ldots X_0)_r$, $Y = (Y_{n_w-1} \ldots Y_0)_r$, $M = (M_{n_w-1} \ldots M_0)_r$ where $0 \leq X, Y < M, 2^{n-1} \leq M < 2^n, r = 2^w$ and $n_w = \lceil n/w \rceil$.

Output: $Z = XY \bmod M$.

1: $Z \Leftarrow 0$
2: **for** $i = n_w - 1$ downto 0 **do**
3:     $Z \Leftarrow Zr + XY_i$
4:     $q_C \Leftarrow \lfloor Z/M \rfloor$
5:     $Z \Leftarrow Z - q_C M$
6: **end for**
7: Return Z.

Furthermore, an integer division with the power of 2 is a simple shift operation in hardware. Since the value of $\hat{q}$ is an estimated value, some correction steps at the end of the modular multiplication algorithm have to be performed. In his thesis, Dhem [8] determines the values of α and β for which the classical modular multiplication, based on Barrett reduction algorithm, needs at most one subtraction at the end of the algorithm. The improved Barrett algorithm [8], uses the following parameters: $\alpha = w + 3$ and $\beta = -2$.

Montgomery's algorithm [15] is the most commonly utilized modular multiplication algorithm today. In contrast to classical modular multiplication, it utilizes right to left divisions. Given an n_w-digit odd modulus M and an integer $U \in \mathbb{Z}_M$, the image or the Montgomery residue of U is defined as $X = UR \bmod M$ where R, the Montgomery radix, is a constant relatively prime to M. If X and Y are, respectively, the images of U and V, the Montgomery multiplication of these two images is defined as:

$$X * Y \triangleq XYR^{-1} \bmod M \ .$$

The result is the image of $UV \bmod M$ and needs to be converted back at the end of the process. For the sake of efficient implementation, one usually uses $R = r^{n_w}$ where $r = 2^w$ is the radix of each digit. Similar to the Barrett multiplication, this algorithm uses a precomputed value $M' = -M^{-1} \bmod r = -M_0^{-1} \bmod r$. The algorithm is shown in Algorithm 2.

2.2 Bipartite Modular Multiplication Method

An algorithm that efficiently combines classical and Montgomery multiplications, both in finite fields of characteristic 2, was first proposed by Potgieter [17] in 2002. Extending this approach to the ring of integers, the bipartite modular multiplication (BMM) was introduced by Kaihara and Takagi in [11]. The method efficiently combines a classical modular multiplication method with Montgomery's modular multiplication algorithm. It splits the operand multiplier into two parts that can be processed separately in parallel, increasing the calculation speed. The calculation is performed using Montgomery residues defined by a modulus M and a Montgomery radix R, $R < M$. Next, we outline the main idea of the BMM method.

Algorithm 2. Montgomery modular multiplication algorithm.

Input: $X = (X_{n_w-1} \ldots X_0)_r$, $Y = (Y_{n_w-1} \ldots Y_0)_r$, $M = (M_{n_w-1} \ldots M_0)_r$, $M' = -M_0^{-1} \bmod r$ where $0 \leq X, Y < M$, $2^{n-1} \leq M < 2^n$, $r = 2^w$, gcd(M,r)=1 and $n_w = \lceil n/w \rceil$.

Output: $Z = XYr^{-n_w} \bmod M$.

1: $Z \Leftarrow 0$
2: **for** $i = 0$ to $n_w - 1$ **do**
3:     $Z \Leftarrow Z + XY_i$
4:     $q_M \Leftarrow (Z \bmod r)M' \bmod r$
5:     $Z \Leftarrow (Z + q_M M)/r$
6: **end for**
7: **if** $Z \geq M$ **then**
8:     $Z \Leftarrow Z - M$
9: **end if**
10: Return Z.

Let the modulus M be an n_w-digit integer, where the radix of each digit is $r = 2^w$ and let $R = r^k$ where $0 < k < n_w$. Consider the multiplier Y to be split into two parts Y_H and Y_L so that $Y = Y_H R + Y_L$. Then, the Montgomery multiplication modulo M of the integers X and Y can be computed as follows:

$$
\begin{aligned}
X * Y &= XYR^{-1} \bmod M \\
&= X(Y_H R + Y_L)R^{-1} \bmod M \\
&= \big((XY_H \bmod M) + (XY_L R^{-1} \bmod M)\big) \bmod M \ .
\end{aligned}
$$

The left term of the last equation, $XY_H \bmod M$, can be calculated using the classical modular multiplication that processes the upper part of the split multiplier Y_H. The right term, $XY_L R^{-1} \bmod M$, can be calculated using the Montgomery algorithm that processes the lower part of the split multiplier Y_L. Both calculations can be performed in parallel. Since the split operands Y_H and Y_L are shorter in length than Y, the calculations $XY_H \bmod M$ and $XY_L R^{-1} \bmod M$ are performed faster than $XYR^{-1} \bmod M$.

2.3   Related Work

Before introducing related work we note here that for the moduli used in all common ECC cryptosystems, the modular reduction can be done much faster than the one proposed by Barrett or Montgomery. Even without any multiplication. This is the reason behind standardizing generalized Mersenne prime moduli (sums/differences of a few powers of 2) [16,1,21].

The idea of speeding up a modular multiplication by simplifying an intermediate quotient was first presented by Quisquater [18] at the rump session of Eurocrypt '90. The method is similar to the one of Barrett except that the modulus is preprocessed before the modular multiplication in such a way that the evaluation of the intermediate quotient basically comes for free. Preprocessing requires some extra memory and computational time, but the latter is negligible when many modular multiplications are performed using the same modulus.

In [12], Lenstra proposes several ways to generate RSA moduli with any number of predetermined leading (trailing) bits, with the fraction of specified bits only limited by security considerations. He points out that choosing such moduli is beneficial both for storage and computational requirements. Furthermore, Lenstra discusses security issues and concludes that the resulting moduli do not seem to offer less security than regular RSA moduli. Joye [10] enhances the method for generating RSA moduli with a predetermined portion proposed in [12].

3   Speeding Up the Bipartite Modular Multiplication

In both Barrett and Montgomery modular multiplication algorithms, the precomputed values of either modulus reciprocal or modulus inverse are used in order to avoid multiple-precision divisions. However, single-precision multiplications still need to be performed (step 4 of the algorithms above). This especially concerns the hardware implementations, as the multiplication with the precomputed values often occurs within the critical path of the whole design. Section 4 discusses this issue in more detail.

Since the BMM method utilizes both Barrett and Montgomery multiplication algorithms, one needs to precompute both $\mu = \lfloor 2^{n+\alpha}/M \rfloor$ and $M' = -M_0^{-1} \bmod r$. Let us, for now, assume that the precomputed values are both of type 2^γ where $\gamma \in \mathbb{Z}$. By tuning μ and M' to be of this special type, we transform a single-precision multiplication with these values into a simple shift operation in hardware. Therefore, we find a set of moduli for which the precomputed values are both of type 2^γ. A lemma that defines this set is given below:

Lemma 1. *Let* $M = 2^n - \Delta 2^w - 1$ *be an n-bit positive integer in radix* $r = 2^w$ *representation with* $\Delta \in \mathbb{Z}$, $w \in \mathbb{N}$ *and* $w < n$. *Now, let* $\mu = \lfloor 2^{n+\alpha}/M \rfloor$ *where* $\alpha \in \mathbb{N}$ *and* $M' = -M_0^{-1} \bmod r$. *The following statement holds:*

$$\mu = 2^\alpha \wedge M' = 1 \quad \Rightarrow \quad 0 \leq \Delta \leq \left\lfloor \frac{2^n - 2^\alpha - 1}{2^w(2^\alpha + 1)} \right\rfloor .$$

Proof. To prove the theorem, we first rewrite $2^{n+\alpha}$ as:

$$2^{n+\alpha} = M2^\alpha + \Delta 2^{w+\alpha} + 2^\alpha .$$

Now, the reciprocal μ of the modulus M can be written as:

$$\mu = \left\lfloor \frac{2^{n+\alpha}}{M} \right\rfloor = 2^\alpha + \left\lfloor \frac{\Delta 2^{w+\alpha} + 2^\alpha}{M} \right\rfloor = 2^\alpha + \left\lfloor \frac{\lambda}{M} \right\rfloor$$

Having that $\mu = 2^\alpha$, the inequality $0 \leq \lambda < M$ must hold. By solving the left part of inequality ($\lambda \geq 0$) we get:

$$\Delta \geq -2^{-w} . \tag{1}$$

Similar, for the right part of inequality ($\lambda < M$) we get:

$$\Delta < \frac{2^n - 2^\alpha - 1}{2^w \left(2^\alpha + 1\right)} \quad . \tag{2}$$

From the condition $M' = -M_0^{-1} \bmod r = 1$ it follows that $M \equiv -1 \bmod r$. This is true for all $\Delta \in \mathbb{Z}$. Finally, a condition that the modulus M is an n-bit integer ($2^{n-1} \leq M < 2^n$) makes the last condition for Δ:

$$-2^{-w} < \Delta \leq 2^{n-w-1} - 2^{-w} \quad . \tag{3}$$

Now, from the inequalities (1), (2), (3) and the fact that $\Delta \in \mathbb{Z}$, follows the final condition for Δ:

$$0 \leq \Delta \leq \left\lfloor \frac{2^n - 2^\alpha - 1}{2^w \left(2^\alpha + 1\right)} \right\rfloor \quad . \qquad \blacksquare$$

The previous theorem defines a set of moduli for which both conditions $\mu = 2^\alpha$ and $M' = 1$ are true. As mentioned earlier, to minimize the number of correction steps in the improved Barrett algorithm [8], we choose $\alpha = w + 3$. Finally, the proposed set is defined as:

$$\mathrm{S} : M = 2^n - \Delta 2^w - 1 \text{ where } 0 \leq \Delta \leq \left\lfloor \frac{2^n - 2^{w+3} - 1}{2^w \left(2^{w+3} + 1\right)} \right\rfloor \quad .$$

Figure 1 further illustrates the properties of the proposed set. As can be seen, the w least significant bits and the $w + 3$ most significant bits are fixed to be all 1's while the other $n - 2w - 3$ bits can be randomly chosen.

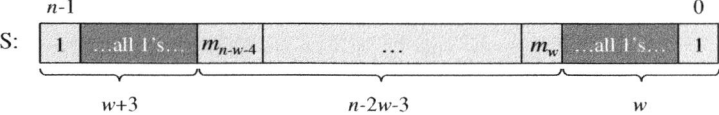

Fig. 1. Binary representation of the proposed set

The evaluation of the intermediate quotient for the improved Barrett algorithm, $\hat{q}$, now becomes equal to:

$$\hat{q} = \left\lfloor \frac{\left\lfloor \frac{Z}{2^{n+\beta}} \right\rfloor \mu}{2^{\alpha - \beta}} \right\rfloor = \left\lfloor \frac{\left\lfloor \frac{Z}{2^{n+\beta}} \right\rfloor 2^\alpha}{2^{\alpha - \beta}} \right\rfloor = \left\lfloor \left\lfloor \frac{Z}{2^{n+\beta}} \right\rfloor 2^\beta \right\rfloor \quad .$$

For $\beta \leq 0$, the previous equation becomes simplified and equivalent to:

$$\hat{q} = \left\lfloor \frac{Z}{2^n} \right\rfloor \quad .$$

Since $M' = 1$, the intermediate quotient for the Montgomery multiplication also gets simplified:

$$q_M = Z \bmod r \quad .$$

Finally, the bipartite modular multiplication for the proposed set of moduli is given in Algorithm 3. After the final addition is performed, one more correction step might be necessary since $0 \leq Z_H + Z_L < 2M$.

Algorithm 3. BMM algorithm for the proposed set of moduli.

Input: $X = (X_{n_w-1} \dots X_0)_r$, $Y = (Y_{n_w-1} \dots Y_0)_r = Y_H r^k + Y_L$, $M = (M_{n_w-1} \dots M_0)_r \in S$, where $0 \leq X, Y < M, r = 2^w, 0 < k < n_w$ and $n_w = \lceil n/w \rceil$.

Output: $Z = XYr^{-k} \bmod M$.

1: $Z_H \Leftarrow 0$
2: **for** $i = n_w - 1$ downto k **do**
3:     $Z_H \Leftarrow Z_H r + XY_i$
4:     $\hat{q} \Leftarrow \lfloor Z_H/2^n \rfloor$
5:     $Z_H \Leftarrow Z_H - \hat{q}M$
6: **end for**
7: **if** $Z_H \geq M$ **then**
8:     $Z_H \Leftarrow Z_H - M$
9: **end if**

1: $Z_L \Leftarrow 0$
2: **for** $i = 0$ to $k - 1$ **do**
3:     $Z_L \Leftarrow Z_L + XY_i$
4:     $q_M \Leftarrow Z_L \bmod r$
5:     $Z_L \Leftarrow (Z_L + q_M M)/r$
6: **end for**
7: **if** $Z_L \geq M$ **then**
8:     $Z_L \Leftarrow Z_L - M$
9: **end if**

Return $Z \Leftarrow Z_H + Z_L$

---

4   Hardware Implementation

To verify our approach in practice, we implement a set of multipliers that are based on our proposal and compare them with the multipliers that support the original BMM algorithm. Obviously, the mission of the BMM algorithm is to utilize the parallel computation and hence, increase the speed of the modular multiplication. Therefore, in order to compare different designs with the same input size, we define a relative throughput as

$$T_r = \frac{1}{N t_{cp}}$$

where t_{cp} is a critical path delay and N is a number of clock cycles. The total throughput is then obtained as $T = BT_r$, where B is the number of bits processed in $1/T_r$ time.

To maximize the throughput, one obviously needs to decrease both N and t_{cp}. Typically, there are plenty of trade-offs to explore in order to make an optimal (in this case fastest) design. To make an objective comparison, we distinguish between designs that aim at the shortest critical path and the ones that achieve the minimum number of clock cycles. We address each of them separately, in the coming subsections.

4.1   Optimization Goal: Shortest Critical Path

A modular multiplier based on the BMM algorithm, depicted in Fig. 2, consists of four multiple-precision multipliers ($\pi_{H1}, \pi_{H2}, \pi_{L1}, \pi_{L2}$). Apart from the multipliers, the architecture contains some additional adders (Σ_L, Σ_H and Σ). The multiple-precision multipliers are implemented in a digit-serial manner which typically provides a good trade-off between area and speed. The multipliers π_{H1} and π_{H2} assemble together the Barrett modular multiplier that processes the most significant half of Y (that is Y_H). Similarly, the multipliers π_{L1} and π_{L2} form the Montgomery modular multiplier that processes the least significant half of Y (that is Y_L). The results of both multipliers are finally added together, resulting in $Z = XYr^{-k} \bmod M$. The parameters k and α are chosen such that the execution speed is maximized and the number of correction steps is minimized: $k = \lfloor n_w/2 \rfloor$ and $\alpha = w + 3$.

A choice of the specific architecture is based on the following criteria. The two levels of parallelism are exploited such that the number of clock cycles needed for one modular multiplication is minimized. First, the BMM algorithm itself is constructed such that the Barrett part and the Montgomery part of the multiplier work independently, in parallel. Second, the multiple-precision multipliers π_{H1} and π_{H2} in the Barrett part, and π_{L1} and π_{L2} in the Montgomery part operate with the independent data such that they run in parallel and speed-up the whole multiplication process. The critical path is minimized and consists of one multiplexer, a single-precision multiplier and an adder (bold line, Fig. 2).

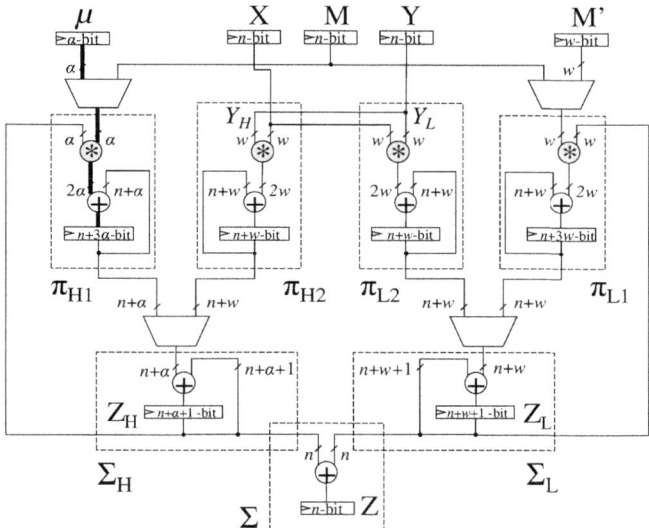

Fig. 2. Datapath of the modular multiplier with the shortest critical path based on BMM method

In order to avoid any ambiguity we provide a graph in Fig. 3 which shows the exact timing schedule of separate blocks inside the multiplier. With i ($0 \le i < k$) we denote the current iteration of the algorithm. Each iteration consists of $n_w + 3$ clock cycles except the first iteration that lasts for $n_w + 1$ cycles.

4.2   Optimization Goal: Minimum Number of Clock Cycles

In order to minimize the number of clock cycles needed for one modular multiplication, the architecture from Fig. 2 is modified as depicted in Fig. 4. Two single-precision multipliers (π_{H3} and π_{L3}), consisting only of a pure combinatorial logic, are added without requiring any clock cycles for calculating their products.

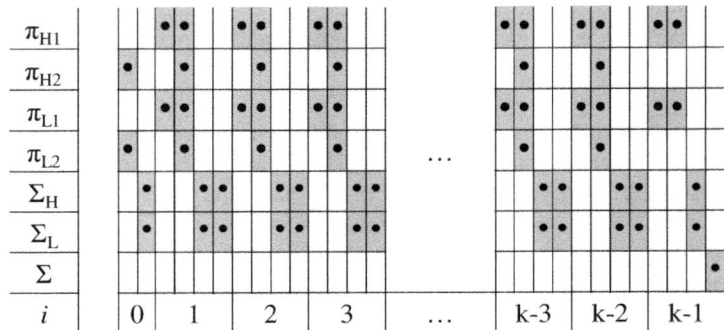

Fig. 3. Timing schedule of the BMM multiplier with the shortest critical path

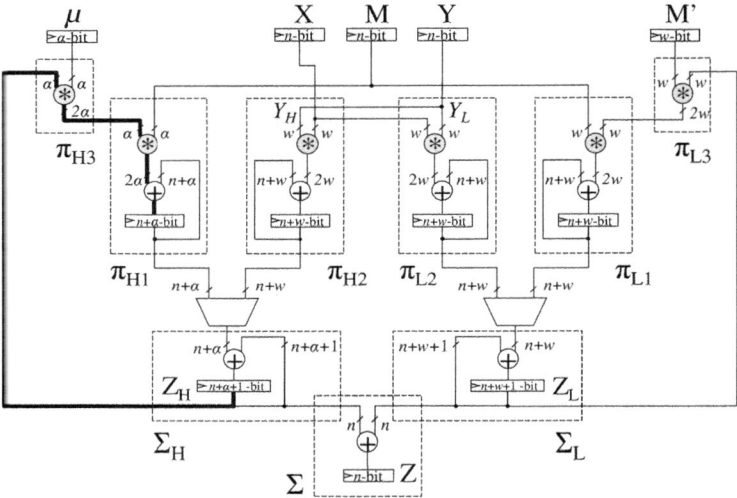

Fig. 4. Datapath of the modular multiplier with the minimized number of clock cycles based on BMM method

We again provide a graph in Fig. 5 which shows the timing schedule of the multiplier. Each iteration now consists of $n_w + 2$ clock cycles except the first that lasts for $n_w + 1$ cycles.

The critical path of the whole design occurs from the output of the register Z_H to the input of the temporary register in π_{H1}, passing through two single-precision multipliers and one adder (bold line).

4.3 Proposed Multiplier

An architecture of the modular multiplier based on the BMM method with the moduli from the proposed set (see Algorithm 3) is shown in Fig. 6. The most important difference is that there are no multiplications with the precomputed values and hence, the

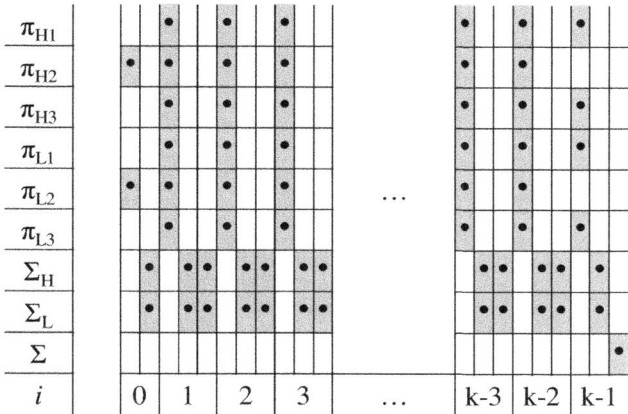

Fig. 5. Timing schedule of the BMM multiplier with the minimized number of clock cycles

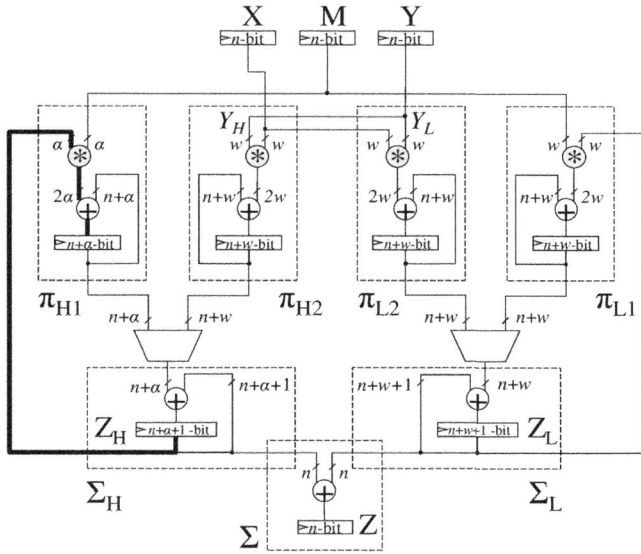

Fig. 6. Datapath of the modular multiplier based on BMM method with the modulus from the proposed set

critical path contains one single-precision multiplier and one adder only. A full timing schedule of the multiplier is given in Fig. 7. The number of cycles remains the same as in the architecture from Fig. 4 while the critical path reduces.

4.4 Results

To show this in practice, we have synthesized 192-bit, 512-bit and 1024-bit multipliers, each with the digit size of 16, 32 and 64 bits. The designs were synthesized using UMC

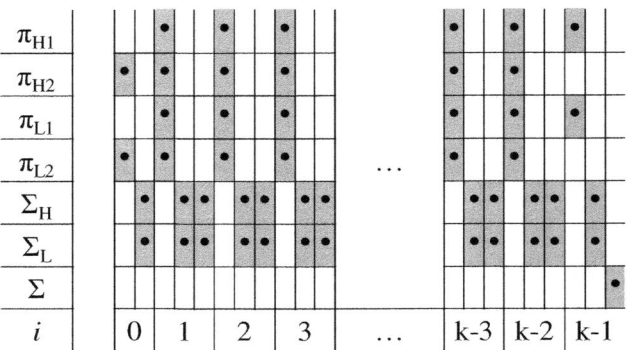

Fig. 7. Timing schedule of the proposed BMM multiplier

0.13 μm CMOS High-Speed standard cell library with Synopsys Design Vision version C-2009.06-SP3. The results are given in Table 1.

Observing the implementation results, we conclude that our proposed design outperforms the standard BMM design with the shortest critical path for at most 18 %. A design that is based on standard BMM with the minimum number of clock cycles is outperformed by at most 67 %. Furthermore, our design consumes less area than all its counterparts.

5   On the Security of the Proposed Set

In this section we analyze the security implications of choosing primes in the proposed set for use in ECC/HECC and in RSA.

In the current state of the art, the security of ECC/HECC over finite fields GF(p^m) only depends on the extension degree m of the field [2]. Therefore, the security does not depend on the precise structure of the prime p. This is illustrated by the particular choices for p that have been made in several standards such as SEC [21], NIST [16], ANSI [1]. In particular, the following primes have been proposed: $p_{192} = 2^{192} - 2^{64} - 1$, $p_{224} = 2^{224} - 2^{96} + 1$, $p_{256} = 2^{256} - 2^{224} + 2^{192} + 2^{96} - 1$, $p_{384} = 2^{384} - 2^{128} - 2^{96} + 2^{32} - 1$, and $p_{521} = 2^{521} - 1$. It is easy to verify that for $w \leq 28$ all primes except p_{224} are in S. In conclusion: choosing a prime of prescribed structure has no influence on the security of ECC/HECC.

The case of RSA requires a more detailed analysis than ECC/HECC. First, we assume that the modulus N is chosen from the proposed set. This is a special case of the security analysis given in [12] followed by the conclusion that the resulting moduli do not seem to offer less security than regular RSA moduli.

Next, we assume that the primes p and q that constitute the modulus $N = pq$ are both chosen in the set S. To analyze the security implications of the restricted choice of p and q, we first make a trivial observation. The number of n-bit primes in the set S for $n > 259 + 2w$ is large enough such that the exhaustive listing of these sets is impossible, since a maximum of $2w + 3$ bits are fixed.

Table 1. Synthesis results for the hardware architectures of 192-bit, 512-bit and 1024-bit modular multipliers

Design	n [bit]	w [bit]	Area [kGE]	t_{cp} [ns]	N	T_r [MHz]
		16	48.20	2.94	178	1.91
	192	32	85.66	4.46	52	4.31
		64	212.40	7.29	16	8.57
		16	96.31	3.17	1118	0.28
Fig. 2	512	32	134.10	4.79	302	0.69
		64	259.84	7.49	86	1.55
		16	177.93	3.33	4286	0.07
	1024	32	208.59	5.19	1118	0.17
		64	356.37	7.46	302	0.44
		16	50.17	4.34	167	1.38
	192	32	84.25	6.81	47	3.12
		64	220.73	12.16	14	5.87
		16	97.44	4.28	1087	0.21
Fig. 4	512	32	127.33	6.95	287	0.50
		64	271.54	12.43	79	1.02
		16	169.49	4.48	4223	0.05
	1024	32	198.01	6.91	1087	0.13
		64	341.59	12.22	287	0.29
		16	44.65	2.92	167	2.05
	192	32	83.14	4.18	47	5.09
		64	204.73	7.25	14	9.85
		16	95.23	3.10	1087	0.30
Fig. 6	512	32	137.41	4.37	287	0.80
		64	247.35	7.45	79	1.70
		16	183.01	3.16	4223	0.07
	1024	32	211.07	4.71	1087	0.20
		64	346.40	7.45	287	0.47

The security analysis then corresponds to attacks on RSA with partially known factorization. This problem has been analyzed extensively in the literature and the first results come from Rivest and Shamir [19] in 1985. They describe an algorithm that factors N in polynomial time if $2/3$ of the bits of p or q are known. In 1995, Coppersmith [5] improves this bound to $3/5$.

Today's best attacks all rely on variants of Coppersmith's method published in 1996 [7,6]. A good overview of these algorithms is given in [13]. The best results in this area are as follows. Let N be an n bit number, which is a product of two $n/2$-bit primes. If half of the bits of either p or q (or both) are known, then N can be factored in polynomial time. If less than half of the bits are known, say $n/4 - \varepsilon$ bits, then the best algorithm simply guesses ε bits and then applies the polynomial time algorithm, leading to a running time exponential in ε. In practice, the values of w (typically $w \leq 64$) and n ($n \geq 1024$) are always chosen such that our proposed moduli remain secure against Coppersmith's factorization algorithm, since at most $2w + 3$ bits of p and q are known.

Finally, we consider a similar approach extended to moduli of the form $N = p^r q$ where p and q have the same bit-size. This extension was proposed by Boneh, Durfee and Howgrave-Graham [4]. Assuming that p and q are of the same bit-size, one needs a $1/(r-1)$-fraction of the most significant bits of p in order to factor N in polynomial time. In other words, for the case $r = 1$, we need half of the bits, whereas for e.g. $r = 2$ we need only a third of the most significant bits of p. These results show that the primes $p, q \in S$, assembling an RSA modulus of the form $N = p^r q$, should be used with care. This is especially true when r is large. Note that if $r \approx \log p$, the latter factoring method factors N in polynomial time for any general primes $p, q \in \mathbb{N}$.

6   Conclusion

A set of moduli for which the performance of bipartite modular multiplication considerably increases is proposed in this work. The size of the set is determined by the digit-size and the length of the modulus. Since the security level of ECC/HECC does not depend at all on the precise structure of the prime p, our proposed set is safe to be used for constructing underlying fields in elliptic curves cryptography. The case of RSA is also discussed with a conclusion that if used with care ($w \leq 64$ and $n \geq 1024$) our proposed set does not decrease the security of RSA.

Additionally, we propose an architecture for a modular multiplier that is based on our method. The results show that, concerning the speed, our proposed architecture outperforms the modular multiplier based on standard BMM method with no additional area overhead.

Acknowledgment

Work supported in part by the IAP Programme P6/26 BCRYPT of the Belgian State, by FWO project G.0300.07, by the European Commission under contract number ICT-2007-216676 ECRYPT NoE phaseII ,and by K.U.Leuven-BOF (OT/06/40).

References

1. ANSI. ANSI X9.62 The Elliptic Curve Digital Signature Algorithm (ECDSA),
   http://www.ansi.org
2. Avanzi, R.M., Cohen, H., Doche, C., Frey, G., Lange, T., Nguyen, K., Vercauteren, F.: Handbook of Elliptic and Hyperelliptic Curve Cryptography. CRC Press, Boca Raton (2005)
3. Barrett, P.: Implementing the Rivest Shamir and Adleman Public Key Encryption Algorithm on a Standard Digital Signal Processor. In: Odlyzko, A.M. (ed.) CRYPTO 1986. LNCS, vol. 263, pp. 311–323. Springer, Heidelberg (1987)
4. Boneh, D., Durfee, G., Howgrave-Graham, N.: Factoring $N = p^r q$ for Large r. In: Wiener, M. (ed.) CRYPTO 1999. LNCS, vol. 1666, pp. 326–337. Springer, Heidelberg (1999)
5. Coppersmith, D.: Factoring with a Hint, IBM Research Report RC 19905 (1995)
6. Coppersmith, D.: Finding a small root of a bivariate integer equation; factoring with high bits known. In: Maurer, U.M. (ed.) EUROCRYPT 1996. LNCS, vol. 1070, pp. 178–189. Springer, Heidelberg (1996)

7. Coppersmith, D.: Small Solutions to Polynomial Equations, and Low Exponent Vulnerabilities. Journal of Cryptology 10(4), 233–260 (1996)
8. Dhem, J.-F.: Design of an Efficient Public-Key Cryptographic Library for RISC-based Smart Cards. PhD Thesis (1998)
9. Diffie, W., Hellman, M.E.: New Directions in Cryptography. IEEE Transactions on Information Theory 22, 644–654 (1976)
10. Joye, M.: RSA Moduli with a Predetermined Portion: Techniques and Applications. Information Security Practice and Experience, 116–130 (2008)
11. Kaihara, M.E., Takagi, N.: Bipartite Modular Multiplication. In: Rao, J.R., Sunar, B. (eds.) CHES 2005. LNCS, vol. 3659, pp. 201–210. Springer, Heidelberg (2005)
12. Lenstra, A.: Generating RSA Moduli with a Predetermined Portion. In: Ohta, K., Pei, D. (eds.) ASIACRYPT 1998. LNCS, vol. 1514, pp. 1–10. Springer, Heidelberg (1998)
13. May, A.: Using LLL-Reduction for Solving RSA and Factorization Problems: A Survey (2007), http://www.informatik.tu-darmstadt.de/KP/publications/07/111.pdf
14. Menezes, A., van Oorschot, P., Vanstone, S.: Handbook of Applied Cryptography. CRC Press, Boca Raton (1997)
15. Montgomery, P.: Modular Multiplication without Trial Division. Mathematics of Computation 44(170), 519–521 (1985)
16. National Institute of Standards and Technology. FIPS 186-2: Digital Signature Standard (January 2000)
17. Potgieter, M.J., van Dyk, B.J.: Two Hardware Implementations of the Group Operations Necessary for Implementing an Elliptic Curve Cryptosystem over a Characteristic Two Finite Field. In: IEEE AFRICON. 6th Africon Conference in Africa, pp. 187–192 (2002)
18. Quisquater, J.-J.: Encoding System According to the So-Called RSA Method, by Means of a Microcontroller and Arrangement Implementing this System, US Patent #5,166,978 (1992)
19. Rivest, R.L., Shamir, A.: Efficient Factoring Based on Partial Information. In: Pichler, F. (ed.) EUROCRYPT 1985. LNCS, vol. 219, pp. 31–34. Springer, Heidelberg (1986)
20. Rivest, R.L., Shamir, A., Adleman, L.: A Method for Obtaining Digital Signatures and Public-Key Cryptosystems. Communications of the ACM 21(2), 120–126 (1978)
21. Standards for Efficient Cryptography. SEC2: Recommended Elliptic Curve Domain Parameters (2010), http://www.secg.org

Constructing Tower Extensions of Finite Fields for Implementation of Pairing-Based Cryptography*

Naomi Benger and Michael Scott

School of Computing
Dublin City University
Ballymun, Dublin 9, Ireland
{nbenger,mike}@computing.dcu.ie

Abstract. A cryptographic pairing evaluates as an element of a finite extension field, and the evaluation itself involves a considerable amount of extension field arithmetic. It is recognised that organising the extension field as a "tower" of subfield extensions has many advantages. Here we consider criteria that apply when choosing the best towering construction, and the associated choice of irreducible polynomials for the implementation of pairing-based cryptosystems. We introduce a method for automatically constructing efficient towers for more classes of finite fields than previous methods, some of which allow faster arithmetic.

We also show that for some families of pairing-friendly elliptic curves defined over $\mathbb{F}_p$ there are a large number of instances for which an efficient tower extension $\mathbb{F}_{p^k}$ is given immediately if the parameter defining the prime characteristic of the field satisfies a few easily checked equivalences.

Keywords: Extension Fields, Pairing implementation, pairing-based cryptosystems, Euler's Conjectures.

1 Introduction

When considering the software implementation of a cryptographic scheme such as RSA, or schemes based on the discrete logarithm problem, an implementation can be written which performs reasonably efficiently for any level of security. For example, an RSA implementation with a 1024-bit modulus can easily be modified to use a 4096-bit modulus, maybe by just changing a single parameter within the program. The same applies to elliptic curve cryptography where a generic implementation will perform reasonably well for a curve with a subgroup of points of size 160-bits, 192-bits or 256-bits. Of course an implementation specially tailored for, and hard-wired to, a particular level of security will perform somewhat better, but not spectacularly so.

The situation for pairing-based cryptography is fundamentally different. An efficient implementation at the 80-bit level of security using the Tate pairing on

---

* Research supported by the Claude Shannon Institute, Science Foundation Ireland Grant 06/MI/006.

M.A. Hasan and T. Helleseth (Eds.): WAIFI 2010, LNCS 6087, pp. 180–195, 2010.

a Cocks-Pinch pairing-friendly curve [10] will be completely different from an implementation at the 128-bit level using the R-ate [16] pairing on a BN curve [6] and very little code will be reusable between the two implementations. In this situation the development and maintenance of good quality pairing code becomes difficult and there is a compelling case for the development of some kind of automatic tool – a *cryptographic compiler* – which can generate good quality code for each case [9].

When using pairing-based protocols, it is necessary to perform arithmetic in fields of the form $\mathbb{F}_{q^k}$, for moderate values of k, so it is important that the field is represented in such a way that the arithmetic can be performed as efficiently as possible. It is this aspect of the implementation of pairing-based protocols which is the focus of this paper. The first contribution of this work is to prove a result which gives a method of checking if a binomial defined over an extension field is irreducible by testing a single element in the base field. This result gives a new method which complements the existing method and gives a means for automatically constructing efficient towers of extensions of finite fields in the cases for which the existing method can not be used or do not give the most efficient algorithms. The resulting constructions are efficient and the usefulness of these results will be shown by the specific application to pairing-based cryptography. The second contribution of this work is to give some constant constructions for the tower extensions for classes of families of pairing-friendly curves.

The remainder of the paper is organised as follows: in §2 the motivation for the work in this paper will be reinforced. In §3 the specific context will be presented. Some existing ideas for constructing tower extensions are briefly explained in §4. A general result to use in the construction of tower extensions for general fields is given in §5 which is applied to the context of PBC in §6. In §6.2 Euclid's conjectures will be presented and used to give concrete tower constructions for some specific families of pairing-friendly curves. In §7 the selection of appropriate polynomials for implementation will be discussed. In §8 we draw some conclusions.

2   Extension Fields

Consider the implementation of the extension field $\mathbb{F}_{p^k}$. The natural representation of the elements of this field is as polynomials of degree $k - 1$, $\mathbb{F}_{p^k} = \mathbb{F}_p[x]/f(x)\mathbb{F}_p[x]$, where $f(x)$ is an irreducible polynomial in $\mathbb{F}_p[x]$ of degree k. For efficiency reasons some effort might be made to choose $f(x)$ to have a minimal number of terms and small coefficients. For example, for the field $\mathbb{F}_{p^2}$, where p is a prime and $p \equiv 3 \mod 4$, a good choice for $f(x)$ would be $x^2 + 1$, and elements can be represented as $ax + b$, with $a, b \in \mathbb{F}_p$. For the case $p \equiv 5 \mod 8$, a good choice for $f(x)$ would be $x^2 - 2$. For the final case $p \equiv 1 \mod 8$ there is no immediately obvious way to choose a suitable irreducible binomial, but for some small value i which is a quadratic non-residue in $\mathbb{F}_p$, $x^2 - i$ would be appropriate.

In some settings the value of the extension degree k might be much greater than 2, in which case the direct polynomial representation becomes more arithmetically complex. For elliptic curve cryptography implemented over "Optimal Extension Fields", (OEFs) as suggested by Bailey and Paar [3], extensions as high as $\mathbb{F}_{p^{30}}$ are considered; in pairing-based cryptosystems, an extension degree of up to 50 is reasonable [10]. OEFs are usually defined as extensions with respect to a small single-word pseudo-mersenne prime. The extension fields that arise in the context of efficient implementations of pairing-based cryptography, however, are rather different.

If the extension degree is a parameter of the implementation then the potentially uncomfortable situation arises where, if the extension degree changes, an optimal implementation must be re-written again, largely "from scratch". The alternative seems to be to use generic polynomial code to construct the extension field, making the implementation slow and bulky. A nice compromise that applies when the extension k is smooth (that is has only small factors) is to use a "tower" of extensions, where one layer builds on top of the last, and ideally where each sub-extension is quite small. For example, $\mathbb{F}_{p^{12}}$ could be implemented as a quadratic extension, of a cubic extension, of a very efficiently implemented (and reusable) quadratic extension field $\mathbb{F}_{p^2}$, as implemented by Devegili et al. [8].

This idea of using a tower of extensions was suggested by Baktir and Sunar [19] as a better way of implementing OEFs, and in the process of doing this they discovered that the resulting simpler implementation resulted in an asymptotically improved method for performing field inversion. The point is that it is relatively easy to implement quadratic and cubic extensions efficiently, whereas the complexity of implementing generic methods over large extensions might result in the inadvertent use of sub-optimal methods.

It is also proposed in the IEEE draft standard "P1363.3: Standard for Identity-Based Cryptographic Techniques using Pairings" that extensions of odd primes are constructed using a tower of extensions created using irreducible binomials at each stage [1].

Clearly it is advantageous to use this towering method when implementing a pairing-based protocol. One issue remains: finding the best tower for a particular value of k. Obviously, for different values of k, we will need to use different towers; a very reasonable approach in the context of Pairing-Based Cryptography (PBC) would be to fix the tower for a particular k. This will be made clear in §6.

The construction does not only depend on k however, but also on p, the characteristic of the base field. There is an existing method for constructing such towers given by Koblitz and Menezes in [15] which can only be used for some p with specific properties, so relying on this method alone places unnecessary restrictions on the parameters of a pairing-friendly curve. Given that pairing-friendly elliptic curves are quite rare, it is clear that we should aim to reduce the number of constraints on the parameters that may compromise the efficiency of the implementation.

Motivating this work is our ambition to contribute to a "cryptographic compiler" [9], that is, a compiler which when given as input the parameters for a pairing-friendly curve, should be able to automatically generate the optimal pairing code, including the optimal field arithmetic implementation.

3   Pairings and Pairing-Friendly Elliptic Curves

The Tate pairing of two linearly independent points P and Q on an elliptic curve $E(\mathbb{F}_{q^k})$, denoted $e(P,Q)$, evaluates as an element of the extension field $\mathbb{F}_{q^k}$. If P is of prime order r, then the pairing evaluates as an element of order r. Here we focus on the case of non-supersingular elliptic curves over prime fields, that is, $q = p$. In practice it is common to choose P as a point on the elliptic curve over the base field, $E(\mathbb{F}_p)$. As is well known, the number of points on this elliptic curve is $p + 1 - t$, where $\mid t \mid \leq 2\sqrt{p}$ (Hasse bound) is the trace of the Frobenius endomorphism [23].

The Tate pairing is only of interest if it is calculated on a "pairing-friendly" elliptic curve. This pairing-friendliness entails that $r \mid (p^k - 1)$ for some reasonably small value of k, that is, the rth roots of unity in $\overline{\mathbb{F}}_p$ are contained in $\mathbb{F}_{p^k}$, the codomain of the pairing. To find the actual parameters of the curve, however, it is also required that the integer $4p - t^2$ (always positive as a consequence of the Hasse condition), has a relatively small non-square part D (the CM discriminant), that is it factors as Dv^2 for small D. Such curves can then be found using the method of complex multiplication (CM) [7].

For the Tate pairing the point Q is commonly represented as a point over some twist $E'(\mathbb{F}_{p^{k/d}})$, where $d \mid k$, as opposed to being on the curve defined over the full extension field $E(\mathbb{F}_{p^k})$. When k is even, the quadratic twist $(d = 2)$ can always be used, when the pairing-friendly curve has a CM discriminant of $D = 1$ and $4 \mid k$, the quartic twist $(d = 4)$ can be used, if $D = 3$, $3 \mid k$ and k is odd, cubic twists $(d = 3)$ can be used and when the CM discriminant is $D = 3$ and $6 \mid k$, the sextic twist $(d = 6)$ can be used. It is preferable to use the highest order twist available, as this leads to a faster more compact implementation [13].

Variants of the Tate pairing have recently been discovered (the ate pairing [13], and the R-ate pairing [16]) that are more efficient in some cases, but which require the roles of P and Q to be reversed. This makes it even more important to use the highest order twist available as a significant part of the pairing calculation is a point multiplication of the first parameter (now Q), which is more expensive than in the Tate pairing.

In their taxonomy of pairing-friendly curves [10], Freeman, Scott and Teske, following a recommendation from Koblitz and Menezes [15, §8.3], particularly recommend curves for which the embedding degree k is of the form $k = 2^i \cdot 3^j$ for $i, j \geq 0$. Here we further restrict that $i \geq 1$, $j \geq 0$ as an even value for k facilitates the important "denominator elimination" optimization for the pairing calculation [4]. In each case we prefer curves which support the maximal twist.

4   Existing Ideas for Constructing General Towers

Let p be an odd prime, and let $n, m > 0$ be integers. the most obvious way to construct the tower of sub-extensions of the field $\mathbb{F}_{p^{nm}}$ over $\mathbb{F}_{p^n}$ would be to use a binomial $x^m - \alpha$ which is irreducible over $\mathbb{F}_{p^n}$ and successively adjoin roots of the previously adjoined root until the tower has been constructed (we refer to this as the 'general method'). We are able to test $x^m - \alpha$ for irreducibility using the following theorem:

Theorem 1. *[18, Theorem 3.75] Let $m \geq 2$ be an integer and $\alpha \in \mathbb{F}_{p^n}^{\times}$. Then the binomial $x^m - \alpha$ is irreducible in $\mathbb{F}_{p^n}[x]$ if and only if the following two conditions are satisfied:*

1. *each prime factor of m divides the order e of $\alpha \in \mathbb{F}_{p^n}^{\times}$, but not $(p^n - 1)/e$;*
2. *If $m \equiv 0 \mod 4$ then $p^n \equiv 1 \mod 4$.*

The *order* of $\gamma \in \mathbb{F}_{p^n}$ is the smallest positive integer e such that $\gamma^e = 1$ in $\mathbb{F}_{p^n}$ and the order is a divisor of $p^n - 1$.

By Theorem 1 we see that the general method above works for all $m \not\equiv 0 \mod 4$. When $m \equiv 0 \mod 4$, this method works if $p^n \equiv 1 \mod 4$.

Given the constraints outlined in §3, it is clear that the tower of extensions used in pairing-based cryptography can be built using a sequence of cubic and quadratic sub-extensions. This was recognised by Koblitz and Menezes in [15]. They called a field $\mathbb{F}_{p^k}$ *pairing-friendly* (not to be confused with a pairing friendly elliptic curve) if $p \equiv 1 \mod 12$ and k is of the form $k = 2^i 3^j$, in which case by [15, Theorem 2] (which is derived from Theorem 1 above) the polynomial $x^k - \alpha$ is irreducible over $\mathbb{F}_p$ if α neither a square nor a cube in $\mathbb{F}_p$. The extension can be constructed using the general method by simply adjoining a cube or square root of some small such α and then successively adjoining a cube or square root of the previously adjoined root until the tower has been constructed. If $j = 0$ then it is sufficient that $p \equiv 1 \mod 4$ and that α be a quadratic non-residue in $\mathbb{F}_p$. This result gives us an easy method for building towers over pairing-friendly fields: simply find an element $\alpha \in \mathbb{F}_p$ which is a quadratic and (when necessary) cubic non-residue and adjoin successive cube and square roots of α to $\mathbb{F}_p$.

There is one major issue remaining, the strict condition that $p \equiv 1 \mod 12$ to give a pairing-friendly field. When searching for pairing-friendly curves of a suitable size there are typically other criteria that we wish to meet (for example, it is preferred that the Hamming weight of the variable that controls the Miller loop in the pairing calculation should be as small as possible [8]). Having to skip a nice curve just because $p \equiv 3 \mod 4$ seems unnecessarily restrictive. Since the publication of [15], new families of pairing-friendly elliptic curves have been discovered which the results of [15] could not have taken into account. In particular, the KSS curves with embedding degree $k = 18$ [14] are good for implementation given the many optimisations possible using these curves. The condition that $p \equiv 1 \mod 12$ here is completely unnecessary as this condition arises from condition 2 of Theorem 1 which is not applicable when $k = 18$.

Given the many applications of pairings in cryptography and the fact that the parameters of a pairing-based protocol are already subject to quite strict constraints, it is clear that there is a necessity for a method to construct towers for fields which would not be considered pairing-friendly (in the sense of Koblitz and Menezes) but would otherwise be favourable for implementation of a pairing-based protocol. The term 'pairing-friendly' field is slightly misleading, as there are families of pairing-friendly elliptic curves attractive for implementation which are defined over fields which do not necessarily satisfy $p \equiv 1 \mod 12$. In a sense, the pairing-friendly fields of [15] are the fields, in the context of pairings, over which it is easy to build the towers. We instead refer to these fields as *towering-friendly* as this gives a more accurate description of these fields – the towers over such fields are easily constructed. This definition is not specific to pairings, but in this setting we would like to use towering-friendly fields for the most efficient implementation possible.

Definition 2. A *towering-friendly* field is a field of the form $\mathbb{F}_{q^m}$, where q is a prime power, for which all prime divisors of m also divide $q - 1$.

In essence, towering-friendly fields are fields for which the tower of sub-extensions can be easily (and most efficiently) constructed; that is, using binomials. The OEFs of Bailey and Paar [3] are by definition towering-friendly fields with characteristic a prime of a special form. The fields said to be pairing-friendly by Koblitz and Menezes are indeed towering-friendly, but these are not the only towering-friendly fields which occur in the context of pairing-based cryptography.

5   General Tower Construction Method

Considering first the general case where p is an odd prime, $n > 0$ and $m > 1$ are integers and we want to construct the tower of sub-extensions of the towering-friendly finite field $\mathbb{F}_{p^{nm}}$ over $\mathbb{F}_{p^n}$. The general method uses a binomial $x^m - \alpha$ which is irreducible in $\mathbb{F}_{p^n}[x]$ and successively adjoins roots of the previously adjoined root until the tower has been constructed, as in [19]. By Theorem 1 the only restriction on α is that α should not be a qth power in $\mathbb{F}_{p^n}$ for any prime divisor q of m. This method works for all m, $m \not\equiv 0 \mod 4$. When $m \equiv 0 \mod 4$, this method will work if $p^n \equiv 1 \mod 4$ (which is always true for even n).

The two issues to address now are:

- we need a method to build a tower when $m \equiv 0 \mod 4$ and $p^n \equiv 3 \mod 4$;
- we need to find a suitable irreducible binomial $x^m - \alpha \in \mathbb{F}_{p^n}[x]$ to construct the tower.

The first problem has a relatively simple solution. We construct first a quadratic extension of $\mathbb{F}_{p^n}$, $\mathbb{F}_{p^{2n}}$, which we will refer to as a *base tower*, using a binomial. We now have $p^{2n} \equiv 1 \mod 4$ so we can use the general method to build the rest of the tower above $\mathbb{F}_{p^{2n}}$ using a binomial $x^{m/2} - \alpha$, where $\alpha \in \mathbb{F}_{p^{2n}}$ (not in $\mathbb{F}_{p^n}$). In the particular case of $n = 1$ this can be done by simply adjoining a square root of -1. This idea is a generalisation of the approach taken by Barreto and Naehrig in [6] to construct the field $\mathbb{F}_{p^{12}}$ over $\mathbb{F}_p$. They first implement an efficient quadratic extension over the base field, and then look for irreducible polynomials of the form $x^6 - \alpha$, where $\alpha \in \mathbb{F}_{p^2}/\mathbb{F}_p$ is neither a square nor a cube.

Remark 3. *The idea of a base tower can be generalised: Suppose $\mathbb{F}_{p^{nm}}$ over $\mathbb{F}_{p^n}$ is not a towering-friendly field. Write $m = m_1 m_2$ such that $\gcd(p^n - 1, m_2) = 1$*

and all the primes dividing m_1 divide $p^n - 1$. If all the primes dividing m_2 divide $p^{2m_1} - 1$ then the tower of $\mathbb{F}_{p^{nm}}$ over $\mathbb{F}_{p^n}$ can be constructed in two parts using the general method. First $\mathbb{F}_{p^{nm_1}}$ over $\mathbb{F}_{p^n}$ is constructed using a binomial, this is the base-tower. Then $\mathbb{F}_{p^{nm}} = \mathbb{F}_{p^{nm_1 m_2}}$ over $\mathbb{F}_{p^{nm_1}}$ is constructed using a binomial defined over $\mathbb{F}_{p^{nm_1}}$ (not over any subfield of $\mathbb{F}_{p^{nm_1}}$). This method can be implemented recursively to achieve an efficient tower for a non-towering-friendly extension.

As to the problem of finding a suitable α for constructing the tower (and also the base tower when necessary), Theorem 1 provides a means for determining whether a given binomial is irreducible, but it does not give an efficient method for constructing the towers: taking random small elements then computing their order in the extension field and verifying that the conditions hold is quite cumbersome, the order could be quite large and this could require a lot of extension field computation for a single element. Using Theorem 1, however, we are able to prove a theorem which results in a simpler method for checking the irreducibility of a polynomial $x^m - \alpha$ in certain cases and hence a more practical method for finding irreducible polynomials to construct the towering-friendly field extensions, particularly in the context of PBC.

We first recall some definitions and properties which will be used in the following theorems and proof: Let $\gamma \in \mathbb{F}_{p^n}$. The *Norm* of $\mathbb{F}_{p^n}$ over $\mathbb{F}_p$ of γ, denoted $N_{\mathbb{F}_{p^n}/\mathbb{F}_p}(\gamma)$, is the product of all its conjugates,

$$N_{\mathbb{F}_{p^n}/\mathbb{F}_p}(\gamma) = \prod_{i=0}^{n-1} (\gamma)^{p^i} \in \mathbb{F}_p.$$

The norm is multiplicative, that is, for $\gamma_1, \gamma_2 \in \mathbb{F}_{p^n}$,

$$N_{\mathbb{F}_{p^n}/\mathbb{F}_p}(\gamma_1 \cdot \gamma_2) = N_{\mathbb{F}_{p^n}/\mathbb{F}_p}(\gamma_1) \cdot N_{\mathbb{F}_{p^n}/\mathbb{F}_p}(\gamma_2)$$

and so for any $\ell \in \mathbb{Z}^+$ we have $N_{\mathbb{F}_{p^n}/\mathbb{F}_p}(\gamma^\ell) = N_{\mathbb{F}_{p^n}/\mathbb{F}_p}(\gamma)^\ell$.

Theorem 4. *Let $m > 1$, $n > 0$ be integers, p an odd prime and $\alpha \in \mathbb{F}_{p^n}^\times$. The binomial $x^m - \alpha$ is irreducible in $\mathbb{F}_{p^n}[x]$ if the following two conditions are satisfied:*

1. *Each prime factor q of m divides $p - 1$ and $N_{\mathbb{F}_{p^n}/\mathbb{F}_p}(\alpha) \in \mathbb{F}_p$ is not a qth residue in $\mathbb{F}_p$;*
2. *If $m \equiv 0 \mod 4$ then $p^n \equiv 1 \mod 4$.*

Proof. To prove this theorem, we show that condition 1 of Theorem 4 implies condition 1 of Theorem 1. We assume that condition 1 of Theorem 4 is true. Let e denote the order of α in $\mathbb{F}_{p^n}$ and q denote a prime divisor of m.

Suppose that $q \mid (p^n - 1)/e$. This implies that $e \mid (p^n - 1)/q$ and so α is a qth power in $\mathbb{F}_{p^n}$. Let $\delta \in \mathbb{F}_{p^n}$ be such that $\delta^q = \alpha$. Taking the norm of α we see that $N_{\mathbb{F}_{p^n}/\mathbb{F}_p}(\alpha) = N_{\mathbb{F}_{p^n}/\mathbb{F}_p}(\delta^q) = N_{\mathbb{F}_{p^n}/\mathbb{F}_p}(\delta)^q$ where $N_{\mathbb{F}_{p^n}/\mathbb{F}_p}(\delta) \in \mathbb{F}_p$ and thus $N_{\mathbb{F}_{p^n}/\mathbb{F}_p}(\alpha)$ is a qth residue in $\mathbb{F}_p$, a contradiction, so $q \nmid (p^n - 1)/e$.

We have also assumed that $q \mid (p^n - 1)$ and since $q \nmid (p^n - 1)/e$ it is clear that $q \mid e$ and so condition 1 of theorem 4 is satisfied.

Using Theorem 4 we are able to verify the irreducibility of a binomial $x^m - \alpha$ over an extension field $\mathbb{F}_{p^n}[x]$, where α is an element of $\mathbb{F}_{p^n}$, by checking the properties of just one particular element of the base field, namely the norm of $\mathbb{F}_{p^n}$ over $\mathbb{F}_p$ of α – a much simpler task than computing the order of an element in $\mathbb{F}_{p^n}$. Theorem 4 can be used in all cases for which the prime divisors of m also divide $p - 1$ to automatically generate towers of extensions over all towering-friendly fields to build an efficient tower of extensions for the extension field $\mathbb{F}_{p^{nm}}$. As already mentioned, if condition 2 of Theorem 1 is not satisfied, the towers can still be easily constructed by first constructing a base tower, a quadratic extension, then using the theorem to construct the tower over the base tower.

We now illustrate the usefulness of Theorem 4 by adapting it to the context of PBC as outlined in §3.

6   Towers in Pairing-Based Cryptography

Given the constraints outlined in §3, it is clear that the tower of extensions can be built as a sequence of quadratic and cubic sub-extensions. There is some freedom as to the best way to order the extensions. The choice here may be influenced by whether or not it is intended to compress the value of the pairing [21,12]. This compressed value can then be further efficiently exponentiated in its compressed form by using Lucas or XTR based methods for times 2 and times 3 compression respectively. This is facilitated by terminating with a quadratic or a cubic extension respectively.

Consider for example the BN curves [6], which have an embedding degree of 12 and which support the sextic twist $t = 6$. In this case $E(\mathbb{F}_{p^2})$ arithmetic must be supported, and so it makes sense that the tower should start with a quadratic extension over the base field. This can be followed by a cubic extension and then a quadratic, or indeed the other way around. Assuming that the highest possible compression should be supported, the tower of choice in this case is $1 - 2 - 4 - 12$. This particular tower construction is given as an example by the IEEE draft standard [1]. Starting with a quadratic extension where possible is preferred (in case a base tower is needed). Taking all these constraints into account, in Table 1 we make the towering recommendations for the curves recommended in [10].

The ρ-value is given by $\frac{\log(p)}{\log(r)}$ for p the characteristic of the field over which the curve is defined and r the cardinality of the group of points on the elliptic curve.

There have been some advances in arithmetic performance in $\mathbb{F}_{p^k}$ based on the final extension being a quadratic extension [2]. Such towers can also be constructed using our method.

6.1   Tower Construction for PBC

From the definition of towering-friendly fields we are only able to distinguish on a specific case-to-case basis if a general extension field is a towering-friendly field.

Table 1. Suggested Towers for Curves with Efficient Arithmetic

k	ρ	D	Twist d	Construction	Tower
4	2	1	4	FST [10]	1-2-4
6	2	3	6	FST [10]	1-2-6
8	1.5	1	4	KSS [14]	1-2-4-8
12	1	3	6	BN [6]	1-2-4-12
16	1.25	1	4	KSS [14]	1-2-4-8-16
18	1.333	3	6	KSS [14]	1-3-6-18
24	1.25	3	6	BLS [5]	1-2-4-8-24
32	1.125	1	4	KSS [14]	1-2-4-8-16-32
36	1.167	3	6	KSS [14]	1-2-6-12-36
48	1.125	3	6	BLS [5]	1-2-4-8-16-48

In the PBC setting we have a little more information. We are able to determine information about some of the parameters for particular curves in advance by making some observations. We see from the following discussion that all the fields $\mathbb{F}_{p^k}$ arising when using the families of pairing-friendly curves in Table 1 are towering-friendly.

Elliptic curves with CM discriminant $D = 1$. Elliptic curves from Table 1 with CM discriminant $D = 1$ have equations of the form $E : y^2 = x^3 + Ax$. We know that these curves are not supersingular (which is the case for curves with such equations defined over a prime field with characteristic $p \equiv 3 \mod 4$ [7]) and so $p \equiv 1 \mod 4$. This means that the field is towering-friendly as all $D = 1$ cases in Table 1 have $k = 2^n$ so the Koblitz-Menezes strategy appears to be optimal. Indeed, in the case of $p \equiv 5 \mod 8$ we can always choose $\alpha = 2$, which leads to fast reduction. An implementation can simply tower up quadratically, by adjoining the square root of the last adjoined element to build the next extension at each step.

Elliptic curves with CM discriminant $D = 3$. For elliptic curves with CM discriminant $D = 3$, p will not always be a pairing-friendly prime in the sense of the Koblitz and Menezes definition, but we do have some information which will aid us in the construction of the towers over $\mathbb{F}_p$. Given that the CM discriminant $D = 3$, we know that the elliptic curve must have an equation of the form $E : y^2 = x^3 + B$. If $p \equiv 2 \mod 3$ then such a curve would be supersingular [23] and so $p \equiv 1 \mod 3$ must be true. We see then that all the fields resulting from this construction are towering-friendly.

For the KSS $k = 18$ curves and FST $k = 6$ curves we are able to use the general method in every case without a base tower (as $k \not\equiv 0 \mod 4$ and both 2 and 3 divide $p - 1$). We simply adjoin successive cubic and quadratic roots of some cubic and quadratic non-residue $\alpha \in \mathbb{F}_p$ in the recommended order.

For all other families of curves, if the prime $p \not\equiv 1 \mod 4$ then we need to use a base tower to construct the tower. One advantage in this case is that we know $p \equiv 3 \mod 4$ and so the base tower $\mathbb{F}_{p^2}$ over $\mathbb{F}_p$ can be efficiently constructed by adjoining a square root of -1. This may actually be more efficient than an implementation using a pairing-friendly field as the arithmetic in $\mathbb{F}_p(\sqrt{-1})$ can be performed faster than in $\mathbb{F}_p(\sqrt{\tau})$ for some other quadratic non-residue $\tau \in \mathbb{F}_p$ [11]. The following Corollary (drawing on ideas from Barreto and Naehrig in [6]) gives a method for finding an appropriate value α such that the polynomial $x^m - \alpha$ is irreducible over a finite field of the form $\mathbb{F}_{p^2} = \mathbb{F}_p(\sqrt{-1})$.

Corollary 5. *The polynomial $x^m - (a \pm b\sqrt{-1})$ is irreducible over $\mathbb{F}_{p^2}$, for $m = 2^i 3^j$, $i, j > 0$, if $a^2 + b^2$ is neither a square nor a cube in $\mathbb{F}_p$.*

Proof. For any element $a \pm b\sqrt{-1}$, $N_{\mathbb{F}_{p^2}/\mathbb{F}_p}(a \pm b\sqrt{-1}) = (a + b\sqrt{-1})(a - b\sqrt{-1}) = a^2 + b^2$. The integer m is of the form $2^i 3^j$ and so by Theorem 4 if $a^2 + b^2$ is neither a quadratic nor a cubic residue modulo p, then $x^m - (a \pm b\sqrt{-1})$ is irreducible over $\mathbb{F}_{p^2}$.

This Corollary is basically Theorem 4 in the case $p \equiv 3 \mod 4$, $n = 2$ and $m = k/2$, this is the case of most concern in PBC. Using this corollary, in order to construct the tower, small values of a and b can be tested until a combination is found such that $a^2 + b^2$ is neither a square nor a cube in $\mathbb{F}_p$. This process only requires a few cubic and quadratic non-residue tests to be performed on elements of the base field. Small values for a and b can be found to help improve efficiency.

As $\frac{1}{2}$ of the non-zero elements of $\mathbb{F}_p$ are non-squares and $\frac{2}{3}$ of the non-zero elements are non-cubes, such an element must exist; in fact, on heuristic grounds it is expected that $\frac{1}{3}$ of the elements will be neither squares nor cubes, which the experimental evidence supports [6].

Given a little more information about p, which is easily found, we are able to give some more specific constructions.

Construction 6. *For approximately $2/3$ of the primes $p \equiv 3$ modulo 8 the polynomial $x^m - (1 + \sqrt{-1})$ is irreducible in $\mathbb{F}_{p^2}[x]$ for $m = 2^i 3^j$, $i, j > 0$.*

Proof. In this case $a^2 + b^2 = 2$. The polynomial will be irreducible if 2 is neither a square nor a cube modulo p. We know that 2 is a quadratic non-residue modulo p when $p \equiv 3 \mod 8$. The only remaining condition is that 2 is not a cube modulo p.

All primes $p \equiv 1 \mod 3$ can be written in the form $p = 3u^2 + v^2$. As Euler conjectured (proved by Gauss [17]) 2 is a cubic residue modulo p if and only if $3 \mid u$. Instinctively we would presume that this occurs $1/3$ of the time. There is currently no proof concerning the number of primes in a quadratic sequence but this is supported by experimental results. So 2 is a cubic non-residue modulo p for approximately $2/3$ of the values of p.

When $p \equiv 7 \mod 8$ the following corollary may be useful:

Construction 7. *For approximately $2/3$ of the primes $p \equiv 2$ or 3 modulo 5 the polynomial $x^m - (2 + \sqrt{-1})$ is irreducible in $\mathbb{F}_{p^2}[x]$ for $m = 2^i 3^j$, $i, j > 0$.*[1]

---

[1] In this case, the polynomial $x^m - (1 + 2\sqrt{-1})$ is also irreducible.

Proof. The values of a and b in Corollary 5 in this case are 2 and 1 respectively, so $a^2 + b^2 = 5$. The polynomial will be irreducible if 5 is neither a square nor a cube modulo p. When $p \equiv 2$ or 3 modulo 5 we know that 5 is a quadratic non-residue modulo p and so the only condition left is that 5 should not be a cube in $\mathbb{F}_p$. With p written in the form $p = 3u^2 + v^2$, we know that 5 is a cube if $15 \mid c$, or $3 \mid a$ and $5 \mid b$, or $15 \mid (a \pm b)$, or $15 \mid (a \pm 2b)$ [17]. Again, there is currently no proof concerning the number of primes in a quadratic sequence but as supported by experimental results we expect that this occurs $1/3$ of the time. So 5 is a cubic non-residue modulo p for approximately $2/3$ of the values of p.

The result of Constructions 6 and 7 is that for around $2/3$ of the fields not considered pairing-friendly we have a more automatic and often more efficient implementation than is possible for pairing-friendly fields.

6.2   Using Euler's Conjectures

For primes which are equivalent to 2 mod 3 it is easily shown that every element is a cubic residue modulo p. For primes which are 1 mod 3 Fermat showed that p can be written as the sum $p = a^2 + 3b^2$ for some integers a and b. Euler conjectured (and Gauss proved) that using this form we can easily determine if some small elements are cubic residues [17]:

1. 2 is a cubic residue $\Leftrightarrow 3 \mid b$.
2. 3 is a cubic residue $\Leftrightarrow 9 \mid b$; or $9 \mid (a \pm b)$.
3. 5 is a cubic residue $\Leftrightarrow 15 \mid b$; or $3 \mid b$ and $5 \mid a$; or $15 \mid (a \pm b)$; or $3 \mid (2a \pm b)$.
4. 6 is a cubic residue $\Leftrightarrow 9 \mid b$; or $9 \mid (a \pm 2b)$.
5. 7 is a cubic residue $\Leftrightarrow 21 \mid b$; or $3 \mid b$ and $7 \mid a$; or $21 \mid (a \pm b)$; or $7 \mid (a \pm 4b)$; or $7 \mid (2a \pm b)$.

These conjectures can be used once p has been constructed to decide if constructions 6 or 7 can be used. For some cases we have this information already.

BN Towers. The prime characteristic p of the field over which a BN curve is defined is parameterised by the polynomial $p(x) = 36x^4 + 36x^3 + 24x^2 + 6x + 1$; an appropriate value x_0 is chosen to give $p = p(x_0)$. It was noticed by Shirase [22] that this parameterisation can be written in the form $p(x) = a(x)^2 + 3b(x)^2$ thus giving us more information about the towers we can construct for certain values of x_0 without having to perform the quadratic and cubic residue tests modulo p. We have $a(x) = 6x^2 + 3x + 1$ and $b(x) = x$. With this additional information, we now see that we are able to use Theorem 4 to put conditions on the values of x_0, which, when satisfied, give an immediate construction for the tower of fields of degree 12 over BN primes.

Considering first BN primes $p \equiv 3$ mod 4 we know that $x_0 \equiv \pm 1$ mod 4 and that we have a towering friendly field which requires a base tower $\mathbb{F}_{p^2}$ which can be constructed by adjoining $\sqrt{-1}$ to $\mathbb{F}_p$. We now need to find an element $a + b\sqrt{-1} \in \mathbb{F}_{p^2}$ such that $x^6 - (a + b\sqrt{-1})$ is irreducible to construct the remaining extensions. From Corollary 5 we know that $x^6 - (a + b\sqrt{-1})$ is irreducible if

$a^2 + b^2$ is neither a square nor a cube in $\mathbb{F}_p$. We know from the conjecture 1 that if $x_0 \equiv \pm 1 \mod 3$ then 2 is a cubic non-residue modulo p. For 2 to be a non-quadratic residue also we need $p \equiv 3 \mod 8$, this implies that $x_0 \equiv 3 \mod 4$. Together, these two constraints give the following:

- If $x_0 \equiv 7$ or $11 \mod 12$ then $x^6 - (1 + \sqrt{-1})$ is irreducible over $\mathbb{F}_{p^2} = \mathbb{F}_p(\sqrt{-1})$.

In [22] the same conclusion is drawn, but using a much more elaborate method. We see that this result supports the claim in Construction 6 as 2/3 of the possible values of x_0 (for $p \equiv 3 \mod 8$) give a p for which 2 is a quadratic non-residue.

Using Theorem 4 we are also able to classify more constructions than those given in [22]. Using a similar method as above:

- If x_0 is odd and $x_0 \equiv 1, 3, 7, 11, 12$ or $13 \mod 15$ then $x^6 - (1 + 2\sqrt{-1})$ is irreducible over $\mathbb{F}_{p^2} = \mathbb{F}_p(\sqrt{-1})$.

Using Euler's conjectures it is also straight forward to set construction for BN primes $p \equiv 1 \mod 4$ not needing a base tower.

- If $x_0 \not\equiv 0 \mod 3$ and $x_0 \equiv 2, 6 \mod 8$ then $x^{12} - 2$ is irreducible;
- If $x_0 \equiv 1, 3, 7, 11, 12$ or $13 \mod 15$ then $x^{12} - 5$ is irreducible;
- If $x_0 \not\equiv 0, 2$ or $4 \mod 9$ and $x_0/2$ is odd then $x^{12} - 6$ is irreducible.

BN curves are quite plentiful and easy to find. Using BN curves in pairing-based protocols means that we need an efficient implementation of $\mathbb{F}_{p^{12}}$ and also of $\mathbb{F}_{p^2}$ as we would use a degree 6 twist. It may be favourable to choose $x_0 \equiv 1 \mod 2$ and x_0 satisfying one of the equivalences above so that $\mathbb{F}_{p^2}$ can be constructed as $\mathbb{F}_p(\sqrt{-1})$ and the tower for $\mathbb{F}_{p^{12}}$ can be constructed using one of Constructions 6 or 7, though these fields would not have originally been considered pairing-friendly. Given that BN curves are so plentiful, this restriction would not impede finding curves appropriate for use.

KSS Towers. When $k = 18$ the parameterisation of $p(x)$ can also be written in the form $a(x)^2 + 3b(x)^2 = p(x)$ where $a(x)$ and $b(x)$ have integer coefficients. In these cases we are also able to give the tower construction if the value x_0 satisfies some easily checked conditions.

KSS $k = 18$. The polynomial parameterisation of p for a KSS $k = 18$ curve is given by

$$p(x) = (x^8 + 5x^7 + 7x^6 + 37x^5 + 188x^4 + 259x^3 + 343x^2 + 1763x + 2401)/21.$$

We also know that $x \equiv 14 \mod 42$ so substituting $x = 42x' + 14$ we obtain the equation

$$p(x') =$$
$$461078666496x'^8 + 1284433428096x'^7 + 1564374047040x'^6 + 1088278335648x'^5 +$$
$$473078255328x'^4 + 131624074008x'^3 + 22896702948x'^2 + 2277529014x' + 99213811.$$

Using Euclid's algorithm and interpolation we find

$$a(x') = 444528x'^4 + 629748x'^3 + 333396x'^2 + 78321x' + 6908,$$

and

$$b(x') = 296352x'^4 + 407484x'^3 + 209916x'^2 + 48091x' + 4143,$$

such that $a(x')^2 + 3b(x')^2 = p(x')$. Using Euler's Conjectures we see that:

- If $x_0' \equiv 1, 4, 5, 8 \mod 12$ then $x^{18} - 2$ is irreducible over $\mathbb{F}_p$;
- If $x_0' \not\equiv 2, 3, 4 \mod 9$ then $x^{18} - 3$ is irreducible over $\mathbb{F}_p$;
- If $x_0' \equiv 7, 9, 12, 14 \mod 15$ then $x^{18} - 5$ is irreducible over $\mathbb{F}_p$;
- If $x_0' \equiv a \mod 42$ then $x^{18} - 6$ is irreducible over $\mathbb{F}_p$,

where $a = \{2, 3, 4, 9, 10, 11, 12, 13, 18, 20, 21, 22, 27, 28, 30, 31, 35, 36, 37, 38,$
$38, 40, 44, 45, 46, 48, 49, 53, 54, 55, 56, 57, 58, 62, 63, 64, 65, 66\}$;

- If $x_0' \equiv 2 \mod 7$ then $x^{18} - 7$ is irreducible over $\mathbb{F}_p$.

7   Twists and Choosing α

When choosing a particular value of α to construct the tower we may find that there are more than one potential values we could use. In this case we must decide which value α is best for implementation. This is illustrated in the following example.

Example 1. The value $x_0 = 4008804000000009_{16}$ generates suitable parameters for a BN curve. Using this x_0 we see that $p \equiv 3 \mod 4$ and we first need a base tower $\mathbb{F}_{p^2} = \mathbb{F}_p(\sqrt{-1})$ before we use the general construction method. We see also that $x_0 \equiv 3 \mod 15$ and x_0 is odd so, as shown in section 6.2, we know immediately that 5 is a cubic and quadratic non-residue in $\mathbb{F}_p$ and so $x^6 - (1 + 2\sqrt{-1})$ is irreducible over $\mathbb{F}_{p^2} = \mathbb{F}_p(\sqrt{-1})$. Using the same reasoning, however, we also know that $x^6 - (2 + 1\sqrt{-1})$, $x^6 - (2 - 1\sqrt{-1})$, $x^6 - (-2 - 1\sqrt{-1})$ and $x^6 - (-2 + 1\sqrt{-1})$ are all irreducible over $\mathbb{F}_{p^2} = \mathbb{F}_p(\sqrt{-1})$. Using this particular value of x_0 we also see that $a^2 + b^2$ is neither a square nor a cube for the (unordered and unsigned) pairs $(a, b) = (1, 3), (1, 5), (2, 3)$ as well as for $(1, 2)$. This example raises an important question:

How do we decide which value will be the best for implementation?

A simple analysis indicates that the optimal choice is the one which minimises $\omega(a) + \omega(b)$, where $\omega(n)$ is the number of additions required to perform a multiplication by n. There is another important point to take into account when choosing α and that is the construction of the twists of the elliptic curve used when computing the pairing.

In §3 it was mentioned that twists are used to improve the efficiency of the pairing computation. To construct a twist of degree d and the isomorphism from the twist to the curve we need an element $i \in \mathbb{F}_{p^{k/d}}$ which is a qth non-residue for all divisors q of k/d. Clearly, for the tower construction we already have such an element. In fact, it would make sense to use the same element to define the twist as we use to construct the tower; though we will have to be slightly more careful in our selection of the element α. An elliptic curve with a twist of degree d actually has $\phi(d)$ twists of degree d, with different numbers of points. The twists used for the curves specified above are of degrees $d = 4$ or 6, both having $\phi(6) = \phi(4) = 2$ possible twists.

For $E(F_p) : y^2 = x^3 + Ax$, the quartic twists are given by $E_1'(\mathbb{F}_{p^{k/t}}) : y^2 = x^3 + Ax/i$ and $E_2'(\mathbb{F}_{p^{k/t}}) : y^2 = x^3 + Ax/i^3$, the twist used for the pairing is the twist with the correct number of points. The respective isomorphisms are given as [20]:

$$E_1' \rightarrow E : (x, y) \rightarrow (i^{1/2}x, i^{3/4}y)$$

and

$$E_2' \rightarrow E : (x, y) \rightarrow (i^{1/2}x/i, i^{1/4}y/i).$$

Similarly, for $E(F_p) : y^2 = x^3 + B$, the sextic twists are given by $E_1'(\mathbb{F}_{p^{k/t}}) : y^2 = x^3 + B/i$ and $E_2'(\mathbb{F}_{p^{k/t}}) : y^2 = x^3 + B/i^5$, the twists must then be tested to find the one with the correct number of points. The respective isomorphisms are given as:

$$E_1' \rightarrow E : (x, y) \rightarrow (i^{1/3}x, i^{1/2}y)$$

and

$$E_2' \rightarrow E : (x, y) \rightarrow (i^{2/3}x/i, i^{1/2}y/i).$$

We see here how important it is to choose the element i to be of the simplest form as the isomorphism will be effected. If we select α such that $i = \alpha^{(1/e)}$, where $e = k/d$, then the isomorphism is basically a free computation [8]. If the curve defined choosing $i = \alpha^{(1/e)}$ does not give the correct number of points, then we must take $i = \alpha^{(3/e)}$ if E' is a quartic twist or $i = \alpha^{(5/e)}$ if E' is a sextic twist. In these cases the isomorphism will be slightly more expensive. This is also discussed in [13].

To summarise, when selecting the element α to define the tower, both $\omega(\alpha)$ and the structure of the twist should be taken into account.

8   Conclusion

In this paper we proved a theorem which leads to a method to determine if a binomial defined over an extension field is irreducible by performing a few tests on one element of the base field. This results in an efficient method of construction for fields which occur in pairing-based cryptography and which were not originally considered to be "pairing-friendly" and could not be constructed using general method discussed in [15]. Using Theorem 5 along with the general construction method we are now able to automatically construct towers of extensions for the implementation of the finite fields used in pairing-based cryptography by performing a few cubic and quadratic non-residue tests on elements of $\mathbb{F}_p$. The resulting constructions are efficient and can contribute to the development of a cryptographic compiler specialised for pairing-based cryptography as described in [9]. We have used our results, Euclid's conjectures and an observation by Shirase [22] to give immediate constructions for a large class of towering-friendly fields used with BN curves. Using Euclid's conjectures we have also given an immediate construction for a large group of towering-friendly fields used with KSS $k = 18$ curves. We are confident that these methods can be extended to other families of pairing-friendly elliptic curves and other embedding degrees to generate automatic tower structures for these curves.

Acknowledgements

The authors thank Rob Granger for his insightful comments and encouraging discussions and thank Paulo Barreto for his helpful comments. We would also like to thank the anonymous reviewers for their constructive comments.

References

1. IEEE P1363.3: Standard for identity-based cryptographic techniques using pairings. Draft 3: Section 5.3.2,
   http://grouper.ieee.org/groups/1363/IBC/index.html
2. Arène, C., Lange, T., Naehrig, M., Ritzenthaler, C.: Faster computation of the Tate pairing. Cryptology ePrint Archive, Report 2009/155 (2009),
   http://eprint.iacr.org/
3. Bailey, D., Paar, C.: Optimal extension fields for fast arithmetic in public-key algorithms. In: Krawczyk, H. (ed.) CRYPTO 1998. LNCS, vol. 1462, pp. 472–485. Springer, Heidelberg (1998)
4. Barreto, P.S.L.M., Kim, H.Y., Lynn, B., Scott, M.: Efficient algorithms for pairing-based cryptosystems. In: Yung, M. (ed.) CRYPTO 2002. LNCS, vol. 2442, pp. 354–368. Springer, Heidelberg (2002)
5. Barreto, P.S.L.M., Lynn, B., Scott, M.: Constructing elliptic curves with prescribed embedding degrees. In: Cimato, S., Galdi, C., Persiano, G. (eds.) SCN 2002. LNCS, vol. 2576, pp. 263–273. Springer, Heidelberg (2003)
6. Barreto, P.S.L.M., Naehrig, M.: Pairing-friendly elliptic curves of prime order. In: Preneel, B., Tavares, S. (eds.) SAC 2005. LNCS, vol. 3897, pp. 319–331. Springer, Heidelberg (2006)
7. Cohen, H., Frey, G. (eds.): Handbook of elliptic and hyperelliptic curve cryptography. CRC Press, Boca Raton (2005)
8. Devegili, A.J., Scott, M., Dahab, R.: Implementing cryptographic pairings over Barreto-Naehrig curves. In: Takagi, T., Okamoto, T., Okamoto, E., Okamoto, T. (eds.) Pairing 2007. LNCS, vol. 4575, pp. 197–207. Springer, Heidelberg (2007)
9. Dominguez Perez, L.J., Scott, M.: Automatic generation of optimised cryptographic pairing functions. In: SPEED-CC Workshop Record– Software Performance Enhancement for Encryption and Decryption and Cryptographic Compilers, vol. 1, pp. 55–71 (2009)

10. Freeman, D., Scott, M., Teske, E.: A taxonomy of pairing-friendly elliptic curves. Journal of Cryptology 23 (2010)
11. Galbraith, S., Lin, X., Scott, M.: Endomorphisms for faster elliptic curve cryptography on a large class of curves. In: Joux, A. (ed.) EUROCRYPT 2009. LNCS, vol. 5479, pp. 518–535. Springer, Heidelberg (2010)
12. Granger, R., Page, D., Stam, M.: On small characteristic algebraic tori in pairing based cryptography. LMS Journal of Computation and Mathematics 9, 64–85 (2006)
13. Hess, F., Smart, N., Vercauteren, F.: The eta pairing revisited. IEEE Trans. Information Theory 52, 4595–4602 (2006)
14. Kachisa, E., Schaefer, E., Scott, M.: Constructing Brezing-Weng pairing friendly elliptic curves using elements in the cyclotomic field. In: Galbraith, S.D., Paterson, K.G. (eds.) Pairing 2008. LNCS, vol. 5209, pp. 126–135. Springer, Heidelberg (2008)
15. Koblitz, N., Menezes, A.: Pairing-based cryptography at high security levels. In: Smart, N.P. (ed.) Cryptography and Coding 2005. LNCS, vol. 3796, pp. 13–36. Springer, Heidelberg (2005)
16. Lee, E., Lee, H., Park, C.: Efficient and generalized pairing computation on abelian varieties. IEEE Trans. Information Theory 55, 1793–1803 (2009)
17. Lemmermeyer, F.: Reciprocity Laws: From Euler to Eisenstein. Springer Monographs in Mathematics. Springer, Heidelberg (2000)
18. Lidl, R., Niederreiter, H.: Finite Fields, 2nd edn. Encyclopedia of Mathematics and its Applications, vol. 20. Cambridge University Press, Cambridge (1997)
19. Baktır, S., Sunar, B.: Optimal tower fields. IEEE Transactions on Computers 53(10), 1231–1243 (2004)
20. Scott, M.: A note on twists for pairing friendly curves, ftp://ftp.computing.dcu.ie/pub/resources/crypto/twists.pdf
21. Scott, M., Barreto, P.: Compressed pairings. In: Franklin, M. (ed.) CRYPTO 2004. LNCS, vol. 3152, pp. 140–156. Springer, Heidelberg (2004), http://eprint.iacr.org/2004/032/
22. Shirase, M.: Universally constructing 12-th degree extension field for ate pairing. Cryptology ePrint Archive, Report 2009/623 (2009), http://eprint.iacr.org/
23. Silverman, J.H.: The Arithmetic of Elliptic Curves. Graduate Texts in Mathematics, vol. 106. Springer, New York (1986)

Delaying Mismatched Field Multiplications in Pairing Computations*

Craig Costello, Colin Boyd,
Juan Manuel Gonzalez Nieto, and Kenneth Koon-Ho Wong

Information Security Institute
Queensland University of Technology, GPO Box 2434,
Brisbane QLD 4001, Australia
{craig.costello,c.boyd,j.gonzaleznieto,kk.wong}@qut.edu.au

Abstract. Miller's algorithm for computing pairings involves performing multiplications between elements that belong to different finite fields. Namely, elements in the full extension field $\mathbb{F}_{p^k}$ are multiplied by elements contained in proper subfields $\mathbb{F}_{p^{k/d}}$, and by elements in the base field $\mathbb{F}_p$. We show that significant speedups in pairing computations can be achieved by delaying these "mismatched" multiplications for an optimal number of iterations. Importantly, we show that our technique can be easily integrated into traditional pairing algorithms; implementers can exploit the computational savings herein by applying only minor changes to existing pairing code.

Keywords: Pairings, Miller's algorithm, finite field arithmetic, Tate pairing, ate pairing.

1   Introduction

For the past decade, the public-key cryptographic community has witnessed an avalanche of novel and exciting protocols based on bilinear pairings; the major trigger being the discovery of identity-based encryption by Boneh and Franklin [8]. The algorithm for computing these pairings, initially proposed by Miller in the mid 1980's [27], was originally too slow for pairing-based protocols to be practically competitive with their RSA and Diffie-Hellman type rivals, and much research has since been invested towards speeding up Miller's algorithm. Consequently, Miller's algorithm has come a long way from its original form, to the point where many possible enhancements have now been fully optimized [2,3,29,4,31,19].

The progress in the field of pairing improvements has seemingly steadied to a pace where a real world programmer could rest assured that their currently optimized (or close to optimized) implementation will most likely remain

---

* The first author acknowledges funding from the Queensland Government Smart State PhD Scholarship. This work has been supported in part by the Australian Research Council through Discovery Project DP0666065.

within arms length of the state-of-the-art implementation, for at least a couple of years. Nevertheless, so long as optimizations continue to be introduced [5,12,10], old pairing code could potentially become outdated quite quickly. The importance of code reusability and integrability might be the difference between an implementer continuing to modify and update their existing code in light of the latest breakthroughs, or shying away from such improvements because of the difficulty in integrating them.

It was shown very recently [10] that it is possible to avoid much of the costly, full extension field arithmetic encountered in pairing computations over large prime fields by replacing a multiplication in the full extension field with more minor multiplications in its proper subfields, decreasing the overall complexity of Miller's algorithm by over 30% in some cases. In these instances, an implementation not encompassing these techniques would perform substantially slower compared to one that does. However, a programmer wishing to implement the methods of avoiding extension field arithmetic in [10] would be facing the task of re-writing most (if not all) of their pairing code from scratch, having to employ new and potentially cumbersome explicit formulas.

In this paper we provide an alternative to the technique in [10] that offers much higher integrability into traditional pairing algorithms and existing pairing code. The idea used herein is the same as that used by Granger, Page and Stam [18, §6], who employ *loop unrolling* to combine two Miller iterations at a time, achieving fewer overall field operations in the case of characteristic three pairing implementations. In this paper we apply this same loop unrolling technique to pairings computed over large prime fields, by analyzing the cost of combining n Miller iterations at a time, and choosing the optimal value of n for various embedding degrees. Unlike the method in [10], our method requires no new explicit formulas for elliptic curve point operations and Miller line computations. Our aim is to inject a new and conceptually simple subroutine into Miller's algorithm that optimizes the field arithmetic occuring between elements of finite fields with different extension degree, with a parallel goal of minimizing the change imposed on existing pairing code.

The rest of this paper is organized as follows. In Section 2 we set notations and give a background on the computation of pairings. In Section 3 we discuss the proposed technique, before analyzing its computational complexity in Section 4. We provide necessary implementation details in Section 5, before providing parameters to optimize its implementation in Section 6, where we also draw comparisons against the traditional version of Miller's algorithm.

2   Preliminaries

Implementing pairings in cryptography most commonly requires the definition of two linearly independent groups, $\mathbb{G}_1$ and $\mathbb{G}_2$, of large prime order r, contained on an elliptic curve E which is defined over a finite field $\mathbb{F}_q$ of large prime characteristic p. Herein, we choose to deal with the most common case of prime fields, so that in fact we have $q = p$. Let π_p be the p-power Frobenius endomorphism on E. In general, the most popular choices for $\mathbb{G}_1$ and $\mathbb{G}_2$ are the two eigenspaces of π_p, restricted to the r-torsion $E[r]$ of E, so that $\mathbb{G}_1 = E[r] \cap \ker(\pi_p - [1])$ and $\mathbb{G}_2 = E[r] \cap \ker(\pi_p - [p])$. Let k be the smallest integer such that $r \mid p^k - 1$; a direct consequence of this is that the field $\mathbb{F}_{p^k}$ is the smallest extension of $\mathbb{F}_p$ that contains all of the points in $E[r]$, so that $\mathbb{F}_{p^k}$ houses both $\mathbb{G}_1$ and $\mathbb{G}_2$ in their entirety. We refer to k the embedding degree, because computing the pairing of any two linearly independent points in $E[r]$ results in an element of an order-r subgroup of the finite field $\mathbb{F}_{p^k}$, i.e. the pairing embeds the points of $E[r]$ into the k-degree extension of $\mathbb{F}_p$. We use $\mathbb{G}_T$ to denote this order-r subgroup of $\mathbb{F}_{p^k}$, since this is the target group of the pairing map. For $k > 1$, the points in $\mathbb{G}_1$ are completely defined over the base field $\mathbb{F}_p$, whilst the points in $\mathbb{G}_2$ are defined over the larger field $\mathbb{F}_{p^k}$.

We assume that our pairing is defined by the Tate methodology rather than the Weil methodology (see [19]), since the Weil pairing has been phased out in practice due to its inefficient computation. The Tate methodology computes a bilinear pairing, e, of two linearly independent points $R, S \in E[r]$, as

$$e(R, S) = f_{m,R}(S)^{(p^k-1)/r}, \tag{1}$$

where $f_{n,R}$ is a function with divisor $\operatorname{div}(f_{m,R}) = m(R) - ([m]R) - (m-1)(\mathcal{O})$, with $\mathcal{O}$ being the neutral element on E. We refer to the function $f_{m,R}$ as the Miller function, since it is computed using Miller's algorithm. This algorithm uses relations between divisors of functions to build $f_{m,R}$ in $\log_2(m)$ iterations in a double-and-add like fashion, as summarized in Algorithm 1.

Pairings that fit into the Tate methodology can be naturally divided into two categories: *Miller-lite* pairings which take $R \in \mathbb{G}_1$ and $S \in \mathbb{G}_2$ and *Miller-full* pairings which take $R \in \mathbb{G}_2$ and $S \in \mathbb{G}_1$. That is, $e_{\text{lite}} : \mathbb{G}_1 \times \mathbb{G}_2 \mapsto \mathbb{G}_T$, whilst $e_{\text{full}} : \mathbb{G}_2 \times \mathbb{G}_1 \mapsto \mathbb{G}_T$. The Tate pairing and the twisted ate pairing [20] are examples of Miller-lite pairing, whilst the ate pairing [20] and its derivatives (the ate$_i$ pairing [26], the R-ate pairing [24], etc) sit under the umbrella of Miller-full pairings. Efficient pairing implementations make use of the twisted curve E' to define a group $\mathbb{G}_2' \in E'$ that is isomorphic to $\mathbb{G}_2 \in E$, but whose elements are contained in a much smaller subfield $\mathbb{F}_{p^e} \subset \mathbb{F}_{p^k}$, where $e = k/d$ and d is the degree of the twist. We let $\psi : E' \to E$ denote the twisting isomorphism from E' to E, so that $\psi(\mathbb{G}_2) = \mathbb{G}_1$. The bulk of the operations encountered in an iteration of Miller's algorithm are computed using the coordinates of R or its image R' under ψ^{-1}, so that Miller-lite pairings benefit from the majority of operations being performed over $\mathbb{G}_1$, which is defined over the base field $\mathbb{F}_p$. Alternatively, Miller-full pairings spend the majority of computations operating on coordinates that are defined over the larger extension field $\mathbb{F}_{p^e}$. Such computations are more costly over extension fields, however Miller-full pairings are usually more efficient than Miller-lite pairings in practice [20,12], because they enjoy a much smaller loop parameter m, meaning that Miller's algorithm requires significantly less iterations.

Any extension fields of $\mathbb{F}_p$ that are required in the pairing computation are best constructed using towers of field extensions. The general method to

---

Algorithm 1. Miller's double-and-add Algorithm

Input: R, S, $m = (m_{l-1}...m_1, m_0)_2$.
Output: $f_{m,R}(S) \leftarrow f$.

1: $T \leftarrow R$, $f \leftarrow 1$.
2: **for** $i = l - 2$ to 0 **do**
3:     $T \leftarrow [2]T$.
4:     Compute a function g, which has divisor $\operatorname{div}(g) = 2(T) - (2T) - (\mathcal{O})$.
5:     Compute $g = g(S)$ (evaluate g at the coordinates of S).
6:     $f \leftarrow f^2 \cdot g$.
7:     **if** $m_i \neq 0$ **then**
8:         $T \leftarrow T + R$.
9:         Compute a function g, which has divisor $\operatorname{div}(g) = (T) + (R) - (T + R) - (\mathcal{O})$.
10:         Compute $g = g(S)$ (evaluate g at the coordinates of S).
11:         $f \leftarrow f \cdot g$.
12:     **end if**
13: **end for**
14: **return** f.

---

construct towers of extension fields in pairing-based cryptography is originally due to Koblitz and Menezes [23], who introduced the notion of *pairing-friendly fields*, where the embedding degree is chosen to be of the form $k = 2^i 3^j$, and the characteristic of the field $\mathbb{F}_p$ is chosen to be $p \equiv 1 \bmod 12$. These conditions allow us to easily build a tower of extensions up to $\mathbb{F}_{p^k}$ using a sequence of $z = i + j$ cubic and quadratic sub-extensions, where the defining polynomial for each of the $d_{i,j}$-degree sub-extensions are actually binomial of the form $x^{d_{i,j}} - \alpha$. Such quadratic and cubic binomials facilitate fast arithmetic over extension fields. Very recently, Benger and Scott [5] broadened the definition of pairing-friendly fields to present the more general notion of *towering-friendly fields*, which are fields of the form $\mathbb{F}_{q^m}$ (q not necessarily prime itself) for which all prime divisors of m also divide $q - 1$, showing that efficient tower constructions can also be achieved without satisfying the more restrictive condition of $p \equiv 1 \bmod 12$ for characteristic p fields.

For elliptic curves, there are only four twist degrees possible: $d = 2$ quadratic twists, $d = 3$ cubic twists, $d = 4$ quartic twists and $d = 6$ sextic twists. In both Miller-lite and Miller-full pairings, it is advantageous to choose the Weierstrass curve model (of the form $y^2 = x^3 + ax + b$) which supports the maximal twist degree d, such that $d \mid k$. Cubic and sextic twists are only possible when $a = 0$, quartic twists when $b = 0$, and quadratic twists impose no condition on the curve constants, although it is usually advantageous to set either a or b to be zero for computational efficiency anyway [1,12].

For quadratic and cubic twists, $\mathbb{F}_{p^k}$ is the direct (quadratic or cubic) sub-extension of the field $\mathbb{F}_{p^e}$. For quartic and sextic extensions, however, we must first extend $\mathbb{F}_{p^e}$ to an intermediate field $\mathbb{F}_{p^h}$, where $\mathbb{F}_{p^e} \subset \mathbb{F}_{p^h} \subset \mathbb{F}_{p^k}$, and the field extensions are formed by taking $\mathbb{F}_{p^h} = \mathbb{F}_{p^e}(\alpha)$ and $\mathbb{F}_{p^k} = \mathbb{F}_{p^h}(\beta)$. We denote the degree of the extensions as $\delta_\alpha = [\mathbb{F}_{p^h} : \mathbb{F}_{p^e}] = h/e$ and $\delta_\beta = [\mathbb{F}_{p^k} : \mathbb{F}_{p^h}] = k/h$, in agreement with $[\mathbb{F}_{p^k} : \mathbb{F}_{p^e}] = \delta_\alpha \delta_\beta = k/e = d$. For all twists, we have that an element of the full extension field, say the Miller function $f \in \mathbb{F}_{p^k}$, takes the form

$$f = \sum_{j=0}^{\delta_\beta - 1} \left(\sum_{i=0}^{\delta_\alpha - 1} f_{j,i} \cdot \alpha^i \right) \cdot \beta^j, \tag{2}$$

where each of the $f_{j,i}$ are contained in $\mathbb{F}_{p^e}$. For quadratic twists we must take $(\delta_\alpha, \delta_\beta) = (2,1)$ and for cubic twists we must take $(\delta_\alpha, \delta_\beta) = (3,1)$. For both quartic and sextic twists, Benger and Scott [5] suggest that the most efficient tower is constructed with $\delta_\alpha = 2$, so that quartic twists should take $(\delta_\alpha, \delta_\beta) = (2,2)$, and sextic twists should take $(\delta_\alpha, \delta_\beta) = (2,3)$. The nature of the tower for the fields that lie between $\mathbb{F}_p$ and $\mathbb{F}_{p^e}$ do not play a role in this work, so we pay no attention to these details, but point the interested reader to [5].

The general twist of a short Weierstrass curve is written as $E'(\mathbb{F}_{p^e}) : y^2 = x^3 + az^4 x + bz^6$, where the isomorphism $\psi : E' \to E$ is defined as $\psi(x', y') = (z^2 x', z^3 y')$. For quartic twists when $b = 0$, we choose $z^4 \in \mathbb{F}_{p^e}$ such that $z^2 \in \mathbb{F}_{p^{k/2}} \notin \mathbb{F}_{p^e}$ and $z \in \mathbb{F}_{p^k} \notin \mathbb{F}_{p^{k/2}}$, so that we can set $\alpha = z^2$ and $\beta = z^3$, resulting in a twisting isomorphism $\psi(x', y') = (\alpha x', \beta y')$ that allows twisted coordinates to be easily integrated with general field elements taking the form of f in (2). Similarly, for sextic twists when $a = 0$, we choose $z^6 \in \mathbb{F}_{p^e}$ such that $z^3 \in \mathbb{F}_{p^{k/3}} \notin \mathbb{F}_{p^e}$ and $z^2 \in \mathbb{F}_{p^{k/2}} \notin \mathbb{F}_{p^e}$, so that we can set $\alpha = z^3$ and $\beta = z^2$, and the twisting isomorphism conveniently becomes $\psi(x', y') = (\beta x', \alpha y')$.

We follow the general trend of reporting results for even k [3,23,5], since such embedding degrees support the denominator elimination optimization [2]. Thus, any k we consider which is divisible by 3 will also be divisible by 6 and admit a sextic twist, so that we do not need to consider cubic twists. Computationally speaking, the treatment of cubic twists is quite different to the other even degree twists and tends to be awkward anyway [12], so curves with odd embedding degree divisible by 3 are generally not chosen in practice, although Lin et al. [25] show that choosing $k = 9$ can be competitive at some security levels.

Remark 1 (A notation for counting costs). This paper is largely concerned with the computational cost of field operations, so we employ a notation that allows us to easily narrate such costs alongside the associated algebra. We use $\mathsf{cost}\big[L \leftarrow J\big]$ to denote the computational cost of computing the set $L = \{L_1, ..., L_i\}$ from the already computed (or available) set $J = \{J_1, ..., J_j\}$. If the best way to compute the set L from the set J is to compute the intermediate set $K = \{K_1, ..., K_j\}$, then we can clearly split the cost, so that $\mathsf{cost}\big[L \leftarrow J\big] = \mathsf{cost}\big[L \leftarrow K\big] + \mathsf{cost}\big[K \leftarrow J\big]$, under the assumption that there does not exist a cheaper way to compute L from J which does not require the computation of K. When referring to the cost of computing the set L without assuming any prior computations, we simply use $\mathsf{cost}\big[L\big]$.

Remark 2 (The squaring vs. multiplication ratio). Our cost analyses are primarily concerned with field multiplications and field squarings and we choose not pay any attention to the much cheaper cost of field additions, although the algorithms presented herein aim to minimize all field operations. We use $\mathbf{m}_i$ and $\mathbf{s}_i$ to represent the respective costs of a multiplication and a squaring in the field $\mathbb{F}_{p^i}$. Since the ratio of the complexity of a field squaring to a field multiplication is specific to the implementation, we leave the discussion general until Section 6 by using the parameter Ω, which denotes the $\mathbf{s} : \mathbf{m}$ ratio. That is, $\mathbf{s} = \Omega \mathbf{m}$, where $0 << \Omega \leq 1$. For example, Bernstein [6] achieves $\Omega = 0.68$ and Hisil [21] reports $\Omega = 0.72$, while the EFD [7] presents results based on the more commonly accepted $\Omega = 0.8$ and $\Omega = 1$ values.

Remark 3 (Non-specific field definitions). In the pairing $e(R, S)$, the respective fields that R and S belong to are different depending on whether the pairing is a Miller-lite or Miller-full pairing. In a Miller-lite pairing computed as $e(R, \psi(S'))$, we take $R \in \mathbb{F}_p$ and $S' \in \mathbb{F}_{p^e}$, whilst a Miller-full pairing computed as $e(\psi^{-1}(R), \psi^{-1}(S))$ (see [12]) has $R' \in \mathbb{F}_{p^e}$ and $S \in \mathbb{F}_p$. Ignoring the twisting elements α and β, then the first and second arguments in a Miller-lite pairing are from $\mathbb{F}_p$ and $\mathbb{F}_{p^e}$ respectively, whilst the same arguments in a Miller-full pairing are from $\mathbb{F}_{p^e}$ and $\mathbb{F}_p$ respectively. In sections 3 and 4, we cover both cases simultaneously by saying that the first argument R belongs to $\mathbb{F}_{p^u}$ and the second argument S belongs to $\mathbb{F}_{p^v}$, where it is understood that $(u, v) = (1, e)$ for Miller-lite pairings and $(u, v) = (e, 1)$ for Miller-full pairings. Most importantly, in either case we have that multiplying an element of $\mathbb{F}_{p^u}$ by an element of $\mathbb{F}_{p^v}$ costs $e\mathbf{m}_1$ (cf. [12]).

Remark 4 (Ignoring additions). As is the common trend in papers discussing optimal pairing implementations, we assume that the loop parameter m has low Hamming weight so that additions are sparse in Miller's algorithm. Thus, when discussing any consecutive iterations of the Miller loop, we assume that no such iterations involve additions.

3   Delaying Mismatched Multiplications

We begin this section by illustrating the potential advantage of delaying "mismatched" multiplications, through the use of a toy example. Suppose we have a basic algorithm that involves n iterations, where the i-th iteration simply involves computing an element a_i, and updating the master function A_i as $A_i \leftarrow a_i \cdot A_{i-1}$, where the master function was initialized as A_0. The output of the algorithm would be $A_n = (((...((A_0 \cdot a_1) \cdot a_2)...)))$, which could alternatively be written, or indeed computed as $A_n = A_0 \cdot (\prod_{i=1}^{n} a_i)$, where the product of the a_i's is computed prior to multiplication with A_0. If both A_0 and the a_i's are general elements of the same field, say $\mathbb{F}_{p^c}$, then both methods of computation would require n multiplications in $\mathbb{F}_{p^c}$, as

$$\text{cost}\big[A_n \leftarrow \{a_1, ..., a_n, A_0\}\big] = \sum_{i=1}^{n} \text{cost}\big[A_i \leftarrow \{A_{i-1}, a_i\}\big] = \sum_{i=1}^{n} 1\mathbf{m}_c = n\mathbf{m}_c,$$

or

$$\mathsf{cost}\big[A_n \leftarrow \{a_1, ..., a_n, A_0\}\big] = \mathsf{cost}\big[A_n \leftarrow \{\prod_{i=1}^{n} a_i, A_0\}\big] +$$

$$\mathsf{cost}\big[\prod_{i=1}^{n} a_i \leftarrow \{a_1, ..., a_n\}\big] = 1\mathbf{m}_c + (n-1)\mathbf{m}_c = n\mathbf{m}_c.$$

However, suppose again that the a_i values are general elements of the field $\mathbb{F}_{p^c}$, but instead suppose that A_0 is a general element of the degree-w extension field $\mathbb{F}_{p^{cw}}$, of $\mathbb{F}_{p^c}$. For ease of exposition, we assume for now that w is prime so that a general element of $\mathbb{F}_{p^{cw}}$ can be expressed as a $(w-1)$-degree polynomial with coefficients in $\mathbb{F}_{p^c}$. In this case, multiplying each of the a_i values by A_0 would typically involve multiplying each of the w coefficients of A_0 by a_i, costing $w\mathbf{m}_c$ each time. Clearly, it would be advantageous to form the product $\prod_{i=1}^{n} a_i$ prior to a multiplication by A_0, as we show by using the same comparison as before, where

$$\mathsf{cost}\big[A_n \leftarrow \{a_1, ..., a_n, A_0\}\big] = \sum_{i=1}^{n} \mathsf{cost}\big[A_i \leftarrow \{A_{i-1}, a_i\}\big] = \sum_{i=1}^{n} w\mathbf{m}_c = wn\mathbf{m}_c,$$

whilst

$$\mathsf{cost}\big[A_n \leftarrow \{a_1, ..., a_n, A_0\}\big] = \mathsf{cost}\big[A_n \leftarrow \{\prod_{i=1}^{n} a_i, A_0\}\big] +$$

$$\mathsf{cost}\big[\prod_{i=1}^{n} a_i \leftarrow \{a_1, ..., a_n\}\big] = w\mathbf{m}_c + (n-1)\mathbf{m}_c = (w+n-1)\mathbf{m}_c.$$

Forming the product of the a_i elements from the smaller field prior to the multiplication by A_0 gives a count of $(w+n-1)\mathbf{m}_c$, as opposed to the $wn\mathbf{m}_c$ that it costs to multiply a_i by A_0 in each iteration. When $n > 1$ and $w > 1$, it is always the case that $wn > (w+n-1)$, so that forming the product of $n > 1$ elements in the smaller field and *delaying* any multiplications by the element in the larger field is always advantageous. The central theme of this paper is applying this idea towards pairing computations, however the story in Miller's algorithm is more complicated than the example above. Firstly, the "mismatched" multiplications above were easy to spot, since we were multiplying general elements from different fields. However, there are other more subtle examples of mismatched multiplications, which we formalize in the following definitions.

Definition 1 (General vs. Special Field Elements). *Let $\omega \in \mathbb{F}_{p^c}$, where $\mathbb{F}_{p^c}$ is constructed as a tower of extensions as $\mathbb{F}_p \subset \mathbb{F}_{p^{c_1}} \subset \mathbb{F}_{p^{c_1 c_2}} \subset ... \subset \mathbb{F}_{p^{c_1 c_2 \cdots c_t}} = \mathbb{F}_{p^c}$, i.e. c_i is the degree of the i-th extension in the tower up to $\mathbb{F}_{p^c}$. Let $C_j = \prod_{i=1}^{j} c_i$ so that we can write the tower as $\mathbb{F}_p \subset \mathbb{F}_{p^{C_1}} \subset \mathbb{F}_{p^{C_2}} \subset ... \subset \mathbb{F}_{p^{C_t}} = \mathbb{F}_{p^c}$. Let $\#\omega(\mathbb{F}_{p^{C_j}})$ be the number of non-zero coefficients in the* *polynomial representation of ω over the subfield $\mathbb{F}_{p^{C_j}}$. If $\#\omega(\mathbb{F}_{p^{C_j}}) < c/C_j$ for any j where $1 \leq j \leq t$, then we call ω a special element of $\mathbb{F}_{p^c}$, otherwise we call ω a general element of $\mathbb{F}_{p^c}$.*

Definition 2 (Mismatched Multiplications). *Let $\omega \in \mathbb{F}_{p^c}$ and $\hat{\omega} \in \mathbb{F}_{p^{\hat{c}}}$. We call the multiplication between ω and $\hat{\omega}$ mismatched if one of the following two conditions hold:*

(i) $c \neq \hat{c}$.

(ii) $c = \hat{c}$, but at least one of ω and $\hat{\omega}$ are special.

We refer to a mismatched multiplication as a type (i) or type (ii) mismatch, depending on which of the above conditions it breaches.

Equipped with the above definitions, we now focus on searching for mismatched multiplications in Miller's algorithm, with the aim of investigating the possibility and potential advantage of optimizing the delay or the avoidance of such multiplications. We start by taking a close look at the doubling stage of Miller's algorithm which is the combination of steps 3, 4, 5 and 6 of Algorithm 1. Steps 3 and 4 involve doubling the point T (i.e. computing $[2]T$ from T), and computing the coefficients of the associated function g with divisor $\mathrm{div}(g) = 2(T) - (2T) - (\mathcal{O})$. These computations only depend on the coordinates of the point $T = (T_x, T_y) \in \mathbb{F}_{p^u}$, and since T_x and T_y are assumed to be general elements of $\mathbb{F}_{p^u}$, we can safely assume that, in general, none of the field multiplications in steps 3 and 4 are mismatched. Many authors have achieved speed ups in pairing computations by focussing on reducing the combined cost of these two steps [9,13,22,1,11,12], where the cost of encapsulated point doubling (step 3) and line computation (step 4) is generally presented together, in terms of the combined number of field multiplications (m) and squarings (s) encountered, as

$$\mathsf{cost}\big[\{g, [2]T\} \leftarrow T\big] = m\mathbf{m}_u + s\mathbf{s}_u = (m + \Omega s)\mathbf{m}_u. \tag{3}$$

For curves with even embedding degrees, the denominator elimination optimization greatly simplifies the form of the line function g, so that $g = g(x, y)$ always (cf. [12]) takes the form

$$g(x, y) = g_x \cdot x + g_y \cdot y + g_0, \tag{4}$$

where $g_x, g_y, g_0 \in \mathbb{F}_{p^u}$. Step 5 of Algorithm 1 involves evaluating g at the coordinates of $S = (S_x, S_y)$, i.e. multiplying g_x by S_x and g_y by S_y. From Section 2, we know that (unless $e = 1$) R and S are contained in different fields, so that the evaluation of g at S incurs two type (i) mismatched multiplications. Following this, step 6 of Algorithm 1 involves squaring the Miller function $f \in \mathbb{F}_{p^k}$, and multiplying this result by $g(S) \in \mathbb{F}_{p^k}$. Although the point S belongs to the field $\mathbb{F}_{p^k}$, each of its coordinates are actually either very special elements of $\mathbb{F}_{p^k}$, or lie in proper subfields of $\mathbb{F}_{p^k}$. For example, in Section 2 we saw that an implementation employing a quartic twist has $(S_x, S_y) = (\alpha \hat{S}_x, \beta \hat{S}_y)$, where $\hat{S}_x, \hat{S}_y \in \mathbb{F}_{p^v}$, or similarly an implementation using a sextic twist has $(S_x, S_y) = (\beta \hat{S}_x, \alpha \hat{S}_y)$ with $\hat{S}_x, \hat{S}_y \in \mathbb{F}_{p^v}$. In both cases, it is clear by Definition 1 that $g(S)$ is a special element of $\mathbb{F}_{p^k}$, so that the multiplication of the Miller function f by $g(S)$ is, by Definition 2, a type (ii) mismatched multiplication.

We concretize the above discussion with an example, where we assume a sextic twist has been used, so that lines 5 and 6 of Algorithm 1 require that we compute a multiplication between

$$f = (f_{2,1} \cdot \alpha + f_{2,0}) \cdot \beta^2 + (f_{1,1} \cdot \alpha + f_{1,0}) \cdot \beta + (f_{0,1} \cdot \alpha + f_{0,0}) \in \mathbb{F}_{p^k}$$

and

$$g(S_x, S_y) = (g_x \hat{S}_x) \cdot \beta + (g_y \hat{S}_y) \cdot \alpha + g_0 \in \mathbb{F}_{p^k},$$

where the $f_{i,j}$'s and both $g_x \hat{S}_x$ and $g_y \hat{S}_y$ are contained in $\mathbb{F}_{p^e}$ (see Remark 3), and g_0 is contained in $\mathbb{F}_p$ for Miller-lite implementations or $\mathbb{F}_{p^e}$ in Miller full implementations. Since $g_x, g_y \in \mathbb{F}_{p^u}$ and $\hat{S}_x, \hat{S}_y \in \mathbb{F}_{p^v}$, the products formed to create $g(S_x, S_y)$ are type (i) mismatches, whilst the multiplication between f and $g(S)$ is a type (ii) mismatch.

There are two natural questions that now arise: *are these mismatches a problem?* and, if so, *what can we do about them?* We can immediately answer the first question by referring back to the toy example at the beginning of this section, where we saw that delaying the multiplications between elements of different sized fields can be very advantageous, particularly if the difference in the extension degrees is large.

We start the answer to the second question by noting the main complication, in terms of mismatched multiplications, that we encounter in an iteration of Miller's algorithm; that being the simultaneous presence of both type (i) and type (ii) mismatches. Specifically, each iteration of Miller's algorithm involves two type (i) mismatches buried inside a larger type (ii) mismatch. An ideal solution might involve minimizing both mismatches simultaneously, but unfortunately we will soon see that this is not possible; namely, that the type (i) mismatches are somewhat unavoidable in Miller's algorithm. However, the solution we adopt follows quite naturally if we start by *trying* to avoid the type (i) mismatches, as follows. In a particular iteration of Miller's algorithm, it seems that the only way we can avoid the type (i) mismatched multiplications is to delay them until the following iteration: let g and $\tilde{g}$ represent two consecutive g's in two iterations of Miller's algorithm, and suppose we temporarily delay evaluating g as S until the following iteration when $\tilde{g}$ is computed. Instead of evaluating both functions separately at S, we form the product two indeterminate functions, $g(x, y)$ and $\tilde{g}(x, y)$, modulo the curve equation, and call it $G(x, y)$. In fact, g would have been multiplied by f and squared in the previous iteration, so that $G(x, y)$ is actually computed as

$$G(x, y) = g(x, y)^2 \cdot \tilde{g}(x, y) = (g_x \cdot x + g_y \cdot y + g_0)^2 \cdot (\tilde{g}_x \cdot x + \tilde{g}_y \cdot y + \tilde{g}_0)$$

$$= \sum_{i=0}^{4} G_{x^i} \cdot x^i + \left[\sum_{i=0}^{3} G_{x^i y} \cdot x^i \right] \cdot y, \tag{5}$$

where we reduce any higher powers of y via the curve equation. We could then evaluate $G(x, y)$ at S and multiply $G(S)$ by the Miller function f, so that delaying the evaluation of g at S and the multiplication of f by $g(S)$ avoided both types of mismatched multiplications for one iteration. However, at this next iteration, we now have many more type (i) multiplications to deal with. Namely, what would have been 4 type (i) mismatches in total (2 for the evaluation of g at S and likewise for the evaluation of $\hat{g}$ at S), has now become 8 type (i) mismatches (multiplying $G_{x^i y}$ by $x^i y$ and G_{x^i} by x^i above). At a first glance then, this idea seems somewhat counterproductive. However, let us assume for now that we are employing a sextic twist so that $(\delta_\alpha, \delta_\beta) = (2, 3)$ and observe the new function G evaluated at $S = (\hat{S}_x \beta, \hat{S}_y \alpha)$, as

$$G(x, y) = \sum_{i=0}^{4} G_{x^i} \cdot S_x^i \cdot \beta^i + \left[\sum_{i=0}^{3} G_{x^i y} \cdot S_x^i \cdot \beta^i \right] \cdot \alpha = \sum_{j=0}^{\delta_\beta} \left(\sum_{i=0}^{\delta_\alpha} \hat{G}_{j,i} \cdot \alpha^i \right) \cdot \beta^j, \tag{6}$$

where each of the $\hat{G}_{j,i}$ are easily derived combinations of the $G_{x^i y}$ and G_{x^i} terms in (5). Importantly, we now have that $G(x, y)$ has become a general element of $\mathbb{F}_{p^k}$, so that performing the multiplication between f and G will fully exploit a routine written to perform optimized multiplication over $\mathbb{F}_{p^k}$. More importantly, we have only had to perform one full extension field multiplication in two Miller iterations. In short, we delayed the multiplication between f and g until g was built up into G (a product of g's), a general element of $\mathbb{F}_{p^k}$, and in doing so we saved a mismatched multiplication in $\mathbb{F}_{p^k}$. The price we pay for this saving is the increased number of type (i) mismatched multiplications that are required to evaluate G at S, as well as an increased number of standard $\mathbb{F}_{p^u}$ multiplications that are required to form the coefficients of G from g and $\hat{g}$. Our goal becomes clear then; we wish to explore whether it is advantageous to spend extra computations in order to achieve the savings offered by avoiding type (ii) mismatches altogether.

In the following sections, we explore these trade-offs in detail. Specifically, we consider delaying the multiplication between the g's and the Miller function f for an arbitrary number (N) of iterations, a process we refer to as N-delay. We track the computational cost of N-delay and determine the optimum values of N for implementations over a variety of embedding degrees. Before moving to the next section, we make the following remarks.

Remark 5. Since we are forced to accept the presence of type (i) mismatches in pairings, one possible solution to the problem described above would be to write a specialized multiplication routine for the type (ii) mismatched multiplication between the general element f and special element g, of $\mathbb{F}_{p^k}$. However, replacing the full $\mathbb{F}_{p^k}$ multiplication routine (that takes two general field elements as inputs) with such a specialized routine means that, to some extent, we are sacrificing the tricks that speed up general multiplications, such as the Karatsuba and Toom-Cook methods. Such optimizations are the reason we build extension fields up as towers of degree 2 and 3 sub-extensions, so we argue that avoiding these optimizations is potentially counterproductive, instead favoring the N-delay techniques herein.

Remark 6. The discussion in this section (and in the next) essentially describes the technique of *loop unrolling*, which was first introduced into pairing computations by Granger *et al.* [18], who merged iterations to exploit the sparsity of g. Speedups were achieved in [18] by combining two consecutive iterations into one merged iteration, in implementations over fields of characteristic three. This technique was later used by Shirase *et al.* [30] in pairing implementations over binary fields. To the best of the authors knowledge, this paper is the first to describe a general algorithm for loop unrolling which merges any number of iterations over large prime fields.

4   The Cost of N-delay

We let N-delay refer to the process of delaying the multiplication of the Miller function f by consecutive function updates g, N times in a row. We make note that $N = 0$ corresponds to the standard Miller routine which delays zero multiplications between f and g, whilst $N = 1$ corresponds to the Miller routine which delays one multiplication (combines two iterations), and so on, so that in general N can be thought of as the number of times a multiplication by f is delayed, whilst $N + 1$ is the number of iterations that are combined. The aim of this section is to obtain an expression for the computational cost of N-delay, in terms of N, so that we can determine the optimal N value for specific implementations. To do this, we determine the cost of delaying a single, but general iteration. That is, we write a general expression for the product of the n different powers of g's accumulated after n iterations, and use this to determine the cost of updating this to the $n + 1$-th product. We then sum this cost from $n = 0$ to $n = N - 1$ to obtain the entire cost of performing N-delay.

We let $G_n(x, y)$ be the cumulative product of the first n indeterminate $g(x, y)$ functions, reduced modulo the curve equation, as

$$G_n(x, y) = \sum_{i=0}^{A_n} a_i \cdot x^i \cdot y + \sum_{i=0}^{B_n} b_i \cdot x^i = G_{n_a}(x) \cdot y + G_{n_b}(x), \qquad (7)$$

where $G_{n_a}(x) = \sum_{i=0}^{A_n} a_i \cdot x^i$ and $G_{n_b}(x) = \sum_{i=0}^{B_n} b_i \cdot x^i$. Similarly, we define the n-th Miller function update from (4) as

$$g_n(x, y) = g_{n_x} \cdot x + g_{n_y} \cdot y + g_{n_0} = g_{n_a} \cdot y + g_{n_b}(x), \qquad (8)$$

where $g_{n_a} = g_{n_y}$ and $g_{n_b}(x) = g_{n_x} \cdot x + g_{n_0}$. The $(n+1)$-th consecutive Miller iteration would multiply the square of $G_n(x,y)$ by the $(n+1)$-th Miller function update, $g_{n+1}(x,y)$, as

$$
\begin{aligned}
G_{n+1}(x,y) = &G_n^2(x,y) \cdot g_{n+1}(x,y) \quad = \quad \left(G_{n_a}(x) \cdot y + G_{n_b}(x)\right)^2 \cdot g_{n+1}(x,y) \\
= &\left(G_{n_a}(x)^2 C(x) + 2 G_{n_a}(x) G_{n_b}(x) \cdot y + G_{n_b}(x)^2\right) \cdot g_{n+1}(x,y) \\
= &\left(g_{n+1_a} h_1(x) + g_{n+1_b}(x) h_2(x)\right) \cdot y + \left(g_{n+1_b}(x) h_1(x) + g_{n+1_a} h_3(x)\right) \\
= &\sum_{i=0}^{A_{(n+1)}} \hat{a}_i \cdot x^i \cdot y + \sum_{i=0}^{B_{(n+1)}} \hat{b}_i \cdot x^i = G_{n+1_a}(x) \cdot y + G_{n+1_b}(x), \quad (9)
\end{aligned}
$$

where $h_1(x) = G_{n_a}(x)^2 C(x) + G_{n_b}(x)^2$, $h_2(x) = 2 G_{n_a}(x) G_{n_b}(x)$, $h_3(x) = 2 G_{n_a}(x) G_{n_b}(x) C(x)$, and y^2 was replaced with $C(x) = x^3 + ax + b$. Paying close attention to (9) allows us to determine the cost of obtaining G_{n+1} from G_n. We make the following observations.

– **Observation 1.** To determine the values of A_{n+1} and B_{n+1}, (9) reveals that

$$
\begin{aligned}
A_{n+1} &= \text{Max}\{\deg(g_{n+1_a}) + \deg(h_1), \deg(g_{n+1_b}) + \deg(h_2)\} \\
&= \text{Max}\{2A_n + 3, 2B_n, A_n + B_n + 1\},
\end{aligned}
$$

and similarly

$$
\begin{aligned}
B_{n+1} &= \text{Max}\{\deg(g_{n+1_b}) + \deg(h_1), \deg(g_{n+1_a}) + \deg(h_3)\} \\
&= \text{Max}\{2A_n + 4, 2B_n + 1, A_n + B_n + 3\}.
\end{aligned}
$$

Since $(A_0, B_0) = (0, 1)$, we always have that $(A_{n+1}, B_{n+1}) = (2A_n + 3, 2A_n + 4)$, from which it follows that

$$
(A_n, B_n) = (3(2^n - 1), 3(2^n - 1) + 1). \quad (10)
$$

– **Observation 2.** The three necessary terms $G_{n_a}^2 = (\sum_{i=0}^{A_n} a_i \cdot x^i)^2$, $G_{n_b}^2 = (\sum_{i=0}^{B_n} b_i \cdot x^i)^2$ and $2 G_{n_a} G_{n_b} = 2(\sum_{i=0}^{A_n} a_i \cdot x^i)(\sum_{i=0}^{B_n} b_i \cdot x^i)$ can be computed using only field squarings as follows. Each of the a_i^2 terms in $G_{n_a}^2$ can be computed first and used to compute (via a squaring) the remaining terms of the form $2 a_i a_j$ in $G_{n_a}^2$, where $i \neq j$. In total, there are $\sum_{i=0}^{A_n} \sum_{j=0}^{i} = (A_n + 1)(A_n + 2)/2$ different $a_i a_j$ combinations contributing to $G_{n_a}^2$, so that $\text{cost}[G_{n_a}^2 \leftarrow G_{n_a}] = [(A_n + 1)(A_n + 2)/2]s_u$. Identically, we have that $\text{cost}[G_{n_b}^2 \leftarrow G_{n_b}] = [(B_n + 1)(B_n + 2)/2]s_u$. Lastly, each of the terms of the form $2 a_i b_j$ in $2 G_{n_a} G_{n_b}$ can be computed at the cost of a squaring using the previously computed a_i^2 and b_j^2 values. There are $(A_n + 1)(B_n + 1)$ such terms contributing to $2 G_{n_a} G_{n_b}$, so that $\text{cost}[2 G_{n_a} G_{n_b} \leftarrow \{G_{n_a}^2, G_{n_b}^2\}] = (A_n + 1)(B_n + 1)s_u$. Importantly, we use (10) to give

$$\text{cost}\big[\{G_{n_a}^2, 2G_{n_a}G_{n_b}, G_{n_b}^2\} \leftarrow \{G_{n_a}, G_{n_b}\}\big] = \text{cost}\big[G_{n_a}^2 \leftarrow G_{n_a}\big]$$
$$+\text{cost}\big[G_{n_b}^2 \leftarrow G_{n_b}\big] + \text{cost}\big[2G_{n_a}G_{n_b} \leftarrow \{G_{n_a}^2, G_{n_b}^2\}\big]$$
$$= \big[(A_n + 1)(A_n + 2)/2 + (A_n + 1)(B_n + 1) + (B_n + 1)(B_n + 2)/2\big]\mathbf{s}_u$$
$$= [3(3 \cdot 2^n - 1)(2^{n+1} - 1)\varOmega]\mathbf{m}_u. \tag{11}$$

— **Observation 3.** Aside from additions, computing the three polynomials h_1, h_2 and h_3 from $G_{n_a}^2$, $2G_{n_a}G_{n_b}$, and $G_{n_b}^2$ requires multiplications by C only. Since we are ignoring additions and assuming that multiplications by curve constants are negligible, we assume that there is no extra cost associated in these computations. That is,

$$\text{cost}\big[\{h_1, h_2, h_3\} \leftarrow \{G_{n_a}^2, 2G_{n_a}G_{n_b}, G_{n_b}^2\}\big] = 0. \tag{12}$$

— **Observation 4.** The cost of multiplying G_n^2 by g_{n+1} is determined by the cost of the required multiplications of the g_{n_a} and g_{n_b} values, and the polynomials h_1, h_2 and h_3. Since $g_{n_a} = g_{n_y} \in \mathbb{F}_{p^u}$, multiplying a d-degree polynomial by g_{n_a} requires $d + 1$ multiplications in $\mathbb{F}_{p^u}$, and since $g_{n_b} = g_{n_x} \cdot x + g_{n_0}$ has $g_{n_x}, g_{n_0} \in \mathbb{F}_{p^u}$, multiplying a d-degree polynomial by g_{n_b} requires $2(d + 1)$ $\mathbb{F}_{p^u}$-multiplications. There are four of these types of multiplications required in (9).

$$(i): \quad \text{cost}\big[g_{n+1_a} \cdot h_1 \leftarrow \{g_{n+1_a}, h_1\}\big] \quad = \quad (\deg(h_1) + 1)\mathbf{m}_u$$
$$(ii): \quad \text{cost}\big[g_{n+1_b} \cdot h_2 \leftarrow \{g_{n+1_b}, h_2\}\big] \quad = \quad 2(\deg(h_2) + 1)\mathbf{m}_u$$
$$(iii): \quad \text{cost}\big[g_{n+1_b} \cdot h_1 \leftarrow \{g_{n+1_b}, h_1\}\big] \quad = \quad 2(\deg(h_1) + 1)\mathbf{m}_u$$
$$(iv): \quad \text{cost}\big[g_{n+1_a} \cdot h_3 \leftarrow \{g_{n+1_a}, h_3\}\big] \quad = \quad (\deg(h_2) + 1)\mathbf{m}_u$$

In the case of (iv), since $h_3 = h_2 \cdot C$, we save 3 multiplications by multiplying g_{n+1_a} and h_2 prior to multiplying by C. Thus, the total cost of obtaining G_{n+1} given G_n^2 and g_{n+1} is the combined costs of (i), (ii), (iii) and (iv) above, which is

$$\text{cost}\big[G_{n+1} \leftarrow \{G_n^2, g_{n+1}\}\big] = (3 \cdot \deg(h_1) + 2 \cdot \deg(h_2) + \deg(h_3) + 6)\mathbf{m}_u$$
$$= (3(2A_n + 3) + 3(A_n + B_n) + 3)\mathbf{m}_u$$
$$= (9A_n + 3B_n + 15)\mathbf{m}_u = (36(2^n - 1) + 18)\mathbf{m}_u, \tag{13}$$

The cost of computing G_{n+1} from G_n. We now collect all of the costs calculated in (3), (11), (12) and (13) to determine the cost of computing G_{n+1} from G_n, as

$$\text{cost}\big[G_{n+1} \leftarrow G_n\big] = \text{cost}\big[G_{n+1} \leftarrow \{G_n^2, g_{n+1}\}\big] + \text{cost}\big[g_{n+1}\big] + \text{cost}\big[G_n^2 \leftarrow G_n\big]$$
$$= \text{cost}\big[G_{n+1} \leftarrow \{G_n^2, g_{n+1}\}\big] + \text{cost}\big[g_{n+1}\big] + \text{cost}\big[\{h_1, h_2, h_3\} \leftarrow \{G_{n_a}, G_{n_b}\}\big]$$
$$= \text{cost}\big[G_{n+1} \leftarrow \{G_n^2, g_{n+1}\}\big] + \text{cost}\big[g_{n+1}\big] + \text{cost}\big[\{h_1, h_2, h_3\}$$
$$\leftarrow \{G_{n_a}^2, 2G_{n_a}G_{n_b}, G_{n_b}^2\}\big] + \text{cost}\big[\{G_{n_a}^2, 2G_{n_a}G_{n_b}, G_{n_b}^2\} \leftarrow \{G_{n_a}, G_{n_b}\}\big]$$
$$= \big[(36(2^n - 1) + 18) + (m + \varOmega s) + 3(3 \cdot 2^n - 1)(2^{n+1} - 1)\varOmega\big]\mathbf{m}_u, \tag{14}$$

The total cost of N-delay. Display (14) allows us to determine the number of $\mathbb{F}_{p^u}$ multiplications required to compute $G_N(x, y)$ from scratch, as follows.

$$\text{cost}\big[G_N(x, y)\big] = \text{cost}\big[G_0\big] + \sum_{n=0}^{N-1} \text{cost}\big[G_{n+1} \leftarrow G_n\big]$$

$$= (m + s\Omega)\mathbf{m}_u + \sum_{n=0}^{N-1}(36(2^n - 1) + 21) + (m + \Omega s) + (18(2^n - 1) + 6)\Omega\mathbf{m}_u$$

$$= [(N + 1)(m + s\Omega) + 3N(\Omega - 6) + 3(2^N - 1)((2^{N+1} - 3)\Omega + 12)]\mathbf{m}_u. \quad (15)$$

We note that the above cost also incorporates the cost of transforming the point T into $[2^{N+1}]T$, as these costs are accounted for in the multiples of $(m + s\Omega)\mathbf{m}_u$ (see (3)). The other computations we need to consider in an iteration involving N-delay are those that occur when evaluating $G_N(x, y)$ at the point $S = (S_x, S_y)$. Setting $n = N$ into (7) reveals that N-delay will require the precomputation of the set $\{S_x^i, \ i = 1...B_N\}$, and the set $\{S_x^i \cdot S_y, \ i = 0...A_N\}$, each of which will be multiplied by an element in $\mathbb{F}_{p^u}$. From Remark 3, we have that such a multiplication costs $e\mathbf{m}_1$, and since there are $A_N + B_N + 1$ such elements, we have that

$$\text{cost}\big[G_N(S) \leftarrow G_N(x, y)\big] = [A_N + B_N + 1]e\mathbf{m}_1 = [6(2^N - 1) + 2]e\mathbf{m}_1. \quad (16)$$

We combine (15) and (16) to obtain the total cost of N-delay as

$$\text{cost}\big[G_N(S)\big] = \text{cost}\big[G_N(S) \leftarrow G_N(x, y)\big] + \text{cost}\big[G_N(x, y)\big]$$
$$= \big[6(2^N - 1) + 2\big]e\mathbf{m}_1 + \big[(N + 1)(m + s\Omega) + 3N(\Omega - 6)$$
$$+ 3(2^N - 1)((2^{N+1} - 3)\Omega + 12)\big]\mathbf{m}_u + (1 + (N + 1)\Omega)\mathbf{m}_k, \quad (17)$$

where the $(1 + (N + 1)\Omega)\mathbf{m}_k$ accounts for the $(N + 1)$ squarings of the Miller function f, as well as the full field multiplication of f with $G_N(S)$ that occurs after N-delay.

5   Implementing N-delay

The advantage of employing N-delay over the technique in [10] is the ease at which a standard implementation of Miller's algorithm can be updated to incorporate N-delay. The routines for the point doublings/additions and encapsulated line computations that are used in the standard version of Miller's algorithm are the same routines used in N-delay, so that this (existing) code is not altered when employing N-delay. We refer to these two standard subroutines as `MillerDBL`, which performs steps 3 and 4 in Algorithm 1, and `MillerADD`, which performs steps 8 and 9 in Algorithm 1, both of which are the same subroutines we call in Algorithm 2.

Since N-delay performs $N + 1$ squarings in the same iteration, we follow the algorithm description in [10] and write the loop parameter in base 2^{N+1}. Our goal is to incorporate N-delay by injecting a new subroutine into Algorithm 1, and slightly tweaking the original Miller code to account for this alteration. After calling MillerDBL, we call the new subroutine GetNewabArrays, which transforms G_n into G_{n+1}, based on equation (9) and the four observations that followed

---

Algorithm 2. Miller N-delay

Input: R, S, N, $m = (m_{l_N-1}...m_1, m_0)_{2^{N+1}}$, $f_{[w]R}$ for each unique non-zero $w \in \{m_{l_N-1}, ..., m_1, m_0\}$.
Output: $f_{m,R}(S) \leftarrow f$.

1: $T \leftarrow R$, $f \leftarrow 1$.
2: **if** $m_{l_N-1} \neq 1$ **then**
3:     $g_x, g_y, g_0, T] = \text{MillerADD}(T, [m_i]R)$.
4:     $f \leftarrow f \cdot f_{[m_i]R}(S)$.
5: **end if**
6: **for** $i = l_N - 2$ to 0 **do**
7:     Compute $[g_x, g_y, g_0, T] = \text{MillerDBL}(T)$.
8:     $\underline{a}(1) \leftarrow g_y$, $\underline{b}(1) \leftarrow g_0$, $\underline{b}(2) \leftarrow g_x$.
9:     $A_n \leftarrow 0$, $B_n \leftarrow 1$.
10:     **for** $i = 0$ to $N - 1$ by 1 **do**
11:         Compute $[g_x, g_y, g_0, T] = \text{MillerDBL}(T)$.
12:         Compute $\underline{a}, \underline{b}, A_n, B_n = \text{GetNewabArrays}(\underline{a}, \underline{b}, A_n, B_n, g_x, g_y, g_0)$.
13:     **end for**
14:     Evaluate $G = (\underline{a}, \underline{b})$ at S.
15:     $f \leftarrow f^{2^{N+1}} \cdot G$.
16:     **if** $m_i \neq 0$ **then**
17:         Compute $[g_x, g_y, g_0, T] = \text{MillerADD}(T, [m_i]R)$.
18:         $f \leftarrow f \cdot f_{[m_i]R} \cdot (g_x \cdot S_x + g_y \cdot S_y + g_0)$
19:     **end if**
20: **end for**
21: **return** f.

Since non-zero m_i that appear in $m = (m_{l-1}...m_1, m_0)_{2^{N+1}}$ can now take values up to $2^{N+1} - 1$, Algorithm 2 must account for the additions of $[m_i]R$ to the point T. We follow the technique in [10] and adjust the step accordingly, by including the precomputed function $f_{[m_i]R}$, with divisor $\text{div}(f_{[m_i]R}) = m_i(R) - ([m_i]R) - (m_i - 1)(\mathcal{O})$, into the addition product on line 18 of Algorithm 2.

6 Optimal N-delay

This section makes use of (17) to determine the value of N which gives the lowest operation count for all even embedding degrees less than $k = 50$. To obtain the m and s values described in (3) that are required in (17), we couple the recommendations for optimal curve construction in [16] with the fastest applicable explicit formulas for $D = 1, 3$ curves that admit high-degree twists in [12]. For curves admitting only quadratic twists, we opt for the CM

Table 1. Optimal N values for Miller-lite pairings on different embedding degrees $k \leq 50$

k	D	m, s	$\mathbb{F}_{p^u} \subseteq \mathbb{F}_{p^e} \subset \mathbb{F}_{p^k}$	$\Omega = 1$ (s = m)		$\Omega = 0.8$ (s = 0.8 m)	
				$N = 0$ count	Optimal N count	$N = 0$	Optimal N count
2	3	2, 7	$\mathbb{F}_p = \mathbb{F}_p \subset \mathbb{F}_{p^2}$	17	−	15	−
4	1	2, 8	$\mathbb{F}_p = \mathbb{F}_p \subset \mathbb{F}_{p^4}$	30	−	26.6	−
6	3	2, 7	$\mathbb{F}_p = \mathbb{F}_p \subset \mathbb{F}_{p^6}$	41	−	36.6	−
8	1	2, 8	$\mathbb{F}_p \subset \mathbb{F}_{p^2} \subset \mathbb{F}_{p^8}$	68	−	61	−
10	some	1, 11	$\mathbb{F}_p \subset \mathbb{F}_{p^5} \subset \mathbb{F}_{p^{10}}$	100	−	90	−
12	3	2, 7	$\mathbb{F}_p \subset \mathbb{F}_{p^2} \subset \mathbb{F}_{p^{12}}$	103	1 96.5	92.6	1 85.5
14	3	2, 7	$\mathbb{F}_p \subset \mathbb{F}_{p^7} \subset \mathbb{F}_{p^{14}}$	155	1 148	140.4	1 132.8
16	1	2, 8	$\mathbb{F}_p \subset \mathbb{F}_{p^4} \subset \mathbb{F}_{p^{16}}$	180	1 159.5	162.2	1 141.1
18	3	2, 7	$\mathbb{F}_p \subset \mathbb{F}_{p^3} \subset \mathbb{F}_{p^{18}}$	165	1 145.5	148.6	1 128.5
20	1	2, 8	$\mathbb{F}_p \subset \mathbb{F}_{p^{10}} \subset \mathbb{F}_{p^{20}}$	254	1 217.5	229	1 191.9
22	1	2, 8	$\mathbb{F}_p \subset \mathbb{F}_{p^{11}} \subset \mathbb{F}_{p^{22}}$	428	1 363	386.8	1 321.2
24	3	2, 7	$\mathbb{F}_p \subset \mathbb{F}_{p^4} \subset \mathbb{F}_{p^{24}}$	287	1 239.5	258.6	1 210.5
26	3	2, 7	$\mathbb{F}_p \subset \mathbb{F}_{p^{13}} \subset \mathbb{F}_{p^{26}}$	581	1 482.5	525	1 425.9
28	1	2, 8	$\mathbb{F}_p \subset \mathbb{F}_{p^7} \subset \mathbb{F}_{p^{28}}$	420	1 347	378.8	1 305.2
30	3	2, 7	$\mathbb{F}_p \subset \mathbb{F}_{p^{10}} \subset \mathbb{F}_{p^{30}}$	409	1 333.5	368.6	1 292.5
32	1	2, 8	$\mathbb{F}_p \subset \mathbb{F}_{p^8} \subset \mathbb{F}_{p^{32}}$	512	1 418.5	461.8	1 367.7
34	3	2, 7	$\mathbb{F}_p \subset \mathbb{F}_{p^{17}} \subset \mathbb{F}_{p^{34}}$	961	2 775.3	867.8	2 678.7
36	3	2, 7	$\mathbb{F}_p \subset \mathbb{F}_{p^6} \subset \mathbb{F}_{p^{36}}$	471	1 382.5	424.6	1 335.5
38	3	2, 7	$\mathbb{F}_p \subset \mathbb{F}_{p^{19}} \subset \mathbb{F}_{p^{38}}$	1187	2 936.7	1071.6	2 817.9
40	1	2, 8	$\mathbb{F}_p \subset \mathbb{F}_{p^{10}} \subset \mathbb{F}_{p^{40}}$	732	2 585.6	660.2	2 510.5
42	3	2, 7	$\mathbb{F}_p \subset \mathbb{F}_{p^7} \subset \mathbb{F}_{p^{42}}$	683	2 536.7	615.6	2 465.9
44	1	2, 8	$\mathbb{F}_p \subset \mathbb{F}_{p^{11}} \subset \mathbb{F}_{p^{44}}$	1220	2 916.3	1099.6	2 792.5
46	1	2, 8	$\mathbb{F}_p \subset \mathbb{F}_{p^{23}} \subset \mathbb{F}_{p^{46}}$	1712	2 1308.3	1544.8	2 1137.7
48	3	2, 7	$\mathbb{F}_p \subset \mathbb{F}_{p^8} \subset \mathbb{F}_{p^{48}}$	835	2 643.3	752.6	2 557.5
50	3	2, 7	$\mathbb{F}_p \subset \mathbb{F}_{p^{25}} \subset \mathbb{F}_{p^{50}}$	1073	1 881.5	970.2	1 778.1

discriminant D that facilitates the best ρ-value for the particular embedding degree (see [16]). If these curves do not have $D = 1$ or $D = 3$, we use the best operation count for general curves reported in [1] and [22]. For example, the maximal twist for $k = 10$ is a quadratic twist, and since such twists are admitted on any curve, we opt for Freeman's curves [15] with optimum $\rho = 1$, rather than the $D = 1$ or $D = 3$ curves that achieve $\rho = 1.5$. We report the optimal N values for both $\Omega = 0.8$ and $\Omega = 1$, although we make note that lesser values of Ω, such as those stated in Remark 2, would be more likely to favor higher values of N, since lower values of Ω give a greater weight to multiplications in the operation count, and $\mathbb{F}_{p^k}$-multiplications are what N-delay avoids. Since the operation count given by (17) is the total count (in terms of $\mathbb{F}_p$-multiplications) for the equivalent of $N + 1$ double-and-add iterations, the counts presented in Table 1 are given as counts equivalent to one iteration of 0-delay (the standard Miller routine in Algorithm 1), and these counts are obtained by dividing the cost in (17) by $N + 1$. We are reporting results for even embedding degrees that are not necessarily 3-smooth. Thus we must extend the standard method of reporting multiplications in fields of extension degree $k = 2^i 3^j$ as $\mathbf{m}_k = 3^i 5^j$ [23,20,12], this complexity being a result of coupling Karatsuba multiplication with Toom-Cook multiplication, the former allowing us to write $\mathbf{m}_{2c} = 3\mathbf{m}_c$, whilst the latter allows us to write $\mathbf{m}_{3c} = 5\mathbf{m}_c$. Montgomery [28] extended Karatsuba-like multiplication methods to polynomials (or extension degrees) of degrees 5, 6 and 7, achieving $\mathbf{m}_{5c} = 13\mathbf{m}_c$, $\mathbf{m}_{6c} = 17\mathbf{m}_c$ and $\mathbf{m}_{7c} = 22\mathbf{m}_c$ respectively. We note that the degree 6 result is of no use here, since it is more advantageous to build a six degree extension as a combination of quadratic and cubic extensions. For higher prime extension degrees, we use the more general result given by Weimerskirch and Paar [32], who generalize the Karatsuba algorithm to arbitrary w-degree extensions to give $\mathbf{m}_{wc} = [w(w+1)/2]\mathbf{m}_c$. The complexity of multiplications in the field of extension degree $k = 2^{e_2} 3^{e_3} 5^{e_5} 7^{e_7} \cdot \prod p_i^{e_{p_i}}$ are reported in terms of $\mathbb{F}_p$-multiplications as

$$\mathbf{m}_k = \Big[3^{e_2} 5^{e_3} 13^{e_5} 22^{e_7} \cdot \prod_{i=1}^{t} (p_i(p_i + 1)/2)^{e_i}\Big]\mathbf{m}_1, \tag{18}$$

where the p_i are the primes greater than 7 in the prime factorization of k. We use (18) to give a fair and relative comparison across all embedding degrees, not to overlook the substantial speed ups recently achieved by El Mrabet and Negre for particular extension degrees [14].

For Miller-full pairings, $N = 0$ was optimal across all embedding degrees, so we do not report the results here (the standard $N = 0$ operation counts in the Miller-full setting for 3-smooth embedding degrees can be found in [12]). Table 1 shows that Miller-lite pairings on curves with even embedding degrees greater than $k = 10$ will always benefit from N-delay. Although $N > 2$ is never optimal, it is still interesting to see that $N = 2$ is optimal in many instances. Consider equation (6) which showed that, even after one delayed iterate, the product of g's becomes a general field element, i.e. $\#G_1(\mathbb{F}_{p^e}) = k/e$ (refer to Definition 1). Any further delay that occurs after the initial delay will actually involve squaring both f and G_1 (or G_n for $n > 1$) separately, which are both general elements of $\mathbb{F}_{p^k}$, prior to multiplying them. One might intuitively guess that multiplying f and G_1 prior to performing the squaring might be preferred, but clearly this is not the case for embedding degrees where $N = 2$ is optimal. In the case of quadratic twists, this preferred delay is even more surprising since the function updates g are already general elements of $\mathbb{F}_{p^k}$. In agreement with [10], it becomes clear that for Miller-lite pairings where the difference between the fields $\mathbb{F}_{p^u} = \mathbb{F}_p$ and $\mathbb{F}_{p^k}$ is larger than in Miller-full pairings where $\mathbb{F}_{p^u} = \mathbb{F}_{p^e}$, it can be very advantageous to spend many extra computations in $\mathbb{F}_p$ in order to delay one single (and most costly) $\mathbb{F}_{p^k}$-multiplication between f and G.

References

1. Arene, C., Lange, T., Naehrig, M., Ritzenthaler, C.: Faster pairing computation. Cryptology ePrint Archive, Report 2009/155 (2009), http://eprint.iacr.org/
2. Barreto, P.S.L.M., Kim, H.Y., Lynn, B., Scott, M.: Efficient algorithms for pairing-based cryptosystems. In: Yung, M. (ed.) CRYPTO 2002. LNCS, vol. 2442, pp. 354–368. Springer, Heidelberg (2002)
3. Barreto, P.S.L.M., Lynn, B., Scott, M.: Efficient implementation of pairing-based cryptosystems. J. Cryptology 17(4), 321–334 (2004)
4. Barreto, P.S.L.M., Naehrig, M.: Pairing-friendly elliptic curves of prime order. In: Preneel, B., Tavares, S.E. (eds.) SAC 2005. LNCS, vol. 3897, pp. 319–331. Springer, Heidelberg (2006)
5. Benger, N., Scott, M.: Constructing tower extensions for the implementation of pairing-based cryptography. In: Hasan, M.A., Helleseth, T. (eds.) WAIFI 2010. LNCS, vol. 6087, pp. 180–195. Springer, Heidelberg (2010)
6. Bernstein, D.J.: Curve25519: New diffie-hellman speed records. In: Yung, M., Dodis, Y., Kiayias, A., Malkin, T.G. (eds.) PKC 2006. LNCS, vol. 3958, pp. 207–228. Springer, Heidelberg (2006)
7. Bernstein, D.J., Lange, T.: Explicit-formulas database, http://www.hyperelliptic.org/EFD
8. Boneh, D., Franklin, M.K.: Identity-based encryption from the Weil pairing. In: Kilian, J. (ed.) CRYPTO 2001. LNCS, vol. 2139, pp. 213–229. Springer, Heidelberg (2001)
9. Chatterjee, S., Sarkar, P., Barua, R.: Efficient computation of Tate pairing in projective coordinate over general characteristic fields. In: Park, C.-s., Chee, S. (eds.) ICISC 2004. LNCS, vol. 3506, pp. 168–181. Springer, Heidelberg (2005)
10. Costello, C., Boyd, C., Nieto, J.M.G., Wong, K.K.-H.: Avoiding full extension field arithmetic in pairing computations. In: Bernstein, D.J., Lange, T. (eds.) AFRICACRYPT 2010. LNCS, vol. 6055, pp. 203–224. Springer, Heidelberg (2010)
11. Costello, C., Hisil, H., Boyd, C., Nieto, J.M.G., Wong, K.K.-H.: Faster pairings on special weierstrass curves. In: Shacham, H. (ed.) Pairing 2009. LNCS, vol. 5671, pp. 89–101. Springer, Heidelberg (2009)
12. Costello, C., Lange, T., Naehrig, M.: Faster pairing computations on curves with high-degree twists. In: Nguyen, P.Q., Pointcheval, D. (eds.) PKC 2010. LNCS, vol. 6056, pp. 209–223. Springer, Heidelberg (2010)
13. Prem Laxman Das, M., Sarkar, P.: Pairing computation on twisted Edwards form elliptic curves. In: Galbraith, Paterson (eds.) [17], pp. 192–210 (2008)
14. El Mrabet, N., Negre, C.: Finite field multiplication combining AMNS and DFT approach for pairing cryptography. In: Boyd, C., González Nieto, J. (eds.) ACISP 2009. LNCS, vol. 5594, pp. 422–436. Springer, Heidelberg (2009)
15. Freeman, D.: Constructing pairing-friendly elliptic curves with embedding degree 10. In: Hess, F., Pauli, S., Pohst, M. (eds.) ANTS 2006. LNCS, vol. 4076, pp. 452–465. Springer, Heidelberg (2006)
16. Freeman, D., Scott, M., Teske, E.: A taxonomy of pairing-friendly elliptic curves. J. Cryptology 23(2), 224–280 (2010)
17. Galbraith, S.D., Paterson, K.G. (eds.): Pairing 2008. LNCS, vol. 5209. Springer, Heidelberg (2008)
18. Granger, R., Page, D., Stam, M.: On small characteristic algebraic tori in pairing-based cryptography. LMS J. Comput. Math. 9, 64–85 (2006)
19. Hess, F.: Pairing lattices. In: Galbraith, Paterson (eds.) [17], pp. 18–38

20. Hess, F., Smart, N.P., Vercauteren, F.: The eta pairing revisited. IEEE Transactions on Information Theory 52(10), 4595–4602 (2006)

21. Hisil, H.: Elliptic Curves, Group Law, and Efficient Computation. PhD thesis, Queensland University of Technology (2010)

22. Ionica, S., Joux, A.: Another approach to pairing computation in Edwards coordinates. In: Chowdhury, D.R., Rijmen, V., Das, A. (eds.) INDOCRYPT 2008. LNCS, vol. 5365, pp. 400–413. Springer, Heidelberg (2008), http://eprint.iacr.org/2008/292

23. Koblitz, N., Menezes, A.: Pairing-based cryptography at high security levels. In: Smart, N.P. (ed.) Cryptography and Coding 2005. LNCS, vol. 3796, pp. 13–36. Springer, Heidelberg (2005)

24. Lee, E., Lee, H.-S., Park, C.-M.: Efficient and generalized pairing computation on abelian varieties. IEEE Transactions on Information Theory 55(4), 1793–1803 (2009)

25. Lin, X., Zhao, C., Zhang, F., Wang, Y.: Computing the ate pairing on elliptic curves with embedding degree $k = 9$. IEICE Transactions 91-A(9), 2387–2393 (2008)

26. Matsuda, S., Kanayama, N., Hess, F., Okamoto, E.: Optimised versions of the ate and twisted ate pairings. In: Galbraith, S.D. (ed.) Cryptography and Coding 2007. LNCS, vol. 4887, pp. 302–312. Springer, Heidelberg (2007)

27. Miller, V.S.: The Weil pairing, and its efficient calculation. Journal of Cryptology 17, 235–261 (2004)

28. Montgomery, P.L.: Five, six, and seven-term Karatsuba-like formulae. IEEE Trans. Computers 54(3), 362–369 (2005)

29. Scott, M., Barreto, P.S.L.M.: Compressed pairings. In: Franklin, M. (ed.) CRYPTO 2004. LNCS, vol. 3152, pp. 140–156. Springer, Heidelberg (2004)

30. Shirase, M., Takagi, T., Choi, D., Han, D.G., Kim, H.: Efficient computation of Eta pairing over binary field with Vandermonde matrix. ETRI journal 31(2), 129–139 (2009)

31. Vercauteren, F.: Optimal pairings. IEEE Transactions on Information Theory 56(1), 455–461 (2010)

32. Weimerskirch, A., Paar, C.: Generalizations of the Karatsuba algorithm for efficient implementations. Cryptology ePrint Archive, Report 2006/224 (2006), http://eprint.iacr.org/

Regenerating Codes for Distributed Storage Networks

Nihar B. Shah[1], K.V. Rashmi[1], P. Vijay Kumar[1], and Kannan Ramchandran[2]

[1] Indian Institute of Science, Bangalore
[2] University of California, Berkeley
{nihar,rashmikv,vijay}@ece.iisc.ernet.in,
kannanr@eecs.berkeley.edu

Abstract. In a storage system where individual storage nodes are prone to failure, the redundant storage of data in a distributed manner across multiple nodes is a must to ensure reliability. Reed-Solomon codes possess the reconstruction property under which the stored data can be recovered by connecting to any k of the n nodes in the network across which data is dispersed. This property can be shown to lead to vastly improved network reliability over simple replication schemes. Also of interest in such storage systems is the minimization of the repair bandwidth, i.e., the amount of data needed to be downloaded from the network in order to repair a single failed node. Reed-Solomon codes perform poorly here as they require the entire data to be downloaded. Regenerating codes are a new class of codes which minimize the repair bandwidth while retaining the reconstruction property. This paper provides an overview of regenerating codes including a discussion on the explicit construction of optimum codes.

Keywords: Distributed storage, MDS codes, Regenerating codes, Repair bandwidth, Interference alignment.

1 Introduction

In a distributed storage network, information pertaining to a single file of size B symbols is dispersed across nodes in the network in such a manner that an end-user can retrieve the data stored by tapping into neighboring nodes. Under the simplest option, all the data is stored at a single node. This however makes the network highly vulnerable to single-node failures and also increases congestion of the links surrounding the particular node. A second option is to replicate the data across ℓ nodes in the network. Such a system can tolerate $\ell - 1$ failures but needs to store ℓB amount of data. If p is the probability of a single node failure, then the probability of data loss is p^ℓ. The links surrounding the servers may still be congested. A third option is one of using erasure codes. Here, data is split into k fragments, and a maximum distance separable (MDS) code, such as a Reed-Solomon code, is used to generate n fragments which are stored across n nodes in the network. Each fragment represents a single symbol from the finite

M.A. Hasan and T. Helleseth (Eds.): WAIFI 2010, LNCS 6087, pp. 215–223, 2010.

field $\mathbb{F}_q$. The properties of an MDS code permit an end-user to retrieve the data by connecting to any k nodes. Thus the system can tolerate $n - k$ node failures and the total amount of data stored is nB/k. The probability of data loss is given by

$$\sum_{r=n-k+1}^{n} \binom{n}{r} p^r (1 - p)^{n-r}, \tag{1}$$

which can be several orders of magnitude lower than the corresponding probability under the replication option. Also, the links surrounding the storage nodes are less likely to be congested as the data is dispersed to a greater extent across the network.

More formally, let the total amount of data to be stored be B symbols from a finite field $\mathbb{F}_q$ of size q. Data is to be stored across n nodes in the network, such that the entire data can be recovered by connecting to any k nodes. This process of recovering the data from any k nodes is termed as *reconstruction* and is depicted in Figure 1a.

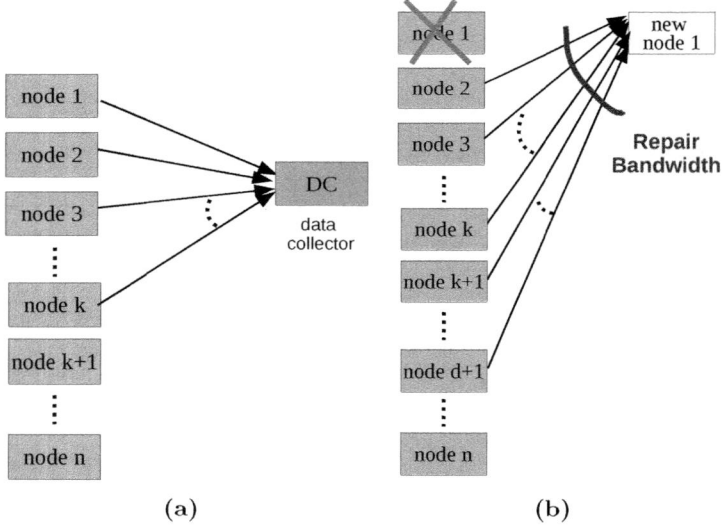

(a)   (b)

Fig. 1. The two main components of a distributed storage system: (a) Reconstruction by a DC and (b) Regeneration of a failed node

Upon failure of an individual node, a self-sustaining data storage network must naturally possess the ability to regenerate (i.e., repair) a failed node. An obvious means to accomplish this is to connect to any k nodes, download the entire data, and extract the data that was stored in the failed node. Figure 2 depicts node regeneration in the RAID 6 storage system which uses a [4, 2]-MDS code. But downloading the entire B units of data to recover the data stored in a single node (which is of the order of B/k) is wasteful and raises the question as

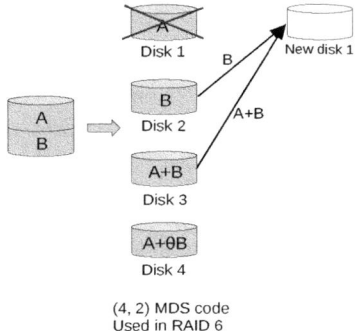

Fig. 2. RAID-6 system with $n = 4$, $k = 2$

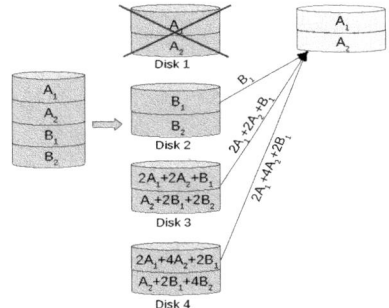

Fig. 3. Repair of a failed node while downloading less than B units

to whether there is a better option. Such an option is provided by the concept of a *regenerating codes*[1].

Traditional erasure codes treat each fragment as a single symbol over the finite field $\mathbb{F}_q$. When individual nodes are only permitted to perform linear operations over $\mathbb{F}_q$, the total amount of data download needed to repair a failed node can be no smaller than the size B of the entire file. Regenerating codes treat the data stored in each node as a vector over a smaller field (which we will however, also denote by $\mathbb{F}_q$). Linear operation over $\mathbb{F}_q$ in this case permits the transfer of a fraction of the data stored at a particular node for the purposes of node repair. This is illustrated in Figure 3 which shows an example regenerating code. The process of repairing a failed node by downloading data from the existing nodes is termed as **regeneration** and the total amount of data download needed for repair is termed the **repair bandwidth**.

We now introduce some parameters (see [1]) associated with a regenerating code. We use α to denote the number of symbols over $\mathbb{F}_q$ stored in each node. Clearly, for reconstruction from any k nodes, one needs

$$\alpha \geq B/k \ . \tag{2}$$

Upon failure of a node, the new node replacing it is permitted to connect to any d existing nodes, while downloading β symbols over $\mathbb{F}_q$ from each of them, making the repair bandwidth equal to $d\beta$. Note that the regeneration is only *functional* in the sense that the only constraint needed to be satisfied is that the new node along with the existing nodes, possess the reconstruction and regeneration properties. This is as opposed to *exact* regeneration, under which duplication of data stored in the failed node is required.

The distributed storage network evolves through successive regeneration over time and this is graphically depicted in Figure 4. In the graph, each storage node with storage capacity α is represented using two nodes: an 'in' node and and an 'out' node with an edge of capacity α linking the two. All edges coming into the node arrive into 'in' and all outgoing edges emanate from 'out'. Data collectors (DC) are represented as sinks connecting to some subset of k nodes in the network.

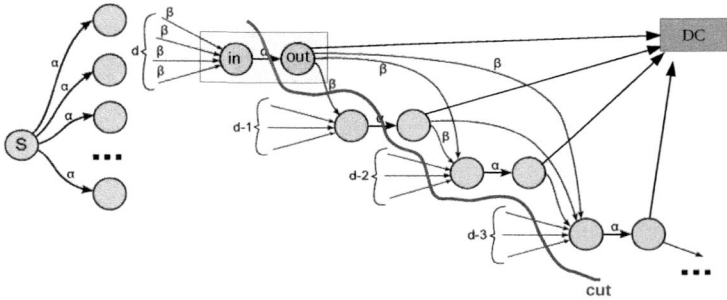

Fig. 4. Illustrating a cut in the network when the sink is a data collector and the system has undergone several rounds of regeneration (from [2])

The max-flow, min-cut bound of network coding can be shown to yield the lower bound:

$$B \leq \sum_{i=0}^{k-1} \min\{\alpha, (d-i)\beta\}. \tag{3}$$

Our interest is in minimizing both the amount of data stored per node as well as the total repair bandwidth, which corresponds respectively to minimizing α and $d\beta$, or equivalently minimizing α and β. This results in a tradeoff between α and β called the 'storage-repair bandwidth tradeoff'. By minimizing first α and then β, one obtains the minimum storage regeneration (MSR) point which is one extreme point of the tradeoff. The MSR point is characterized by

$$\alpha_{\min} = \frac{B}{k} \tag{4}$$

$$d\beta_{\min} = \frac{B}{k(1 - \frac{k-1}{d})} . \tag{5}$$

The other extreme point is the minimum bandwidth regeneration (MBR) point, which is obtained by first minimizing β and then minimizing α. This point is characterized by

$$d\beta_{\min} = \frac{dB}{\sum_{i=0}^{k-1}(d-i)} = \frac{B}{k(1 - \frac{k-1}{2d})} \tag{6}$$

$$\alpha_{\min} = d\beta_{\min} = \frac{B}{k(1 - \frac{k-1}{2d})}. \tag{7}$$

From this it can be seen that regenerating codes potentially offer a savings in repair bandwidth by a factor of k. We illustrate with an example. Let $n = 35$, $k = 5$, $d = 34$, $B = 150$ tera bytes (TB) and $\alpha = 30$ TB. Both RS and regenerating codes permit the entire file to be recovered by connecting to any $k = 5$ nodes. However while the repair bandwidth required by RS codes equals 150 TB, the use of regenerating codes brings this down to to just 34 TB, thereby reducing the repair bandwidth by an approximate factor of $k = 5$.

The authors in [2] prove the existence of codes achieving the tradeoff curve. No explicit constructions however, are given. In our recent work [3–5] we provide explicit code constructions at the MSR and MBR points for various set of parameters. Some of the explicit codes provided are systematic i.e., a set of k out of the n nodes store all source symbols in uncoded form. A data collector connecting to these k systematic nodes can reconstruct the data without need for any further processing of the downloaded symbols. To maintain the systematic nature of the code, we need to *exactly* regenerate a failed node, i.e., the new node replacing the failed node should store data identical to what was stored in the failed node.

In all our constructions, we assume $\beta = 1$ without loss of optimality. This choice gives the code a high degree of flexibility while simultaneously reducing the field size q as well as the complexity of encoding and decoding the code.

2   Explicit Regenerating Code Constructions

Our recent code constructions [3–5] applicable to a variety of situations are are tabulated in Table 1.

In this section, we describe as an example, one of the code constructions listed in Table 1 which operates at an MSR point. At an MSR point, the code stores the minimum possible amount of data possible while enabling reconstruction, which forces the code to be MDS (over a vector alphabet). In addition, the code must be constructed in such a way so as to also minimize the repair bandwidth.

The code construction is linear over $\mathbb{F}_q$, i.e. any symbol stored is a linear combination of the source symbols, and only linear operations at any node are permitted. Define a column vector $\underline{z}$ of length B consisting of the source symbols. Each source symbol can independently take values from $\mathbb{F}_q$. Hence, the B source symbols can be thought of as forming a B-dimensional vector space over $\mathbb{F}_q$.

Since the code is linear, any stored symbol can be written as $\underline{\ell}^t \underline{z}$ for some column vector $\underline{\ell}$. This vector $\underline{\ell}^t$ is termed the *global kernel* associated with that

Table 1. Explicit regenerating code constructions

Operating point	System parameters	Type of regeneration	Minimum field size required
MSR	$d \geq 2k - 1$	Exact regeneration of systematic nodes	$n + d - 2k + 1$
MSR	$d = k + 1$	Approximately exact regeneration of all nodes	n
MBR	$d = n - 1$	Exact regeneration of all nodes	$\binom{n}{2}$
MBR	All	Exact regeneration of systematic nodes	$n - k + d$

symbol, and it is these global kernels that define the code; the actual symbols stored depend on the instantiation of $\underline{z}$. Since a node stores α symbols, it can be considered as storing α vectors of the code, and hence can be represented by a $\alpha \times B$ matrix. We will say that the node *stores* this matrix. Denote the matrix stored in node m as $\mathbf{G}^{(m)}$.

For the regeneration of a failed systematic node, d other nodes contribute a single symbol each over $\mathbb{F}_q$ (since $\beta = 1$ in our constructions). We say that each node *passes a vector* for the regeneration of the failed node. The row vector passed will represent the global kernel of the symbol passed.

One of the key concepts driving code construction is that of *interference alignment*, a concept originally introduced in [6] in the context of communication over wireless channels. There are significant differences though in the manner in which this principle is applied in storage channels.

An Example Code Construction. Since the intent here is to provide a flavor of code construction rather than provide all details, we confine the discussion below to an example that captures all features of the particular method of construction. The general code construction holds for all $d \geq 2k - 1$, and is available in [4]. In the example, we select the parameter set $n = 6$, $d = 5$, $k = 3$ and $B = 9$. Setting $\beta = 1$ we obtain $\alpha = 3$ for the MSR point. We assume that all symbols lie in $\mathbb{F}_7$.

Each node stores a 3×9 matrix. Let the first three nodes be systematic. Hence,

$$
\begin{aligned}
\mathbf{G}^{(1)} &= [I_3 \ 0_3 \ 0_3] \\
\mathbf{G}^{(2)} &= [0_3 \ I_3 \ 0_3] \\
\mathbf{G}^{(3)} &= [0_3 \ 0_3 \ I_3].
\end{aligned} \tag{8}
$$

Thus the first three columns of any global kernel are termed as the *component along systematic node* 1, the next three columns as the component along node 2, and the last three columns as the component along node 3.

Let $\Psi_3 = \begin{bmatrix} \psi_1^{(4)} & \psi_2^{(4)} & \psi_3^{(4)} \\ \psi_1^{(5)} & \psi_2^{(5)} & \psi_3^{(5)} \\ \psi_1^{(6)} & \psi_2^{(6)} & \psi_3^{(6)} \end{bmatrix}$ be a 3×3 Cauchy matrix. A Cauchy matrix has the property that all of its sub-matrices are of full rank. The three non-systematic nodes store the matrices $\mathbf{G}^{(m)}$, $m = 4, 5, 6$, given by

$$\mathbf{G}^{(m)} = \begin{bmatrix} 2\psi_1^{(m)} & 2\psi_2^{(m)} & 2\psi_3^{(m)} & \psi_2^{(m)} & 0 & 0 & \psi_3^{(m)} & 0 & 0 \\ 0 & \psi_1^{(m)} & 0 & 2\psi_1^{(m)} & 2\psi_2^{(m)} & 2\psi_3^{(m)} & 0 & \psi_3^{(m)} & 0 \\ 0 & 0 & \psi_1^{(m)} & 0 & 0 & \psi_2^{(m)} & 2\psi_1^{(m)} & 2\psi_2^{(m)} & 2\psi_3^{(m)} \end{bmatrix} \qquad (9)$$

Regeneration. When a systematic node fails, the new node replacing it downloads one symbol each from all the existing nodes. Since the other two systematic nodes cannot pass any useful information, it is the three vectors from the non-systematic nodes that have to provide all the information about the failed node. In other words, the components along the failed node in the three vectors passed by the non-systematic should be linearly independent. For the regeneration of systematic node l ($\in \{1, 2, 3\}$), each non-systematic node passes its lth row, and we can see that the components along the failed node form the Cauchy matrix Ψ_3 which is non-singular.

Although the other systematic nodes do not contribute any useful information, they perform a very important task – that of interference cancelation. The vectors passed by the non-systematic nodes have non-zero components along the other systematic nodes which need to be removed. These components are termed as *interference*. The interference along any other systematic node can be canceled only by the vector passed by that systematic node. Since each systematic node passes only one vector, the components along a particular existing systematic node in the vectors passed by the non-systematic nodes should be aligned in a single dimension. This is *interference alignment* in the context of regenerating codes.

Our code construction satisfies the above interference alignment conditions. For example, consider regeneration of systematic node 1. Each non-systematic node passes its first row. The first rows of $\mathbf{G}^{(m)}$, $m = 4, 5, 6$, have components along the systematic nodes (nodes 2 and 3) aligned in the direction [1 0 0]. Now, the second and third systematic nodes pass [0 0 0 1 0 0 0 0 0], and [0 0 0 0 0 0 1 0 0] respectively and cancel out the interference leaving behind the matrix [Ψ_3 0_3 0_3]. Since Ψ_3 is invertible, systematic node 1 can be exactly regenerated.

Reconstruction. For reconstruction of the data, the DC can connect to any three nodes in the system. For reconstruction to be possible, the 9×9 matrix formed by juxtaposing the node matrices of these three nodes one below the other should be non-singular.

Reconstruction is trivially satisfied when the data collector connects to all the three systematic nodes. Now consider the case of a DC connecting to the three non-systematic nodes. Let C_1 be the matrix formed by juxtaposing the matrices stored in these three nodes one below the other.

Theorem 1. *The matrix C_1 is full rank.*

Proof. In C_1, group the ith ($i = 1, 2, 3$) rows of all the three nodes together to obtain the matrix C_2. Thus,

$$
C_2 = \begin{bmatrix}
2\psi_1^{(4)} & 2\psi_2^{(4)} & 2\psi_3^{(4)} & \psi_2^{(4)} & 0 & 0 & \psi_3^{(4)} & 0 & 0 \\
2\psi_1^{(5)} & 2\psi_2^{(5)} & 2\psi_3^{(5)} & \psi_2^{(5)} & 0 & 0 & \psi_3^{(5)} & 0 & 0 \\
2\psi_1^{(6)} & 2\psi_2^{(6)} & 2\psi_3^{(6)} & \psi_2^{(6)} & 0 & 0 & \psi_3^{(6)} & 0 & 0 \\
0 & \psi_1^{(4)} & 0 & 2\psi_1^{(4)} & 2\psi_2^{(4)} & 2\psi_3^{(4)} & 0 & \psi_3^{(4)} & 0 \\
0 & \psi_1^{(5)} & 0 & 2\psi_1^{(5)} & 2\psi_2^{(5)} & 2\psi_3^{(5)} & 0 & \psi_3^{(5)} & 0 \\
0 & \psi_1^{(6)} & 0 & 2\psi_1^{(6)} & 2\psi_2^{(6)} & 2\psi_3^{(6)} & 0 & \psi_3^{(6)} & 0 \\
0 & 0 & \psi_1^{(4)} & 0 & 0 & \psi_2^{(4)} & 2\psi_1^{(4)} & 2\psi_2^{(4)} & 2\psi_3^{(4)} \\
0 & 0 & \psi_1^{(5)} & 0 & 0 & \psi_2^{(5)} & 2\psi_1^{(5)} & 2\psi_2^{(5)} & 2\psi_3^{(5)} \\
0 & 0 & \psi_1^{(6)} & 0 & 0 & \psi_2^{(6)} & 2\psi_1^{(6)} & 2\psi_2^{(6)} & 2\psi_3^{(6)}
\end{bmatrix}
\tag{10}
$$

Multiply the 3 groups of 3 rows each by Ψ_3^{-1} to get a matrix C_3 given by

$$
C_3 = \begin{bmatrix}
\Psi_3^{-1} & 0_3 & 0_3 \\
0_3 & \Psi_3^{-1} & 0_3 \\
0_3 & 0_3 & \Psi_3^{-1}
\end{bmatrix} C_2
\tag{11}
$$

$$
= \begin{bmatrix}
2 & 0 & 0 & 0 & 0 & 0 & 0 & 0 & 0 \\
0 & 2 & 0 & 1 & 0 & 0 & 0 & 0 & 0 \\
0 & 0 & 2 & 0 & 0 & 0 & 1 & 0 & 0 \\
0 & 1 & 0 & 2 & 0 & 0 & 0 & 0 & 0 \\
0 & 0 & 0 & 0 & 2 & 0 & 0 & 0 & 0 \\
0 & 0 & 0 & 0 & 0 & 2 & 0 & 1 & 0 \\
0 & 0 & 1 & 0 & 0 & 0 & 2 & 0 & 0 \\
0 & 0 & 0 & 0 & 0 & 1 & 0 & 2 & 0 \\
0 & 0 & 0 & 0 & 0 & 0 & 0 & 0 & 2
\end{bmatrix}
\tag{12}
$$

Further elementary row operations can be used to show that this matrix is full rank over the field $\mathbb{F}_7$, and hence the data collector can recover all the source data.   ∎

Similarly, when the data collector connects to a combination of systematic and non-systematic nodes, it can be shown that the resultant 9×9 matrix is invertible, which enables the data collector to recover all the source data.

References

1. Dimakis, A.G., Godfrey, P.B., Wainwright, M.J., Ramchandran, K.: Network Coding for distributed storage systems. In: Proc. IEEE INFOCOM (May 2007)
2. Wu, Y., Dimakis, A.G., Ramchandran, K.: Deterministic Regenerating codes for Distributed Storage. In: Allerton Conference on Control, Computing and Communication, Urbana-Champaign (September 2007)

3. Rashmi, K.V., Shah, N.B., Kumar, P.V., Ramchandran, K.: Explicit construction of optimal exact regenerating codes for distributed storage. In: Proc. Allerton Conference on Control, Computing and Communication, Urbana-Champaign (September 2009)
4. Shah, N.B., Rashmi, K.V., Kumar, P.V., Ramchandran, K.: Explicit codes minimizing repair bandwidth for distributed storage. In: Proc. IEEE Information Theory Workshop, Cairo (January 2010)
5. Rashmi, K.V., Shah, N.B., Kumar, P.V., Ramchandran, K.: Explicit and Optimal Exact-Regenerating Codes for the Minimum-Bandwidth Point in Distributed Storage. In: Proc. IEEE International Symposium on Information Theory, Austin (June 2010)
6. Cadambe, V.R., Jafar, S.A.: Interference alignment and the degree of freedom for the K user interference channel. IEEE Transactions on Information Theory 54(8), 3425–3441 (2008)

On Rationality of the Intersection Points of a Line with a Plane Quartic

Roger Oyono[1] and Christophe Ritzenthaler[2],[*]

[1] Équipe GAATI, Université de la Polynésie Française, BP 6570 - 98702 Faa'a - Tahiti - Polynésie française
roger.oyono@upf.pf
[2] Institut de Mathématiques de Luminy, UMR 6206 du CNRS, Luminy, Case 907, 13288 Marseille, France
ritzenth@iml.univ-mrs.fr

Abstract. We study the rationality of the intersection points of certain lines and smooth plane quartics C defined over $\mathbb{F}_q$. For $q \geq 127$, we prove the existence of a line such that the intersection points with C are all rational. Using another approach, we further prove the existence of a tangent line with the same property as soon as $\operatorname{char} \mathbb{F}_q \neq 2$ and $q \geq 66^2 + 1$. Finally, we study the probability of the existence of a rational flex on C and exhibit a curious behavior when $\operatorname{char} \mathbb{F}_q = 3$.

Keywords: smooth plane quartics, rationality, intersection points, tangent line, flex.

Subject Classifications: 11G20, 14G05, 14G15, 14H45, 14N10.

1 Introduction

In computational arithmetic geometry, it is an important task to develop an efficient group law for the Jacobian variety of algebraic curves defined over a finite field. One of the most important and recent application of such efficient arithmetic comes from cryptography [18,19,4]. In [7], the authors introduced an efficient algorithm to perform arithmetic in the Jacobian of smooth plane quartics. The presented algorithm depends on the existence of a rational line l intersecting the quartic in rational points only. Moreover, the more special l is (for instance, tangent, tangent at a flex,...), the better is the complexity of the algorithm. Motivated by the above efficiency argument, we prove here the following theorems.

Theorem 1. *Let C be a smooth plane quartic over the finite field $\mathbb{F}_q$ with q elements. If $q \geq 127$, then there exists a line l which intersects C at rational points only.*

[*] The second author acknowledges the financial support of the grant MTM2006-11391 from the Spanish MEC and of the grant ANR-09-BLAN-0020-01 from the French ANR.

M.A. Hasan and T. Helleseth (Eds.): WAIFI 2010, LNCS 6087, pp. 224–237, 2010.

Theorem 2. *Let C be a smooth plane quartic over $\mathbb{F}_q$ and assume that char $\mathbb{F}_q \neq 2$. If $q \geq 66^2 + 1$, then there exists a tangent to C which intersects C at rational points only.*

In [7], the authors gave heuristic arguments and computational evidences that the probability for a plane smooth quartic over a finite field to have a rational flex is about 0.63. In this article, we present a 'proof' which depends on a conjectural analogue over finite fields of a result of [13] on the Galois group G of the 24 flexes of a general quartic. Harris proved that, over $\mathbb{C}$, the group G is as big as possible, namely the symmetric group S_{24}. Unfortunately, Harris's proof uses monodromy arguments which cannot be adapted so easily in positive characteristic. Worse, we surprisingly found that Harris's result is not valid over fields of characteristic 3 and we prove that for this field the Galois group is S_8. This is a consequence of the peculiar fact that, in characteristic 3, a smooth plane quartic C has generically only 8 flexes (with multiplicity 3), which belong to a conic. We suspect that this is the only exceptional case.

The methods used for these three problems are various and can be generalized or adapted to other questions. This was our principal motivation to write down our approaches in the case of quartics. It illustrates also the very unusual behavior of special points and lines in small characteristics.

Coming back to our initial motivation, it appears nowadays unlikely (due to recent progress in index calculus [5]) that smooth plane quartics may be used for building discrete logarithm cryptosystems. However, it is interesting to mention that the complexity analysis of the index calculus attack of [5] uses an asymptotic bound for the number of lines intersecting a smooth plane quartic in four distinct rational points, in the spirit of Theorem 1.

The paper is organized as follows: Section 2 gives a brief overview on the possible geometric intersections of a line and a smooth plane quartic. In Section 3, we give a proof of Theorem 1 using Chebotarev density theorem for covers of curves. In Section 4, we prove Theorem 2 using the tangential correspondence and its associated curve $X_C \subset C \times C$. The crucial point is to prove that X_C is geometrically irreducible in order to apply Hasse-Weil bound for (possibly singular) geometrically irreducible curves. Finally, in Section 5, we study the possible generalization of Theorem 2 to characteristic 2. We cannot prove that X_C is always geometrically irreducible. However if this is true, we can give a bound using intersection theory. We also address the question of the probability of existence of a rational flex and show that in characteristic 3, the flexes are on a conic.

Conventions and notation. In the following, we denote by $(x : y : z)$ the coordinates in $\mathbb{P}^2$, and by (x, y) the coordinates in $\mathbb{A}^2$. Let p be a prime or 0 and $n \geq 1$ an integer. We use the letter K for an arbitrary field of characteristic p and let $k = \mathbb{F}_q$ be a finite field with $p \neq 0$ and $q = p^n$ elements. When C is a smooth geometrically irreducible projective curve, we denote by κ_C its canonical divisor. Operators such as Hom, End or Aut applied to varieties over a field K will always refer to K-rational homomorphisms and endomorphisms.

2   Structure of the Canonical Divisor

In this section, we recall some geometric facts about special points and lines on a plane smooth quartic. Let K be an algebraically closed field of characteristic p and C be a smooth (projective) plane quartic defined over K. The curve C is a non hyperelliptic genus 3 curve which is canonically embedded. Hence the intersection of C with a line l are the positive canonical divisors of C. There are 5 possibilities for the intersection divisor of l and C denoted $(l \cdot C) = P_1 + P_2 + P_3 + P_4$:

case 1. The four points are pairwise distinct. This is the generic position.
case 2. $P_1 = P_2$, then l is tangent to C at P_1.
case 3. $P_1 = P_2 = P_3$. The point P_1 is then called a *flex*. As a linear intersection also represents the canonical divisor κ_C, these points are exactly the ones where a regular differential has a zero of order 3. The curve C has infinitely many flexes if and only if $p = 3$ and C is isomorphic to $x^4 + y^3 z + yz^3 = 0$ which is also isomorphic to the Fermat quartic $x^4 + y^4 + z^4 = 0$ and to the Klein quartic $x^3 y + y^3 z + z^3 x = 0$. This is a *funny curve* in the sense of [14, Ex.IV.2.4] or a *non classical* curve in the sense of [28, p.28]. On the contrary, if C has finitely many flexes, then these points are the Weierstrass points of C and the sum of their weights is 24.
case 4. $P_1 = P_2$ and $P_3 = P_4$. The line l is called a *bitangent* of the curve C and the points P_i *bitangency points*. If $p \neq 2$, then C has exactly 28 bitangents (see for instance [23, Sec.3.3.1]). If $p = 2$, then C has respectively 7, 4, 2, or 1 bitangents, if the 2-rank of its Jacobian is respectively 3, 2, 1 or 0 [27]. Recall that the p-rank γ of an abelian variety A/K is defined by $\#A[p](K) = p^\gamma$.
case 5. $P_1 = P_2 = P_3 = P_4$. The point P_1 is called a *hyperflex*. Generically, such a point does not exist. More precisely, the locus of quartics with at least one hyperflex is of codimension one in the moduli space M_3 (see [29, Prop.4.9,p.29]). If $p = 3$ and C is isomorphic to the Fermat quartic then the number of hyperflexes of C is equal to 28. Otherwise, the weight of a hyperflex is greater than or equal to 2, so there are less than 12 of them [26]. Note that the weight of a hyperflex is exactly 2 when $p > 3$ or 0 and that the weight of a flex which is not a hyperflex is 1 if $p \neq 3$. See also [30] for precisions when $p = 2$ and Section 5.2 when $p = 3$.

A point P can even be more special. Let $P \in C$ be a point and let us denote $\phi_P : C \to |\kappa_C - P| = \mathbb{P}^1$ the degree three map induced by the linear system $|\kappa_C - P|$. If this cover is Galois, such a point P is called a (inner) *Galois point* and we denote $\mathrm{Gal}(C)$ the set of Galois points of C.

Lemma 1 ([8,9,10,11]). *Let C be a smooth plane quartic defined over K. The number of Galois points is at most 4 if $p \neq 3$ and at most 28 if $p = 3$. Moreover, the above bounds are reached, respectively by the curve $yz^3 + x^4 + z^4 = 0$ and the Fermat quartic.*

In the sequel, we will need the first item of the following lemma.

Lemma 2. *Let C be a smooth plane quartic defined over K. There is always a bitangency point which is not a hyperflex unless*

- $p = 3$ *and C is isomorphic to the Fermat quartic,*
- $p = 2$ *and* $\operatorname{Jac} C$ *is supersingular,*
- $p = 2$ *and C is isomorphic to a 2-rank one quartic*

$$(ax^2 + by^2 + cz^2 + dxy)^2 + xy(y^2 + xz) = 0$$

with $ac \neq 0$,
- $p = 2$ *and C is isomorphic to a 2-rank two quartic*

$$(ax^2 + by^2 + cz^2)^2 + xyz(y + z) = 0$$

with $abc \neq 0$ and $b + c \neq 0$.

Proof. According to Section 2 case 5, the number of hyperflexes when C is not isomorphic to the Fermat quartic and $p = 3$, is less than 12. On the other hand, if $p \neq 2$, a curve C has 28 bitangents, and thus there are at least $28 \cdot 2 - 12 \cdot 2 = 32$ bitangency points which are not hyperflexes. However, for the Fermat quartic in characteristic 3, all bitangency points are hyperflexes.

There remains to look at the case $p = 2$ for which we use the classification of [31], [22]. The quartic C falls into four categories according to its number of bitangents:

1. if C has only one bitangent, then C is isomorphic to a model of the form $Q^2 = x(y^3 + x^2 z)$ where $Q = ax^2 + by^2 + cz^2 + dxy + eyz + fzx$ and $c \neq 0$. The unique bitangent $x = 0$ intersects C at points $(x : y : z)$ satisfying $by^2 + cz^2 + eyz = 0$. Therefore, C has a hyperflex if and only if $e = 0$, i.e. C falls into the subfamily $\mathcal{S}$ [22, p.468] of curves whose Jacobian is supersingular. Conversely, any curve in $\mathcal{S}$ has a hyperflex.
2. if C has two bitangents, then C is isomorphic to a model of the form $Q^2 = xy(y^2 + xz)$ where $Q = ax^2 + by^2 + cz^2 + dxy + eyz + fzx$ and $ac \neq 0$. All bitangency points are then hyperflexes if and only if $e = f = 0$.
3. if C has four bitangents, then C is isomorphic to $Q^2 = xyz(y + z)$ with

$$Q = ax^2 + by^2 + cz^2 + dxy + eyz + fzx \text{ and } abc \neq 0, b + c + e \neq 0.$$

   All bitangency points are hyperflexes if and only if $d = e = f = 0$.
4. if C has seven bitangents, then C is isomorphic to $Q^2 = xyz(x + y + z)$ with $Q = ax^2 + by^2 + cz^2 + dxy + eyz + fzx$ and some open conditions on the coefficients [22, p.445]. The bitangents are

$$\{x, y, z, x + y + z, x + y, y + z, x + z\}.$$

   Suppose that the intersection points of $x = 0$, $y = 0$ and $z = 0$ with C are hyperflexes, then $d = e = f = 0$. Moreover $x + y = 0$ gives a hyperflex if and only if $(a + b)y^2 + yz + cz^2$ is a perfect square, which is never possible.   □

3   Proof of Theorem 1

Let $q \geq 127$ be a prime power. Note that, as an easy consequence of Serre-Weil bound, we know that

$$\#C(k) \geq q + 1 - 3 \cdot \lfloor 2\sqrt{q} \rfloor = 62.$$

For the proof, we follow the same strategy as [5, p.604]. The lines intersecting C at P are in bijection with the divisors in the complete linear system $|\kappa_C - P|$. We wish to estimate the number of completely split divisors in this linear system, since such a divisor defines a line solution of Theorem 1. To get the existence of such a divisor, we will use an effective Chebotarev density theorem for function fields, as in [21, Th.1]. This theorem assumes that the cover is Galois but we can reduce to this case thanks to the following lemma.

Lemma 3. *Let K/F be a finite separable extension of function fields over a finite field. Let L be the Galois closure of K/F. A place of F splits completely in K if and only if it splits completely in L.*

Proof. It is clear that, if a place $P \in F$ splits completely in L, it splits completely in K. Conversely, let G be the Galois group of L/F and H be the Galois subgroup of L/K. By construction (see [3, A.V.p.54]), L is the compositum of the conjugates K^σ with $\sigma \in G/H$. If a place $P \in F$ splits completely in K, it splits completely in each of the K^σ. It is then enough to apply [25, Cor.III.8.4] to conclude. □

We consider the separable geometric cover $\phi_P : C \to |\kappa_C - P| = \mathbb{P}^1$ of degree 3 induced by the linear system $|\kappa_C - P|$. We may assume that no rational point in $\mathbb{P}^1$ is ramified for ϕ_P. Otherwise, it is easy to see that the fiber of ϕ_P above this point has only rational points and the line defined by these points intersects the quartic in rational points only. Theorem [21, Th.1] boils down to the following proposition.

Proposition 1. *a) If the cover ϕ_P has a non-trivial automorphism, then the number N of completely split divisors in $|\kappa_C - P|$ satisfies*

$$\left| N - \frac{q+1}{3} \right| \leq 2\sqrt{q} + |D|$$

where $|D| = \sum_{y \in \mathbb{P}^1, ramified} \deg y$.
b) If the cover ϕ_P has a non-trivial $\bar{k}$-automorphism not defined over k, then there are no completely split divisors in $|\kappa_C - P|$.
c) If the cover ϕ_P has no non-trivial $\bar{k}$-automorphism, then the number N of completely split divisors in $|\kappa_C - P|$ satisfies

$$\left| N - \frac{q+1}{6} \right| \leq \sqrt{q} + |D|.$$

Proof (of Theorem 1). To avoid case (b) of Proposition 1, it is enough that P is not a Galois point (and we will avoid case (a) as well). By Lemma 1, we know that the number of such P is less than 28. So let the point $P \in C(k) \backslash (\mathrm{Gal}(C) \cap C(k))$. Then the cover ϕ_P has a completely split divisor if

$$\frac{q+1}{6} > \sqrt{q} + |D|. \tag{1}$$

Using Riemann-Hurwitz formula we get

$$|D| \leq (2 \cdot 3 - 2) - 3 \cdot (0 - 2) = 10.$$

Hence the inequality (1) is satisfied as soon as $q \geq 127$.   □

Remark 1. We do not pretend that our lower bound 127 is optimal. In [2], the converse problem is considered (*i.e.* the existence of a plane (not necessarily smooth) curve with no line solution of Theorem 1) but their bound, 3, is also far from being optimal in the case of quartics. Indeed, by [17], for $q = 32$, there still exists a pointless smooth plane quartic for which of course there is no line satisfying Theorem 1.

4   Proof of Theorem 2

Let C be a smooth plane quartic defined over a field K. Let the map $T : C \to \mathrm{Sym}^2(C)$ be the *tangential correspondence* which sends a point P of C to the divisor $(T_P(C) \cdot C) - 2P$. We associate to T its correspondence curve

$$X_C = \{(P, Q) \in C \times C : Q \in T(P)\}$$

which is defined over K. Our goal is to show that when $K = k = \mathbb{F}_q$ with $q > 66^2$ and $p \neq 2$, then X_C has a rational point, *i.e.* there is $(P, Q) \in C(\mathbb{F}_q)^2$ such that $(T_P(C) \cdot C) = 2P + Q + R$ for some point R, necessarily in $C(\mathbb{F}_q)$. Thus, the tangent $T_P(C)$ is a solution of Theorem 2.

To do so, we first study properties of the geometric covers $X_C \to C$. Let $\pi_i : X_C \to C$, $i = 1, 2$, be the projections on the first and second factor. The morphism π_1 is a 2-cover between these two projective curves.

Lemma 4. *Let K be an algebraically closed field of characteristic p. The projection $\pi_1 : X_C \longrightarrow C$ has the following properties:*

1. *The ramification points of π_1 are the bitangency points of C,*
2. *π_1 is separable,*
3. *The point $(P, Q) \in X_C$ such that P, Q are bitangency points and P is not a hyperflex (i.e. $P \neq Q$) is a regular point if and only if $p \neq 2$,*
4. *If $p \neq 2$, the only possible singular points of X_C are the points (P, P) where P is a hyperflex of C.*

Proof. The first property is an immediate consequence of the definition of a bitangent.

If π_1 is not separable then $p = 2$ and π_1 is purely inseparable. Thus $\#\pi_1^{-1}(P) = 1$ for all $F \in C$, *i.e.* all P are bitangency points. This is impossible since the number of bitangents is finite (less than or equal to 7).

Let $F(x, y, z) = 0$ be an equation of C. Let $Q \neq P$ be a point of C defining a point (P, Q) in $X_C \backslash \Delta$ where Δ is the diagonal of $C \times C$. For such points, it is easy to write local equations as follows. We can suppose that $P = (0 : 0 : 1) = (0, 0)$, $Q = (1 : 0 : 1) = (1, 0)$ and assume that $f(x, y) = F(x, y, 1) = 0$ is an equation of the affine part of C. Then, if we consider the curve Y_C in $\mathbb{A}^4(x, y, z, t)$ defined by

$$
\begin{cases}
f(x, y) & = 0, \\
f(z, t) & = 0, \\
\frac{\partial f}{\partial x}(x, y)(z - x) + \frac{\partial f}{\partial y}(x, y)(t - y) & = 0,
\end{cases}
$$

$Y_C \backslash \Delta$ is an open subvariety of X_C containing (P, Q). The Jacobian matrix at the point $(P, Q) = ((0, 0), (1, 0))$ is equal to

$$
\begin{pmatrix}
\frac{\partial f}{\partial x}(0, 0) & 0 & \frac{\partial^2 f}{\partial x^2}(0, 0) - \frac{\partial f}{\partial x}(0, 0) \\
\frac{\partial f}{\partial y}(0, 0) & 0 & \frac{\partial^2 f}{\partial x \partial y}(0, 0) - \frac{\partial f}{\partial y}(0, 0) \\
0 & \frac{\partial f}{\partial x}(1, 0) & \frac{\partial f}{\partial x}(0, 0) \\
0 & \frac{\partial f}{\partial y}(1, 0) & \frac{\partial f}{\partial y}(0, 0)
\end{pmatrix}.
$$

Now, if P and Q are bitangency points, then the tangent at these points is $y = 0$, so $\frac{\partial f}{\partial x}(0, 0) = \frac{\partial f}{\partial x}(1, 0) = 0$. The only non-trivially zero minor determinant of the matrix is then

$$
\frac{\partial f}{\partial y}(1, 0) \cdot \frac{\partial f}{\partial y}(0, 0) \cdot \frac{\partial^2 f}{\partial x^2}(0, 0).
$$

So $(P, Q) \in X_C$ is not singular if and only if $\frac{\partial^2 f}{\partial x^2}(0, 0) \neq 0$. This can never be the case if $p = 2$, so we now suppose that $p \neq 2$. We can always assume that the point $(0 : 1 : 0) \notin C$ and we write

$$
f(x, y) = x^4 + x^3 h_1(y) + x^2 h_2(y) + x h_3(y) + h_4(y),
$$

where h_i are polynomials (in one variable) over K of degree $\leq i$. Since $y = 0$ is a bitangent at P and Q, we have

$$
f(x, 0) = x^2(x - 1)^2 = x^4 - 2x^3 + x^2,
$$

and thus $h_2(0) = 1$. Now

$$
\frac{\partial^2 f}{\partial x^2}(0, 0) = 2h_2(0) \neq 0.
$$

Finally if $(P, Q) \in X_C$ is not ramified for π_1, it is a smooth point. This proves the last assertion. $\square$

We want to apply the following version of Hasse-Weil bound to the curve X_C.

Proposition 2 ([1]). *Let X be a geometrically irreducible curve of arithmetic genus π_X defined over $\mathbb{F}_q$. Then*

$$|\#X(\mathbb{F}_q) - (q+1)| \le 2\pi_X \sqrt{q}.$$

In particular if $q \ge (2\pi_X)^2$ then X has a rational point.

Hence, to finish the proof, we need to show that X_C is geometrically irreducible and then to compute its arithmetic genus. For the first point, we use the following easy lemma for which we could not find a reference.

Lemma 5. *Let $\phi : X \to Y$ be a separable morphism of degree 2 between two projective curves defined over an algebraically closed field K such that*

1. *Y is smooth and irreducible,*
2. *there exists a point $P_0 \in Y$ such that ϕ is ramified at P_0 and $\phi^{-1}(P_0)$ is not singular.*

Then X is irreducible.

Proof. Let $s : \tilde{X} \to X$ be the normalization of X and $\tilde{\phi} = \phi \circ s : \tilde{X} \to Y$. Due to the second hypothesis, $\tilde{\phi} : \tilde{X} \to Y$ is a separable, ramified 2-cover. Clearly, X is geometrically irreducible if and only if $\tilde{X}$ is.

Let us assume that $\tilde{X}$ is not irreducible. There exist smooth projective curves $\tilde{X}_1$ and $\tilde{X}_2$ such that $\tilde{X} = \tilde{X}_1 \cup \tilde{X}_2$. Then, consider for $i = 1, 2$, $\tilde{\phi}_i = \tilde{\phi}_{|\tilde{X}_i} : \tilde{X}_i \to Y$. Each of these morphisms is of degree 1 and since the curves are projective and smooth, they define an isomorphism between $\tilde{X}_i$ and Y.

Since $P_0 \in Y$ is a ramified point, $\tilde{\phi}_1^{-1}(P_0) = \tilde{\phi}_2^{-1}(P_0)$. It follows that

$$\tilde{\phi}^{-1}(P_0) \in \tilde{X}_1 \cap \tilde{X}_2$$

so that $\tilde{\phi}^{-1}(P_0)$ is singular, which contradicts the hypothesis. $\qquad\square$

Proof (of Theorem 2). Let us first start with C $\bar{k}$-isomorphic to the Fermat quartic in characteristic 3. By Section 2 case 3, all its points are flexes. So if there exists $P \in C(k)$, then the tangent at P cuts C at P and at another unique point which is again rational over k. Now when $q > 23$ and $q \ne 29, 32$, it is proved in [17] that a genus 3 non-hyperelliptic curve over $\mathbb{F}_q$ has always a rational point and the result follows.

We suppose that C is a smooth plane quartic not $\bar{k}$-isomorphic to the Fermat quartic if $p = 3$. As we assumed that $p \ne 2$, by Lemma 5, Lemma 2 and Lemma 4, we conclude that X_C is an geometrically irreducible projective curve. Moreover, if we assume that C has no hyperflex, then X_C is smooth and we can compute its genus g_{X_C} using Riemann-Hurwitz formula for the 2-cover $\pi_1 : X_C \to C$ ramified over the $2 \cdot 28$ bitangency points. In fact,

$$2g_{X_C} - 2 = 2(2 \cdot 3 - 2) + 56,$$

and thus $g_{X_C} = 33$. The family of curves X_C is flat over the locus of smooth plane quartics C by [6, Prop.II.32]. As the arithmetic genus π_{X_C} is constant in flat families [14, Cor.III.9.10] and equal to g_{X_C} for smooth X_C [14, Prop.IV.1.1], we get that $\pi_{X_C} = 33$ for any curve C. We can now use Proposition 2 to get the bound $q > (2 \cdot 33)^2$. $\qquad\square$

5   Further Results and Open Questions

5.1   The Tangent Case in Characteristic 2

It is of course tempting to extend Theorem 2 to characteristic 2. Unfortunately, as we have seen in Lemma 4, all ramified points are singular and we cannot apply Lemma 5 to show that X_C is geometrically irreducible. However, explicit computations with Magma, see

http://iml.univ-mrs.fr/~ritzenth/programme/tangent-char2.mag,

suggest that X_C is still geometrically irreducible (see also Remark 2).

Conjecture 1. Let C be a smooth plane quartic over a field k of characteristic 2. Then the correspondence curve X_C is geometrically irreducible.

Even if we assume the conjecture, we still have to compute the arithmetic genus of X_C. It is not easy in this case, as wild ramification occurs. We therefore suggest another point of view, which can actually be used in any characteristic.

The notation is like in Section 4. In order to emphasize how general the method is, we will denote by $g = 3$ the genus of C and by $d = 4$ its degree. The map T is a correspondence with *valence* $\nu = 2$, *i.e.* the linear equivalence class of $T(P) + \nu P$ is independent of P. Let denote by E (resp. F) a fiber of π_1 (resp. π_2) and Δ the diagonal of $C \times C$. We get, as in the proof of [12, p.285], that X_C is linearly equivalent to $aE + bF - \nu\Delta$ for some $a, b \in \mathbb{Z}$ to be determined. Then, one computes the arithmetic genus π_{X_C} of X_C thanks to the adjunction formula [14, Ex.V.1.3.a]

$$2\pi_{X_C} - 2 = X_C.(X_C + \kappa_{C \times C})$$

where $\kappa_{C \times C}$ is the canonical divisor on $C \times C$. Using that (see for instance [12, p.288], [14, Ex.V.1.6])

$$\begin{cases} E.E = F.F & = 0, \\ E.F = \Delta.E = \Delta.F & = 1, \\ \kappa_{C \times C} & \equiv_{\text{num}} (2g - 2)E + (2g - 2)F, \\ \Delta^2 & = (2 - 2g), \end{cases}$$

we find

$$\pi_{X_C} = ab - 15.$$

Now, we determine the values of a and b. One has

$$X_C.E = b - \nu = \deg \pi_1 = d - 2 = 2$$

so $b = 4$. Also, $X_C.F = a - \nu = \deg \pi_2$. The degree of π_2 is equal to the degree of the dual curve C^* minus 2. By [16, p.786]

$$\deg C^* = \frac{d(d - 1)}{m}$$

where n is the inseparable degree of the dual map $C \to C^*$. In the case of smooth plane quartics, the degree m equals 1 and $a = 12$ except (see [15, Cor.2.4])

- if $p = 3$ where C is geometrically isomorphic to the Fermat quartic and then $m = 3$;
- or if $p = 2$, then $m = 2$ and $a = 6$.

Plugging the values of a and b, we find that

- $\pi_{X_C} = 33$ if $p \neq 2$ and if C is not $\bar{k}$-isomorphic to the Fermat quartic in characteristic 3;
- $\pi_{X_C} = 9$ if $p = 2$.

Hence we get the following proposition.

Proposition 3. *If Conjecture 1 is true, then any smooth plane quartic over $\mathbb{F}_q$ with $q = 2^n$ and $n > 8$ has a tangent which intersects the quartic at rational points only.*

Remark 2. We can prove Conjecture 1 for generic quartics C using the previous background. Indeed, if X_C is not geometrically irreducible, it is the union over $\bar{k}$ of two geometrically irreducible curves X_1 and X_2, birationally equivalent to C, and of arithmetic genus

$$0 \leq \pi_{X_1} = \pi_{X_2} \leq g = 3.$$

Now by [14, Ex.V.1.3.c]

$$9 = \pi_{X_C} = \pi_{X_1} + \pi_{X_2} + X_1.X_2 - 1.$$

Hence $4 \leq X_1.X_2 \leq 10$. But $X_1.X_2$ is greater than or equal to the number of ramification points of $\pi_1 : X_C \to C$ which are the bitangency points of C. When this number is greater than 10 (and generically, for 2-rank 3 quartics with no hyperflex, it is 14), we get that X_C is geometrically irreducible.

5.2 The Case of Flexes

The heuristic results and computations of [7] tend to suggest that a random plane smooth quartic over $\mathbb{F}_q$ has an asymptotic probability of about 0.63 to have at least one rational flex when q tends to infinity. We describe how to turn the heuristic strategy into a proof.

Let $\mathbb{P}^{14}$ be the linear system of all plane quartic curves over a field K and $I_0 = \{(p, l), \ p \in l\} \subset \mathbb{P}^2 \times \mathbb{P}^{2*}$. Let $I_4 \subset \mathbb{P}^{14} \times I_0$ be the locus

$$I_4 = \{(C, (p, l)), \ C \text{ is smooth and } p \text{ is a flex of } C \text{ with tangent line } l\}.$$

Harris proved the following result using monodromy arguments.

Theorem 3 ([13, p.698]). *The Galois group of the cover $I_4 \to \mathbb{P}^{14}$ over $\mathbb{C}$ is the full symmetric group S_{24}.*

Let us assume for a moment that this result is still valid over finite fields. Then, using a general Chebotarev density theorem for function fields like in [24, Th.7], this would mean that the probability of finding a rational flex is within $O(1/\sqrt{q})$ of the probability that a random permutation of 24 letters has a fixed point, which is

$$p_{24} := 1 - \frac{1}{2!} + \frac{1}{3!} - \ldots - \frac{1}{24!} \approx 1 - \exp(-1) \approx 0.63.$$

Unfortunately, it is not easy to transpose Harris's proof over any field. And actually, Harris's result is not true in characteristic 3, as the following proposition implies.

Proposition 4. Let $C : f(x_1, x_2, x_3) = 0$ be a smooth plane quartic defined over an algebraically closed field K of characteristic 3. The flexes of C are the intersection points of C with a certain curve $H_C : h_C = 0$ of degree less than or equal to 2. The curve H_C can be degenerate as in the case of the Fermat quartic where $h_C = 0$.

Proof. We use the method to compute flexes of a plane curve of degree d described in the appendix of [7] (see also [26, Th.0.1]). Indeed, when the characteristic of the field divides $2(d-1)$, one cannot use the usual Hessian and one should proceed as follows.
Let $C : f = 0$ be the generic plane quartic over K

$$\begin{aligned}
f(x, y, z) := {} & a_{00}y^4 + y^3(a_{10}x + a_{01}z) + y^2(a_{20}x^2 + a_{11}xz + a_{02}z^2) \\
& + y(a_{30}x^3 + a_{21}x^2z + a_{12}xz^2 + a_{03}z^3) \\
& + (a_{40}x^4 + a_{31}x^3z + a_{22}x^2z^2 + a_{13}xz^3 + a_{04}z^4),
\end{aligned}$$

We define as in [7]

$$\begin{aligned}
2\bar{h} &= 2f_1f_2f_{12} - f_1^2 f_{22} - f_2^2 f_{11}, \\
&= f_1(f_2f_{12} - f_1f_{22}) + f_2(f_1f_{12} - f_2f_{11})
\end{aligned}$$

where f_i or f_{ij} are the partial derivatives with respect to ith variable (or to ith and jth variables). It is then easy to check via a computer algebra system, see

http://iml.univ-mrs.fr/~ritzenth/programme/flex-char3.mw,

that

$$2\bar{h} - a_{20}f^2 - f \cdot (ax^3 + by^3 + cz^3)z = \tilde{h}_C \cdot z^2,$$

with

$$\begin{aligned}
a &:= a_{40}a_{11} + a_{21}a_{30} - 2a_{20}a_{31}, \\
b &:= a_{10}a_{11} + a_{00}a_{21} - 2a_{20}a_{01}, \\
c &:= 2a_{12}^2 + a_{13}a_{11} + a_{20}a_{04} + a_{03}a_{21} + a_{02}a_{22},
\end{aligned}$$

where $\tilde{h}_C$ is a homogeneous polynomial in $K[x, y, z]$ of degree 6, for which nonzero coefficients appear only for the monomials $x^6, y^6, z^6, x^3y^3, x^3z^3$ and

y^3z^3. Since the map $u \mapsto u^3$ is an isomorphism of K, there is a polynomial $h_C \in K[x, y, z]$ satisfying $\tilde{h}_C = h_C^3$. If we suppose that there is no flex at infinity, the flexes are the intersection points of $\bar{h} = 0$ and $f = 0$, so they are also the intersection points of $h_C = 0$ and $f = 0$ and $h_C = 0$ is the equation of a (possibly degenerate) conic H_C. □

Remark 3. As $\tilde{h}_C = h_C^3$, we see that the weight of a flex which is not a hyperflex is 3 in characteristic 3.

Corollary 1. *The Galois group of the cover $I_4 \to \mathbb{P}^{14}$ over $\bar{\mathbb{F}}_3$ is the full symmetric group S_8.*

Proof. Note that the Galois group G of the cover is the Galois group of the x-coordinate of the 8 flexes of the general quartic. Hence, G is included in S_8. To show that G is exactly S_8, we are going to specialize the general quartic to smooth quartics over finite fields with 8 distinct flexes having different arithmetic patterns. More precisely, to generate S_8 we need to produce (see [20, Lem.4.27]):

- a smooth quartic with 8 Galois conjugate flexes over $\mathbb{F}_3$:
$$2x^4 + 2x^3y + x^3z + 2x^2z^2 + xy^3 + 2xy^2z + y^3z + yz^3 = 0;$$

- a smooth quartic with one rational flex and 7 Galois conjugate flexes over $\mathbb{F}_3$:
$$x^3y + x^2z^2 + 2xy^3 + xy^2z + 2xyz^2 + 2xz^3 + y^4 + 2yz^3 = 0;$$

- a smooth quartic with two quadratic conjugate flexes and 6 rational flexes over $\mathbb{F}_9$:
$$a^6x^4 + ax^3y + a^7x^3z + a^6x^2y^2 + a^2x^2z^2 + a^7xy^3 + a^7xy^2z \\ + xyz^2 + a^5xz^3 + a^5y^4 + a^3y^3z + a^5y^2z^2 + 2yz^3 + a^7z^4 = 0,$$

where $a^2 - a - 1 = 0$. □

Corollary 2. *Let C be a smooth plane quartic over $\mathbb{F}_{3^n}$. The probability that C has a rational flex tends to*

$$p_8 := 1 - \frac{1}{2!} + \frac{1}{3!} - \ldots - \frac{1}{8!} \approx 0.63$$

when n tends to infinity.

Remark 4. The fact that the Galois group G in characteristic 3 is S_8 and not S_{24} was unnoticed in our computations in [7] because $|p_{24} - p_8| \leq 10^{-5}$.

General reduction arguments show that the Galois group G remains S_{24} almost all p. We conjecture that $p = 3$ is the only exceptional case.

Conjecture 2. The Galois group of the cover $I_4 \to \mathbb{P}^{14}$ over $\bar{\mathbb{F}}_p$ is the full symmetric group S_{24} if $p \neq 3$ and S_8 otherwise.

Acknowledgments. We would like to thank Noam Elkies and Pierre Dèbes for their suggestions and references in Section 5.2.

References

1. Aubry, Y., Perret, M.: A Weil theorem for singular curves. In: Pellikaan, P., De Gruyter, V. (eds.) Proceedings of arithmetic, geometry and coding theory, vol. IV, pp. 1–7 (1995)
2. Balico, E., Cossidente, A.: On the number of rational points of hypersurfaces over finite fields. Results Math. 51, 1–4 (2007) (electronic)
3. Bourbaki, N.: Elements of mathematics. Commutative algebra. ch. 5-7 (Éléments de mathématique. Algèbre commutative. Chapitres 5 à 7.) Reprint of the 1985 original. Springer, Berlin (2006)
4. Cantor, D.: Computing in the Jacobian of a hyperelliptic curve. Math. Comp. 48(177), 95–101 (1987)
5. Diem, C., Thomé, E.: Index calculus in class groups of non-hyperelliptic curves of genus three. J. Cryptology 21(4), 593–611 (2008)
6. Eisenbud, D., Harris, J.: The geometry of schemes. Graduate Texts in Mathematics, vol. 197. Springer, New York (2000)
7. Flon, S., Oyono, R., Ritzenthaler, C.: Fast addition on non-hyperelliptic genus 3 curves. In: Algebraic geometry and its applications. Ser. Number Theory Appl., vol. 5, pp. 1–28. World Sci. Publ., Hackensack (2008)
8. Fukasawa, S.: Galois points on quartic curves in characteristic 3. Nihonkai Math. J. 17(2), 103–110 (2006)
9. Fukasawa, S.: On the number of Galois points for a plane curve in positive characteristic. II. Geom. Dedicata 127, 131–137 (2007)
10. Fukasawa, S.: On the number of Galois points for a plane curve in positive characteristic. Comm. Algebra 36(1), 29–36 (2008)
11. Fukasawa, S.: Galois points for a plane curve in arbitrary characteristic. Geom. Dedicata 139, 211–218 (2009)
12. Griffiths, P., Harris, J.: Principles of algebraic geometry. Wiley Classics Library. John Wiley & Sons Inc., New York (1994); Reprint of the 1978 original
13. Harris, J.: Galois groups of enumerative problems. Duke Math. J. 46(4), 685–724 (1979)
14. Hartshorne, R.: Algebraic geometry. Graduate Texts in Mathematics, vol. 52. Springer, New York (1977)
15. Homma, M.: A souped-up version of pardini's theorem and its application to funny curves. Compos. Math. 71, 295–302 (1989)
16. Homma, M.: On duals of smooth plane curves. Proc. Am. Math. Soc. 118, 785–790 (1993)
17. Howe, E.W., Lauter, K.E., Top, J.: Pointless curves of genus three and four. In: Aubry, Y., Lachaud, G. (eds.) Algebra, Geometry, and Coding Theory (AGCT 2003), Société Mathématique de France, Paris. Séminaires et Congrès, vol. 11 (2005)
18. Koblitz, N.: Elliptic curve cryptosystems. Math. Comp. 48(177), 203–209 (1987)
19. Koblitz, N.: Hyperelliptic cryptosystems. J. cryptology 1, 139–150 (1989)
20. Milne, J.: Fields and Galois theory, version 4.21, http://www.jmilne.org/math/CourseNotes/FT.pdf
21. Kumar Murty, V., Scherk, J.: Effective versions of the Chebotarev density theorem for function fields. C. R. Acad. Sci. Paris Sér. I Math. 319(6), 523–528 (1994)
22. Nart, E., Ritzenthaler, C.: Non hyperelliptic curves of genus three over finite fields of characteristic two. J. of Number Theory 116, 443–473 (2006)

23. Ritzenthaler, C.: Problèmes arithmétiques relatifs à certaines familles de courbes sur les corps finis. PhD thesis, Université Paris VII (2003)
24. Serre, J.-P.: Zeta and L functions. In: Arithmetical Algebraic Geometry (Proc. Conf. Purdue Univ., 1963), pp. 82–92. Harper & Row, New York, (1965)
25. Stichtenoth, H.: Algebraic Function Fields and Codes. Lectures Notes in Mathematics, vol. 314. Springer, Heidelberg (1993)
26. Stöhr, K.-O., Voloch, J.F.: Weierstrass points and curves over finite fields. Proc. London Math. Soc. (3) 52(1), 1–19 (1986)
27. Stöhr, K.-O., Voloch, J.F.: A formula for the cartier operator on plane algebraic curves. J. für die Reine und Ang. Math. 377, 49–64 (1987)
28. Torres, F.: The approach of Stöhr-Voloch to the Hasse-Weil bound with applications to optimal curves and plane arcs (2000), http://arxiv.org/abs/math.AG/0011091
29. Vermeulen, A.M.: Weierstrass points of weight two on curves of genus three. PhD thesis, university of Amsterdam, Amsterdam (1983)
30. Viana, P., la Torre, O.-P.: Curves of genus three in characteristic two. Comm. Algebra 33(11), 4291–4302 (2005)
31. Wall, C.T.C.: Quartic curves in characteristic 2. Math. Proc. Cambridge Phil. Soc. 117, 393–414 (1995)

Reflections about a Single Checksum

Ulrich Tamm

German Language Department of Business Informatics,
Marmara University, Istanbul, Turkey

Abstract. A single checksum for codes consisting of n integer components is investigated. In coding theory this is mostly used for single error–correction in unconventional error models. If the errors are such that a single component c_i is distorted to $c_i \pm e_i$, the analysis leads to equivalent group factorizations. We shall present several code constructions for this model, give a short survey on the coding theoretical and mathematical background, and also emphasize applications in cryptography and computer science.

Keywords: single – error correction, perfect codes, group factorization, steganography, distributed computing.

1 Introduction

We are considering the following checksum for a code which consists of all words $(c_1, \ldots, c_n) \in Z$ fulfilling

$$\sum_{i=1}^{n} w_i \cdot c_i = 0 \bmod m, \qquad (1)$$

where $(w_1, \ldots, w_n) \in Z$ is a fixed sequence of weights and n is the length of the code.

Since the checksum is reduced modulo m, of course, the components of the code words may be regarded as letters over an alphabet of size m. Usually, this is indeed the case. However, for some applications, the letters c_i are from a much smaller alphabet.

For instance, in the famous Varshamov – Tenengolts codes [37], which arise for $(w_1, \ldots, w_n) = (1, 2, \ldots, n)$, and $m = n + 1$, the code words are binary. Another case, where the c_i's are chosen from a smaller alphabet will be discussed more detailed in the application in steganography.

Varshamov – Tenengolts codes are able to correct single asymmetric errors [4] and were later also applied by Levenshtein [18] in order to correct single deletions.

Later, Levenshtein and Vinck [19] and Martirossian [22] used the checksum (1) in order to analze single – error correcting codes for further unconventional error models as peak shifts in run–length–limited codes. The effect of a single error is reflected in the behaviour of the syndrome, which should be changed to

M.A. Hasan and T. Helleseth (Eds.): WAIFI 2010, LNCS 6087, pp. 238–249, 2010.
© Springer-Verlag Berlin Heidelberg 2010

a value different from 0 by a linear combination of the weights corresponding to the codeword's coordinates involved in this error.

The proper choice of the weight sequence is crucial for the quality of the code. It strongly depends on the type of error to be corrected.

Vinck and Morita [38] later called the codes obtained via checksum (1) integer codes.

The most important case is the error type of substitution of the letter c_i by c_i'. Then the resulting syndrome is

$$w_1 c_1 + \ldots w_{i-1} c_{i-1} + w_i c_i' + w_{i+1} c_{i+1} + \ldots w_n c_n$$

$$= w_i (c_i' - c_i), \text{ for } i = 1, \ldots, n \tag{2}$$

A single error may involve more than one component. For instance, if the two adjacent letters c_i and c_{i+1} are permuted, then the syndrome will be

$$w_1 c_1 + \ldots w_{i-1} c_{i-1} + w_i c_{i+1} + w_{i+1} c_i + w_{i+2} c_{i+2} + \ldots w_n c_n =$$

$$(w_i - w_{i+1})(c_{i+1} - c_i), \text{ for } i = 1, \ldots, n-1.$$

A very similar syndrome (with n being the number of runs in a run – length limited sequence) occurs in the correction of peak shifts discussed by Levenshtein and Vinck [19].

In order to be able to correct one single error, the syndromes of an integer code have to be pairwisely different. So if the possible distortions, which can be corrected by the integer code, are from an error set $\mathcal{E} \subset Z_m$ and the linear combinations of the weights (for instance w_i for substitutions (2) or $w_i - w_{i+1}$ for permutations) are from a set $\mathcal{H} \subset Z_m$, then we have to assure that

$$e \cdot h \neq e' \cdot h' \text{ for all } e, e' \in \mathcal{E} \text{ and } h, h' \in \mathcal{H}. \tag{3}$$

If, in addition, all elements from the set $Z_m \setminus \{0\}$ occur as a product in (3), then the code is said to be perfect. For a perfect integer code in Z_m the pair $(\mathcal{E}, \mathcal{H})$ is also known as splitting of the additive group Z_m, cf. [30] and we shall also use this notion in the following.

Usually, we choose $m = p$ a prime number, such that we operate in finite fields. Then $\mathcal{E} \cdot \mathcal{H}$ yields a factorization of the multiplicative group Z_p^* (here multiplication of two sets means the set of all possible products of one element in one set with an element of the other set). For the theory of group factorizations we refer to [34], [26].

Integer codes for the error sets $\mathcal{E} = \{\pm 1, \pm 2, \ldots, \pm k\}$ are denoted as k – shift codes or k – shift designs and arise in the study of peak shift correction [19] and of correction of errors in the so – called Stein sphere [7], where a single component is distorted in such a way that the received letter c_i' is of the form $c_i' = c_i + j, j \in \{\pm 1, \pm 2, \ldots, \pm k\}$. Conditions for the existence of perfect k – shift codes have been introduced for $k = 1, 2$ and $k = \frac{m-1}{2}$ in [19] and for the parameters $k = 3$ and $k = 4$ in [24,35].

In [23] the error set $\mathcal{E} = \{\pm 1, \pm a\}$ is discussed. This corresponds to the error model, in which a letter c_i is changed to one of its nearest neighbours on the

$a \times a$ – grid, where a component (x, y) is represented by the number $x + y \cdot a$ This can be described in such a way that the received letter is contained in the set $\{c_i = 1, c_i \pm a\}$.

The error set $\mathcal{E} = \{\pm 1, \pm a, \pm b\}$ was studied in [35] as a special case of the more general $\{\pm 1, \pm a, \ldots, \pm a^r, \pm b, \ldots, \pm b^s\}$ for positive integers r, s.

In Section 2 we shall discuss further applications of the checksum (1) in cryptography and computer science. Especially, it is related to the efficient placement of processors in distributed computing.

In Section 3 we concentrate on error – correction. First, we present a very general method to obtain perfect integer codes for any error set of the form $\mathcal{E} = \{1, a_1, \ldots a_{k-1}\}$ in Z_p, where $p = 2k + 1$ is an odd prime number. Perfect codes for the same error set in Z_m with composite residues m can be derived from them. Our method makes use of the fact that the multiplicative group Z_p^* is cyclic and hence generated by one element $g \in \{2, \ldots, p-2\}$. A perfect code is shown to exist in Z_p exactly if the powers μ_i in the representation $a_i = g^{\mu_i}$ of the elements in the error set fall into the different congruence classes modulo k.

This result is rather obvious. The big advantage compared to previous approaches, however, is that besides the existence an efficient algorithm is available to check this condition, and also to explicitly construct the codes. This will be analyzed for special error sets $\mathcal{E}$.

Finally, in Section 5, we relax the model not further requiring perfectness but a good packing. This may be even a harder task, since the error spheres around a code word are rather nasty. So, finding a perfect code via algebraic methods may be easier than finding a good packing via combinatorial methods.

2   Applications in Computer Science and Cryptography

Lattice tilings: Mathematically, a group factorization obtained from a perfect code with error set $F(k) = \{\pm 1, \pm 2, \ldots, \pm k\}$ corresponds to a tiling of the Euclidean space by a certain star body, the (n, k)–cross. This is a collection of unit n–dimensional cubes, with one cube in the center and a number of k consecutive cubes attached to each of its faces.

More exactly, a lattice tiling of the n–dimensional Euclidean space exists, if $F(k)$ "splits" some abelian group, which for the groups Z_p is just a factorization. We do not go into detail here and refer to [30], [33] for further reading.

Distributed computing: In parallel computing processors may share some resources as memory. This is usually modeled by a graph, where the vertices denote the processors, and an edge between two vertices means that the corresponding processors are connected in the network. Some resources as memories, software modules, or I/O–connections may be expensive and hence only be placed at a subset of the processors [3], [5], [14]. An efficient placement of these resources leads to the concepts of codes in graphs and domination in graphs, e.g., [2], [15], [16], [21]. Usually, a combinatorial treatment is necessary. However, if the underlying graphs have a very regular structure, as a ring or a grid, then checksum (1) can be used to construct an efficient placement. This is essentially equivalent to a good code, where the code words correspond to the processors equiped with the resources and the error spheres around them correspond to the direct neighbours (or neighbours within a certain distance).

Packet loss in internet protocols: Sloane [28] used Varshamov – Tenengolts codes in order to protect internet protocols against packet losses or genome sequences against a deletion of one letter in the sequence. He also analyzed, in which cases it might be better to compute the checksum (1) modulo a number d different from 0.

SEC–DED Codes: If we choose all weights $w_i = 1$, the checksum (1) can, of course, be used to detect a single error, which would result in a sum different from 0. However, it can not be recognized from which component i this error results. Chosing instead the weights $w_i = i$ indeed gives information about the location of the component if the alphabet size m is appropriately large and hence also automatically allows to correct this error. A different way to achieve this goal is a second checksum. If this checksum is carefully chosen, it may also detect a second error. Such SEC–DED codes (single–error correction, double–error detection) are implemented in computer memories, where the errors occcur in a single bit. Combining l bits to an integer modulo $m = 2^l$, integer codes with two checksums have been constructed as SEC–DED codes.

Steganography: In steganography we have a situation complementary to coding theory, where the errors occur at random. Here, sender and receiver agree on a certain set of n positions in which the sender may slightly change the value of the component c_i. In order that these changes will not be detected, there have to be very few changes of very small amplitude. For instance, it may only be allowed to change one component by adding plus or minus 1 to c_i, i.e. $c_i' = c_i \pm 1$. The checksum

$$\sum_{i=1}^{n} i \cdot c_i' \bmod 2n + 1$$

will then be used to decode the corresponding message. Lisonek's idea in [20] was that by appropriate choice of the weights in checksum (1) it is possible to obtain a better perfomance by changing two components by ± 1 and decode the message. In order to assure a succesful decoding, Lisonek chose the weights $w_1, w_2, \ldots, w_m$ from a symmetric sum cover $S = \{0, \pm w_1, \pm w_2, \ldots, \pm w_n\}$, which means that $S + S = Z_m$. This has the effect that the message can be decoded, if all the sums $w_i + w_j$ with $i \neq j$ are different.

  In order to avoid too much overlap, the task hence is to find a large enough m such that $S + S = Z_m$. Lisonek provides a table for small m. Note that it is really required that $i \neq j$. If $i = j$ would be allowed an amplitude 2 would be possible for some i. The symmetry condition as above (including $\pm w_i$) seems to be a new requirement compared to previous calculations, for instance, by Graham and Sloane [10]

Interestingly, this application in steganography, adresses rather the additive structure of the group Z_m, whereas the factorizations important for the analysis of the change of amplitude in only one component, can be analyzed via factorizations, which of course rely on the multiplicative structure of Z_m.

Group factorizations in cryptography: Factorizations of groups are also used to construct cryptosystems, which may replace RSA when, for instance, reliable quantum computers will once be in use [25]. However, the group Z_p important for our error–correcting codes is too simple for such applications and does not allow a one–way function for encoding.

Double error correction: In principle, it would also be possible to use one checksum (1) in order to correct more than one error. For instance, Lisonek's idea for steganography may in some cases be transfered to error–correcting codes (steganography, however, is more related to covering than to packing). Usually, a second check will be carried out, as we saw in the application of SEC–DED codes. These codes, however, only correct bit flips. The analysis of 2–error correcting codes for even the simplest symbol changes is extremely difficult. In [17] one construction is provided. Even more difficult is the correction of errors of distance 2 in the Lee metric [13]. The reason is that this may arise in two ways. Either one component c_i is distorted to $c_i' = c_i \pm 2$ or two components are distorted by $c_i' = c_i \pm 1$ and $c_j' = c_j \pm 1$, This is a combination of the single–error correction as studied by Martirossian [22] and the double–error correction related to the steganographic model discussed by Lisonek.

Further applications: Further applications of the checksum (1) arise in coding for memories with defects [1], [27] and for tilings by certain polyominoes as studied by Golomb [8].

3   A General Construction for Perfect Integer Codes

The analysis can be reduced to groups Z_p where p is a prime number. In this case a splitting $(\mathcal{E}, \mathcal{H})$ corresponds to a factorization $\mathcal{E} \cdot \mathcal{H}$ of the group Z_p^*.

For a composite number $m = p_1^{s_1} \cdots p_r^{s_r}$ a perfect integer code in Z_m can be obtained from the perfect integer codes in Z_{p_i} for the prime factors p_i, $i = 1, \ldots r$, of m. For sets $\mathcal{E}$ of small size it has even been shown that this is the only way to obtain perfect integer codes for composite m [30], [24].

The idea here will be to arrange that the set $\mathcal{H}$ obtained from the weights in the definition of an integer code (1) consists of a subgroup $\mathcal{G}$ of Z_p^* and its translates in the cosets of $\mathcal{G}$. Since Z_p^* is a cyclic group, it is generated by one element g. If $\mathcal{G}$ is a subgroup of Z_p^*, its order must be a divisor of $p - 1$ and $\mathcal{G}$ itself must be generated by a power of g, i. e. for some t dividing $p - 1$

$$\mathcal{G} = \{g^{jt} : j = 0, \ldots, p - 1\}.$$

Theorem 1: Let $\mathcal{E} = \{1, a_1, \ldots, a_{k-1}\}$ be the error set of an integer code, let g be a generator of Z_p^* and let $a_i = g^{\nu_i}$ in Z_p^* for $i = 1, \ldots, k - 1$. Then a perfect *integer code with error set $\mathcal{E}$ exists in Z_p, exactly if for some divisor l of $\frac{p-1}{k}$ the powers ν_i are such that $\nu_i = l\mu_i$ for $i = 0, \dots, k-1$, where the μ_i's fall into the different congruence classes modulo k, i. e.,*

$$\{\mu_1 \ mod \ k, \dots, \mu_{k-1} \ mod \ k\} = \{1, \dots, k-1\}. \tag{4}$$

Proof: There are basically two possible structures for the integer code $\mathcal{H}$ when the error (or splitting) set $\mathcal{E} = \{1, a_1, \dots, a_k\}$.

Type 1: $\mathcal{H}$ is already the subgroup $\mathcal{G}$ in Z_p^*. In this case

$$\mathcal{H} = \{g^{jk} : j = 0, \dots, \frac{p-1}{k}\},$$

where g is a generator of Z_p^*.

Now express the elements of $\mathcal{E}$ as powers of the generator g, namely $a_i = g^{\mu_i}$ for $i = 1, \dots, k-1$. In order to assure that all products $e \cdot h$ ($e \in \mathcal{E}, h \in \mathcal{H}$) are different, one has to guarantee that (with $\mu_0 = 0$) all products $g^{\mu_i} \cdot g^{jk} = g^{jk+\mu_i}$ are different for all possible choices $j = 0, \dots, \frac{p-1}{k}, i = 0, \dots, k-1$. This obviously holds if and only if the μ_i fall into the different congruence classes modulo k, i. e., if

$$\{\mu_0 \ mod \ k, \dots, \mu_{k-1} \ mod \ k\} = \{0, \dots, k-1\}.$$

If, additionally, $|\mathcal{E}| \cdot |\mathcal{H}| = p - 1$, then $\mathcal{H}$ is a perfect integer code.

Type 2: $\mathcal{G} = \{g'^{jk} : j = 0, \dots, \frac{p-1}{kl}\}$, where $g' = g^l$ for some $l \geq 2$. In this case $t = lk$, l being a divisor of $\frac{p-1}{k}$ and the integer code $\mathcal{H}$ will be of the form

$$\mathcal{H} = \bigcup_{\text{coset } C \text{ of } \mathcal{G}} x_C \cdot \mathcal{G},$$

where x_C may be any representative of the coset C.

With the same argumentation as above it can be seen that in order to assure that all products $e \cdot h$ ($e \in \mathcal{E}, h \in \mathcal{H}$) are different, it suffices to show that powers μ_i of the elements of the error set $a_i = g^{l\mu_i}$, $i = 1, \dots, k-1$, fall into the different congruence classes modulo k.

From this theorem the next algorithm is immediate:

Algorithm IntegerCode. (set $\mathcal{E} = \{1, a_1, \dots, a_{k-1}\}$, prime number p)

(1) Find a generator g of Z_p^*
(2) for $i = 1$ to $k - 1$ write $a_i = g^{\nu_i}$ in Z_p^*
(3) for all divisors l of $\frac{p-1}{k}$
    (3a) write $\nu_i = l\mu_i$
    (3b) if $\{\mu_0 \ mod \ k, \dots, \mu_{k-1} \ mod \ k\} = \{0, \dots, k-1\}$ output $(\mathcal{G} = \{(g^l)^{jk}, j = 0, \dots, \frac{p-1}{k}\})$

Remarks
1) Perfect integer codes of type 1 just correspond to the coset splittings in [30], where the elements of $\mathcal{E}$ then are representatives of the cosets of the subgroup $\mathcal{H}$.

2) Computational results from [31], [35] suggest that at least for small sets $\mathcal{E}$ other types of perfect integer codes than those in the above theorem do not occur at all or only occur sporadically.

The above theorem is rather obvious. However, from the algorithmic point of view it has several advantages.

1) The algorithm "Integer Code" is rather efficient. The existence of an integer code in large groups (size 100000, fo instance) can be checked in a few seconds. Especially, in the mathematical literature, only existence results were analyzed without providing an efficient algorithm.

2) The integer code itself is automatically provided - just the set $\mathcal{H}$ from the theorem. This may be further analyzed, for instance, in order to express the integer code in dependence of the elements of $\mathcal{E}$.

3) From the integer code the structure of lattice tilings is clear. This means that the lattice points, in which to place the centers of the star bodies by which one would like to tile the space R^n, can be easily obtained. This may be of interest, for instance, in distributed computing, wher the processors should be placed efficiently in regular graphs such as a grid or a hypecrube.

Let us illustrate these advantages with some basic examples:

The set $\mathcal{E} = \{1, a\}$

1) The condition on the existence of a perfect integer code is that a has an even order modulo p.

2) The set $\mathcal{H}$ then consists of the group $\mathcal{G}$ of the even powers of a in Z_p^* (and its translates in the respective cosets). This was was discussed already in [23] and in [19] and [22] for $a = 2$.

3) If you want to tile a path or a cycle $\{0, \ldots, n-1\}$ by paths of length 2, place the initial vertex of these paths in the even positions. Obviously, this is only possible, if n is even.

4) The generalization to the set $\mathcal{E} = \{1, a, a^2, \ldots, a^{r-1}\}$ is straightforward – the integer code consists of the powers of a^r.

The set $\mathcal{E} = \{1, a, b\}$

1) The conditions on the existence of a perfect integer code in Z_p were derived in [35]

  1 The orders of a and b are both divisible by 3.
  2 Whenever $b^k = a^l$ for some integers k, l, then $k + l \equiv 0 \bmod 3$.

2) The integer code then consists of the subgroup $\mathcal{G} = \{a^i \cdot b^j, i - j \equiv 0 \bmod 3\}$, which is generated by the elements a^3, b^3 and $a \cdot b$, and its translates.

3) In a lattice tiling of R^2 or a tiling of a grid (direct product of two paths or two cycles) by the cross (body containing vertices $(u, v), (u, v+1), (u+1, v)$) place the central vertices (u, v) in positions i, j with $i - j = 0 \bmod 3$. Of course a tiling of the direct product of cycles $Z_k \times Z_l$ is only possible if k and l are both divisible by 3.

4) The generalization to sets $\mathcal{E} = \{1, a, \ldots, a^{r-1}, b, \ldots, b^{s-1}\}$ is possible. The integer code then basically is $\mathcal{H} = \{a^i \cdot b^j, i - j \equiv 0 \bmod (r + s - 1)\}$.

For the above two sets $\mathcal{E}$ the analysis was still possible without Theorem 1. The structure of the integer code still can be obtained using graph theoretic arguments. This argumentation breaks down when the set $\mathcal{E}$ is getting bigger. The following result can easily be derived with Theorem 1. A graph theoretic approach would already be very difficult.

The set $\mathcal{E} = \{1, a, b, c\}$
Theorem 2: *A perfect integer code with error set $\mathcal{E} = \{1, a, b, c\}$ (with a, b and c in the appropriate order) exists in Z_p^*, exactly if the following conditions hold*

1 *In $Z_p*/$ the orders of a and b are divisible by 4 and the order of c is divisible by 2,*
2 *whenever $a^i \cdot b^j \in \mathcal{G}$ then $i + j \equiv 0 \mod 4$,*
3 *whenever $a^i \cdot c^j \in \mathcal{G}$ then $2i + j \equiv 0 \mod 4$,*
4 *whenever $b^i \cdot c^j \in \mathcal{G}$ then $2i + j \equiv 0 \mod 4$.*

The set $\mathcal{H}$ then is determined by the subgroup $\mathcal{G}$ generated by the elements $a^4, b^4, a \cdot b, c^2$.

Proof of Theorem 2: With Theorem 1, a perfect integer code exists, exactly if for some l, the powers in the representations $a = (g^l)^{\mu_1}$, $b = (g^l)^{\mu_2}$, and $c = (g^l)^{\mu_3}$ fall into the three different congruence classes 1, 2, and 3 modulo 4. In order to follow the statement in the theorem, we choose $\mu_1 \equiv 1 \mod 4$, $\mu_2 \equiv 3 \mod 4$, and $\mu_3 \equiv 2 \mod 4$. Since $\mathcal{G}$ consists of all elements of the form g^{4l}, it follows immediately that a^4, b^4, $a \cdot b$, and c^2 and because of the group structure all their products must be contained in $\mathcal{G}$. Further, if $a^i \cdot b^j \in \mathcal{G}$ then obviously $i + j \equiv 0 \mod 4$, if $a^i \cdot c^j \in \mathcal{G}$ then obviously $2i + j \equiv 0 \mod 4$, and if $b^i \cdot c^j \in \mathcal{G}$ then obviously $2i + j \equiv 0 \mod 4$.

The set $\mathcal{E} = \{1, a, b, c, d\}$
When $|\mathcal{E}|$ is getting bigger, conditions similar to those in Theorem 2 must be verified. As in Theorem 2, one has to assure that several products of the elements in $\mathcal{E}$ are contained in the set $\mathcal{H}$. For instance, if $\mathcal{E} = \{1, a, b, c, d\}$ with $a = g^1 \mod k$, $b = g^2 \mod k$, $c = g^3 \mod k$, and $d = g^4 \mod k$, then the elements a^5, b^5, c^5, d^5, $a \cdot d$, and $b \cdot c$ must be contained in $\mathcal{G}$.

Symmetric errors
Most interesting for applications are symmetric errors, in which case the error sets $\mathcal{E}$ under consideration are of the form $\{\pm 1, \pm a_1, \ldots, \pm a_{k-1}\}$. In this case we identify the elements x and $-x$ in Z_p^* and hence consider factorizations by the set $\{1, a_1, \ldots, a_{k-1}\} \in Z_p^*/\{1, -1\}$ ad apply Theorem 1 to this setting.

In [35] and [36] special choices of the parameters a, b, and c relevant for the important shift codes were considered.

1. $\mathcal{E} = \{\pm 1, \pm 2, \pm 3, \pm 4\}$. There are only 9 prime numbers up to 10000 for which a perfect shift code exists, namely $p = 97$, 1873, 2161, 3457, 6577, 6673, 6961, 7297, and 7873. Observe that in this case perfect shift codes of type 1 cannot exist. Since 2 is a quadratic residue for primes $\equiv 1 \mod 8$ it must be an even power of a generator. Hence, the number 4 must be of the form g^{μ_3} with

$\mu_3 \equiv 0 \bmod 4$ and hence μ_3 would be in the same congruence class as $\mu_0 = 0$, the power of the element $1 \in \mathcal{E}$. This means that one has to search for perfect shift codes of type 2 in this case, which seemingly do not occur so frequently. The same holds for any other set containing the element 4, for instance for the set $\{\pm 1, \pm 3, \pm 4, \pm 5\}$, and in many cases also for sets $\mathcal{E}$ containing the number 2 – if not, the number 2 must be chosen as element c in Theorem 2.

2. $\mathcal{E} = \{\pm 1, \pm 2, \pm 3, \pm 5\}$. Perfect integer codes exist for $p = 137$, 953, 1697, 2417, 2533, 2753, 2777, 2897, 4073, 4673, 5153, 5417, 5657, 6113, 6257, 6737, 7193, 7433, 8753, 9257, 9497, 9857.

These primes have been found with the help of Theorem 1 in a few seconds. Further sets $\mathcal{E}$ can be studied similarly.

We also quickly found the following shift codes:

4. $\mathcal{E} = \{\pm 1, \pm 2, \pm 3, \pm 4, \pm 5\}$. Perfect shift codes exist for $p = 421$, 701, 2311, 2861, 3181, 3491, 3931, 4621, 5531, 6121, 7621, 7741, 9001, 9161, 9941.

4   Packings with Error Spheres

Since perfect integer codes seem to be quite sparsely distributed one might relax the conditions and no longer require perfectness but a good packing with the error spheres.

The special error set $\mathcal{E} = \{\pm 1, \pm 2, \ldots, \pm k\}$ is also denoted as $F(k)$ in the literature. We say that $F(k)$ n–packs $\mathcal{G}$ with packing set $\mathcal{H}$ of size n if all products $m \cdot h$ with $m \in F(k), h \in \mathcal{H} \subset \mathcal{G}$ are different. Of course, then $\mathcal{H}$ is a k–shift code of size n.

Let $m(k, n)$ denote the size of the smallest group $\mathcal{G}$ such that a k–shift code of size n exists in $\mathcal{G}$. For a good shift code (or the corresponding packing of the group $\mathcal{G}$) one would expect that $m(k, n)$ is not much bigger than the theoretical lower bound $2nk + 1$, which is obtained for a perfect k–shift code.

Such packings have been considered e.g in [6], [12], [32]. Some applications to Information Theory are discussed in [29] and [11]. The following asymptotical result is known [29].

$$\lim_{k \to \infty} \frac{m(k, n)}{k^2} = 1$$

Motivated by the geometric application of tiling the space with certain star bodies, where n is the dimension of the space, here the parameter n is fixed and k tends to infinity. The result shows that good packings in this case cannot be expected, since $m(k, n)$ is about k^2, which is much bigger than $2kn + 1$ for n small compared to k.

For applications in Coding Theory, however, one would rather fix k and look for code constructions suitable for any n.

In [30] several constructions for packings by the cross $F(k)$ (k–shift codes) are presented. Especially, an almost perfect $(p - 1)$–shift code of size $p + 1$ exists in cyclic groups of order $2p^2$ for an odd prime number p. For instance, if $p = 5$, then Z_{50} is packed by $F(4) = \{\pm 1, \pm 2, \pm 3, \pm 4\}$ with packing set $\mathcal{S} = \{1, 5, 9, 11, 19, 21\}$. Observe that all the products $m \cdot h$ of elements $m \in F(4)$

and $h \in \mathcal{S}$ are different and that only the elements 0 and $p^2 = 25$ in Z_{50} cannot be obtained as such products. Indeed, the general structure of the shift code is

$$\mathcal{S} = \{1, p, 2p \pm 1, 4p \pm 1, \ldots, 2\frac{p-1}{2}p \pm 1\}.$$

Another idea would be to follow the constructions in the previous section no longer requiring perfectness. So one has to arrange that for a generator g the powers $\mu_1 = 0, \ldots, \mu_k$ of $1 = g^{\mu_1}$, $2 = g^{\mu_2}, \ldots$, $k = g^{\mu_k}$ fall into different congruence classes. In this case a close packing wil be obtained as above (if some further divisibilty conditions hold). This way, one can also see, that the packings become worse the fewer residue classes are occupied by the μ_i's.

Some further sporadic constructions in [30] usually proceed by following the orbit of special elements in the group.

For the systematic search of 3–shift codes, this gave us the idea to the following greedy algorithm:

Consider the orbit of the element 3 in the cyclic group Z_l, i.e. the set

$$\mathcal{F} = \{3^s : s = 0, \ldots, \text{ord}(3)\}.$$

Starting with $i = 0$ include the element 3^i in the shift code $\mathcal{S}$ if possible, and set $i \leftarrow i + 1$. When the search in $\mathcal{F}$ is finished, continue with the same procedure in the residue classes $a \cdot \mathcal{F}$, $a \in \mathcal{G}$.

This way, we found some quite good shift codes: For instance, in Z_{40} the 3–shift code $\{1, 4, 5, 7, 9, 17\}$ of size 6 improves the value in Table V-4 on p. 316 in [30], where only an example of a 3–shift code of size 6 in Z_{43} was given. Further, Z_{56} contains the 3–shift code $\{1, 4, 5, 7, 9, 11, 13, 25\}$ of size 8 and in Z_{88} there is a 3–shift code of size 13, namely $\{1, 4, 7, 9, 11, 15, 17, 23, 25, 31, 36, 39, 41\}$.

Observe that all the group orders l here are divisible by 4. In this case the algorithm behaves very nice, since usually almost all odd elements and almost all elements $\equiv 2 \bmod 4$ are included in some sphere around a codeword in the shift code.

A similar construction – following the orbit of the element 2 – allowed us to find the following 4–shift code $\{1, 5, 8, 9, 11, 13, 14, 17, 23, 35, 37, 40\}$ of size 12 in Z_{99}.

The Greedy algorithm does not always perfom very well. Indeed, we have examples where it finds very bad packings. It would be interesting to find conditions under which good packings can be obtained using the Greedy algorithm.

Further, here it was only applied to the sets $F(3)$ and $F(4)$. The reason is that one should follow the orbit of one element 2, say, in the powers of the other one, then 3. For $F(5)$ one also has to consider the powers of the element 5, which becomes much more difficult. It would be interesting to find good packings in this case.

5   Concluding Remarks

A single checksum modulo an integer m was studied. In coding theory this is usually applied to correct unconventional types of errors such as asymmetric errors, deletions and insertions, permutations or peak shifts in run–length–limited codes.

Mathematically, the analysis of perfect codes in this setting is based on group factorizations, which are best carried out if p is a prime number. The problem is of mathematical interest itself, since a factorization (or equivalently a perfect code) corresponds to a lattice tiling of the the Euclidean space by certain star bodies.

Depending on the mathematical approach, several further applications in computer science and cryptography arise. From a lattice tiling of the Euclidean space one may obtain efficient placement of resources in computer networks as grids. From group factorizations, error correcting or steganographic codes can be derived.

To analyze the existence of perfect codes in Z_p we present a method to simply check the powers μ_i of the elements e_i in the set of tolerated errors, when they are represented as powers of a generator, i. e., $e_i = g^{\mu_i}$. This also yields an efficient computational criterion.

Good packings are very difficult to find. For small error sets a greedy algorithm provided some very close packings improving results of a previous table by Stein.

References

1. Belitskaja, E.E., Sidorenko, V.R., Stenström, P.: Testing of Memory with Defects of Fixed Configurations. In: Proceedings of 2nd International Workshop on Algebraic and Combinatorial Coding Theory, Leningrad, pp. 24–28 (1990)
2. Biggs, N.: Perfect Codes in Graphs. J. Combin. Theory Ser. B 15, 289–296 (1973)
3. Chen, H., Tzeng, N.: Efficient Resource Placement in Hypercubes Using Multiple-Adjacency Code. IEEE Trans. Comput. 43, 23–33 (1994)
4. Constantin, S.D., Rao, T.R.N.: On the Theory of Binary Asymmetric Error Correcting Codes. Information and Control 40, 20–26 (1979)
5. Dorbec, P., Mollard, M.: Perfect Codes in Cartesian Products of 2–Paths and Infinite Paths. The Electronic Journal of Combinatorics 12, R65 (2005)
6. Everett, H., Hickerson, D.: Packing and Covering by Translates of Certain Nonconvex Bodies. Proceedings of the American Mathematical Society 75 (1), 87–91 (1979)
7. Golomb, S.: A General Formulation of Error Metrics. IEEE Trans. Inform. Theory 15, 425–426 (1969)
8. Golomb, S.: Polyominoes, 2nd edn. Princeton University Press, Princeton (1994)
9. Golomb, S., Welch, L.R.: Algebraic Coding and the Lee Metric. In: Mann, H.B. (ed.) Error Correcting Codes, pp. 175–194 (1968)
10. Graham, R.L., Sloane, N.J.A.: On Additive Bases and Harmonious Graphs. SIAM J. Algebr. Discr. Math. 1, 382–404 (1980)
11. Hamaker, W., Stein, S.: Combinatorial Packing of R^3 by Certain Error Spheres. IEEE Trans. Inform. Theory 30, 364–368 (1984)
12. Hickerson, D., Stein, S.: Abelian Groups and Packing by Semicrosses. Pacific J. Math. 122, 95–109 (1986)
13. Horak, P.: On Perfect Lee Codes. Discrete Math. 309, 5551–5561 (2009)
14. Jerebic, J., Klavžar, S., Špacapan, S.: Characterizing r-Perfect Codes in Direct Products of Two and Three Cycles. Inform. Process. Lett. 94, 1–6 (2005)
15. Jha, P.K.: Perfect r-Domination in the Kronecker Product of Three Cycles. IEEE Trans. Circuit Systems – I: Fundamental Theory Appl. 49, 89–92 (2002)

16. Jha, P.K.: Perfect r-Domination in the Kronecker product of Two Cycles with an Application to Diagonal Toroidal Mesh. Inform. Process. Lett. 87, 163–168 (2003)
17. Kostadinov, H., Manev, N., Morita, H.: Double ± 1 Error - Correctable Codes and Their Applications to Modulation Schemes. In: Proceedings 11th Int. Workshop Algebraic and Combinatorial Coding Theory, Pamporovo, Bulgaria, pp. 155–160 (2008)
18. Levenshtein, V.I.: Binary Codes with Correction for Deletions and Insertions of the Symbol 1 (in Russian). Problemy Peredachi Informacii 1, 12–25 (1965)
19. Levenshtein, V.I., Vinck, A.J.H.: Perfect (d,k)–Codes Capable of Correcting Single Peak Shifts. IEEE Trans. Inform. Theory 39, 656–662 (1993)
20. Lisonek, P.: Sum Covers in Steganography. In: Proceedings 11th Int. Workshop Algebraic and Combinatorial Coding Theory, Pamporovo, Bulgaria, pp. 186–191 (2008)
21. Livingston, M., Stout, Q.F.: Perfect Dominating Sets. Congr. Numer. 79, 187–203 (1990)
22. Martirosyan, S.: Single – Error Correcting Close Packed and Perfect Codes. In: Proceedings 1st INTAS International Seminar on Coding Theory and Combinatorics, Thahkadzor, Armenia, pp. 90–115 (1996)
23. Morita, H., Geyser, A., van Wijngaarden, A.J.: On Integer Codes Capable of Correcting Single Errors in Two–Dimensional Lattices. In: Proceedings IEEE Int. Symp. Inform. Theory, Yokohama, p. 16 (2003)
24. Munemasa, A.: On Perfect t–Shift Codes in Abelian Groups. Designs, Codes, and Cryptography 5, 253–259 (1995)
25. Qu, M., Vanstone, S.A.: Factorizations of Elementary Abelian p-Groups and their Cryptographic Significance. J. Cryptology 7, 201–212 (1994)
26. Sands, A.D., Szabo, S.: Factoring Groups into Subsets. CRC Press, Boca Raton (2009)
27. Sidorenko, V.: Tilings of the Plane and Codes for Translational Metrics. In: Proceedings IEEE Int. Symp. Inform. Theory, Trondheim, p. 107 (1994)
28. Sloane, N.J.A.: On Single Deletion–Correcting Codes. In: Arasu, K.T., Seress, A. (eds.) Codes and Designs (Ray–Chaudhuri Festschrift), pp. 490–499. de Gruyter, Berlin (2002)
29. Stein, S.: Packing of R^n by Certain Error Spheres. IEEE Trans. Inform. Theory 30, 356–363 (1984)
30. Stein, S.: Tiling, Packing, and Covering by Clusters. Rocky Mountain J. Math. 16, 277–321 (1986)
31. Stein, S.: Splitting Groups of Prime Order. Aequationes Mathematicae 33, 62–71 (1987)
32. Stein, S.: Packing Tripods: Math. Intelligencer 17(2), 37–39 (1995)
33. Stein, S., Szabó, S.: Algebra and Tiling, The Carus Mathematical Monographs, vol. 25. The Mathematical Association of America (1994)
34. Szabó, S.: Topics in Factorization of Abelian Groups. Birkhäuser, Basel (2004)
35. Tamm, U.: Splittings of Cyclic Groups and Perfect Shift Codes. IEEE Trans. Inform. Theory 44, 2003–2009 (1998)
36. Tamm, U.: On Perfect Integer Codes. In: Proceedings IEEE Int. Symp. Inform. Theory, Adelaide, Australia (2005)
37. Varshamov, R.R., Tenengolts, G.M.: One Asymmetric Error Correcting Codes (in Russian). Avtomatika i Telemechanika 26, 288–292 (1965)
38. Vinck, A.J.H., Morita, H.: Codes over the Ring of Integers Modulo m. IEICE Transactions on Fundamentals of Electronics, Communications and Computer Sciences E81–A (10), 2013–2018 (1998)

Efficient Time-Area Scalable ECC Processor Using μ-Coding Technique

Mohamed N. Hassan and Mohammed Benaissa

Department of Electronic & Electrical Engineering,
University of Sheffield, Mappin Street, Sheffield, S1 3JD, UK
{m.nabil,m.benaissa}@shef.ac.uk

Abstract. The work in this paper discusses the feasibility of a low-resource ECC processor implementation over $GF(2^m)$ that supports scalability across a set of standards curves for application in resource constrained environments. A new architecture based on the microcoding technique and targeted to FPGAs is presented for the implementation of a low resource ECC processor design that is scalable to support the 131, 163, 283, 571 bits suite of recommended curves without significant deterioration of the performance. The processor is parameterized for 8, 16, 32-bit data-paths, to quantify the gain in terms of time and area in each case. The implementation results obtained show that the microcode approach results in a lesser area overhead for the ECC point multiplication compared to a full hardware implementation; this makes such approach attractive for numerous applications, where the hardware resources are scarce, as in security in wireless sensor nodes, mobile handsets, and smart cards.

Keywords: Elliptic curve cryptography, binary finite fields, Microcode, FPGA.

1 Introduction

The development of efficient algorithms and architectures for public key systems (PKS) has gathered a significant impetus in the last few years; this is attributed to the increasingly important role of PKS, in many life applications for providing authentication, integrity, confidentiality, and non-repudiation services, while minimizing complexities and vulnerabilities in the management of keys [2]. Elliptic Curve Cryptography (ECC) cryptosystems have emerged recently as an attractive alternative to the well established RSA especially in resource critical applications due to their superior security strength per bit, which accounts for the use of much shorter keys than RSA [2].

The design space for ECC extends across the software and the hardware implementation domains. Fundamentally, hardware methods offer higher speed and bandwidth stability, thus providing more scope for real-time operation. ASIC and FPGA are two distinct environments for implementing cryptographic algorithms in hardware. The FPGA environment is adopted in this work as it enables reconfigurability at low cost and without dissipating the design efficiency. The proposed ECC processor architecture is based on a firmware control to control the execution of the arithmetic operations on the reconfigured hardware. The firmware

M.A. Hasan and T. Helleseth (Eds.): WAIFI 2010, LNCS 6087, pp. 250–268, 2010.
© Springer-Verlag Berlin Heidelberg 2010

control is realized via microcoded instructions stored in a block memory on the FPGA. A Specialized instruction set has been developed to execute the ECC point operation and to support the scalability prerequisite. Such an approach simplifies by far the control of applications that conventionally require substantial computational power like ECC; by emulating or (coding) their behavior as simple instructions called microinstructions, yielding a significant reduction in the circuit complexities and their corresponding timing delays.

The adoption of microcoding in implementing the ECC point multiplication has been reported by a number of researchers. [3] implemented, on 0.18μm CMOS technology, a universal cryptography processor for smart-card applications that incorporates the Data encryption standards (DES), Advanced encryption standards (AES), and ECC curve with 83-bit. [4] introduced a generic ECC processor over $GF(2^m)$ on FPGA; their proposed processor supports standard curves m=163, 193, 233 besides non-standard curves for m ≤ 255. [5] employed the microcode technique to compare the performance gain of using the projective coordinates over affine coordinates and two designs with two different microcodes with the same arithmetic circuit were presented. The highest field length reported in [5] was 473-bit without incorporating the scalability property. In [6], two distinct microcode units are adopted to provide the point operation and finite field arithmetic control units in a processor working over $GF(2^{167})$. However, none of the reported designs addressed the scalability issues.

In this work, we demonstrate that it is possible to provide the required level of field agility by using a limited amount of hardware resources. The overheads associated with the control logic are mitigated by adopting the microcode technique. The microcoding approach facilitates the design of the control unit and makes it easier to develop and debug, yielding a processor's architecture that can provide limited programmability that is enough to adapt to changes in standards.

The presented ECC processor architecture based on microcode control is augmented with small footprint arithmetic units to perform the computation of the ECC point multiplication across many standards curves covering both NIST [7] and SECG [8], namely, 131, 163, 283, 571 with practical performance figures. The design is parameterized for different data widths 8, 16, 32 bits without the need to modify the micro-program (microcode) to quantify the performance gain in each case. Additionally, a baseline design that implements the point operation control as a hardwired control is provided for comparison.

The paper commences with a brief background of ECC and finite field arithmetic in section 2, followed by an introduction to the microcoding technique in section 3. The detailed ECC processor architecture is revealed in section 4. Section 5 provides the implementation results and analysis. Section 6 concludes the paper.

2   Background Theory

Mathematically, an elliptic curve E defined over binary finite field $GF(2^m)$ is the set of points together with a point at infinity called O satisfies the reduced Weierstarss equation on the form :

$$E: y^2+xy=x^3+ ax^2 + b \tag{1}$$

Where, a, b ∈ $GF(2^m)$ and b ≠ 0.

The central operation in any ECC system is called the scalar point multiplication given in (2).

$$Q=K.P \qquad\qquad (2)$$

Where P, Q $\in$ E and K is an integer. The computation of (2) or evenly the scalar point multiplication can be carried out in one way to compute Q given P and K. However, solving (2) in the reverse way to find K given P and Q is believed to be mathematically intractable [2]. Many efficient algorithms in the context of ECC have been proposed to calculate (2). Herein, Lopez-Dahab's method based on Montgomery ladder trick over $GF(2^m)$ outlined in [10] is adopted to compute (2) due to its symmetrical structure, which by far simplifies its emulating using microcode technique, in addition to, its potential to resist timing attacks and power analysis attacks [2]. This method has two affine and projective flavors, we opted the later one to evade the costly modular inversion operation over $GF(2^m)$. The approximate running time for this method is $(6M+5S)\times(m-1) + I +10M+3S$, where, the symbols M, S, I denote modular multiplication, squaring, inversion over $GF(2^m)$ respectively.

2.1 Arithmetic in Binary Finite Fields

In this work, the binary field arithmetic is realized as a polynomial arithmetic modulo irreducible polynomial $p(x)$ as suggested by ECC standards [7, 8]. Finite field multiplications performed using Comba's algorithm [9]. Modular addition is implemented as a bitwise XOR of any two elements defined over $GF(2^m)$. The squaring operation is accomplished through algorithm 2.39 in [2] with word lengths 8, 16, and 32-bits. Modular inversion has the most time and area complexities of all the finite field arithmetic operations. In the context of ECC, two main methods are used to implement the modular inversion, Extended Euclidian algorithm (EEA) and Fermat's little theorem (FLT). Herein, Inversion is implemented using an efficient algorithm 5.6 based on FLT in [2]. FLT uses multiplications and squaring only to compute the inversion. The total number of multiplications is given by $\lfloor \log_2(m-1) \rfloor + h(m-1)-1$, where $h(m-1)$ denotes the Hamming weight, with $(m-1)$ squaring operations.

3   Microcode Design

Customarily, the control unit in any application is implemented as a finite state machine, which may entail a complex logic and long time delays, especially in applications involve highly intensive computations. An old trend has become active recently used as alternative for the hardwired control called *Microcoding*. From the hardware perspective, a microcoding description can be regarded as higher-level abstraction than a finite-state machine [15]. In some context, this is called *firmware* control. The later term reflects the fact that a microcode technique is midway between the hardware and software design space [15].

The microcode (microprogramming) technique is initiated first by M.V. Wilkes in the early 1950s [1]. Wilkes proposed an approach to design a control unit that was systematic and avoided the hardwired control's complexities [15]. The principle advantage of such design strategy is that the circuits that constitute the microcode control are very simple and require tiny pieces of logic. Furthermore, it can be changed without requiring recompilation of the entire processor. This by far facilities the design of the control unit and makes it easier to develop and debug, yielding a quick, cheap and less error-prone solution for applications require complex control circuits like the design in our hand. In security terms, it is of interest for any security application to incorporate the capability to change the higher-level operations (e.g. point doubling and addition algorithm) if better or more secure algorithms for these operations are proposed.

To control a logic or hardware operations in a data-flow e.g. multiplexers, decoders, or arithmetic logic units using microcode approach, a group of bits called a *control word* or *microinstruction* is emanated from the microcode control unit. A *microinstruction* contains controlling bits that represents the micro-operation and bits to determine the next address generation to the sequencer circuit. Hence, the control word can be coded or programmed to initiate the required micro-operations. Conventionally, microinstructions can be grouped in the code memory in what so called a *Routine*. A typical microcode control organization sketched in figure 1 is used to perform this functionality.

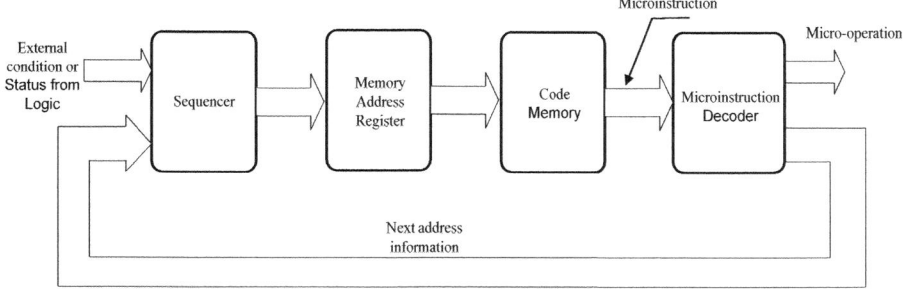

Fig. 1. Microcode control organization

As shown in figure 1, a microcode control consists of a code memory (control store) to hold the microinstructions (each instruction represents one or more micro-operation) [1, 15]. The code memory contains only the instructions that can be emulated and have frequent or non-frequent usage [15]. Meanwhile, the hardware can implement other instructions that cannot be emulated. Thus, the microcode design is a mixture of emulated and hardwired instruction in one platform [1, 15]. The code memory is managed by sequencer logic through a memory address register, which holds the address of the next microinstruction required for execution. The code memory outputs a microinstruction, which will be decoded in a decoder circuit to generate a decoded control word to execute corresponding micro-operations.

4 Processor Architecture

For ECC point multiplication, assigning the tasks between the hardware and microcode regions has to be undertaken efficiently to capitalize the adoption of microcoding technique. The decision to allocate a task to either of the domains is based on the design criterion. In our case, where reducing the controlling circuits' overheads is paramount, in addition to the prerequisite to incorporate the scalability, the decision is made to microcode the controlling of point operations (senior level) (controlling the point doubling - addition and conversion to the affine coordinates) and to implement the finite field arithmetic control (sublevel control) as a finite state machine (hardwired control).

The key motivations for this choice are two folds, first to exploit the symmetric nature of the Montgomery point multiplication, which facilitates the development of the microinstructions in a linear code style with reasonable code memory size to emulate the point operations. Second, microcoding the sublevel control would require specialist support from the hardware resources, which may incur more area penalties, and increase the design complexity.

The top-level of our proposed processor is shown in figure 2. In this work, a traditional Harvard architecture is employed; that way, the processor's data-path is isolated from both the control path and the key path. This architecture not only reduces the overall complexity of the design, but also, maximizes the performance of the overall circuits and enables our scalable ECC processor to be parameterized for different data-path widths without the need to alter the microprogram.

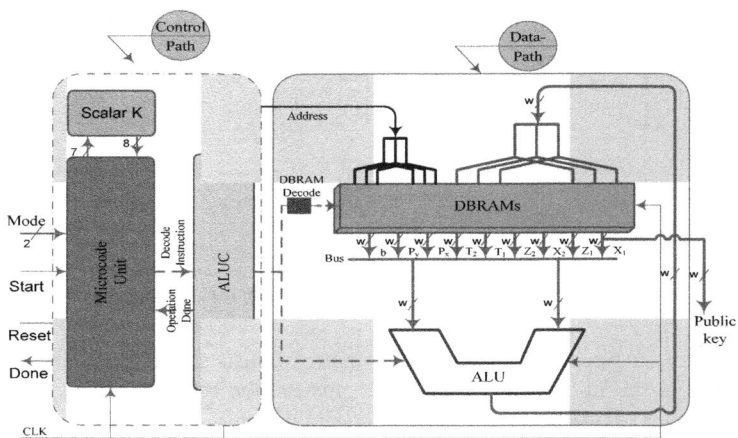

Fig. 2. The top level of the new area-efficient scalable microcode ECC processor

The control in this design is achieved via two levels of hierarchal control using hybrid control approach. The head of the hierarchy is the microcode unit, which controls the point operation execution through commanding the sub-level control to perform the underlying finite field arithmetic. The processor data path contains the arithmetic circuits, switching multiplexers, and the storage blocks (Dual Block memory DBRAM). The sublevel control executes the senior level commands and manages the data-path's traffic between the ALU and storage elements. Mode pins are provided to enable the selection of the required field extension, for instance, when the mode is set to "00", this value is equivalent to the field extension 131-bit, the other values are "01"=163-bit, "10"=283-bit, and "11"=571-bit.

4.1 Instruction Set Design

As shown in figure 3, each instruction has a 16-bit width and contains four fields. Each field has four bits long. The first field is dedicated to hold the micro-operation (Opcode). The other fields hold values according to the category of the microinstruction. Four categories of instructions are used to control the execution of the point group operations. They are finite field arithmetic instructions, branching instructions, logical and arithmetic instructions, in addition to the memory instructions. The finite field arithmetic instructions are MUL, SQR, ADD, these instructions execute the multiplication, squaring and addition operations over $GF(2^m)$ respectively. The reduction operation is interleaved with the multiplication and squaring operations, consequently, no special instruction is needed for this operation. The arithmetic instructions require three fields to represent the addresses of the source operands and one address for the destination operand.

The branching instructions JMPZ, JMPNZ, JMP are used to perform a conditional or unconditional branching to control the flow of the point operations and the conversion from the projective to the affine coordinates. The branching instructions require two fields to hold the branching address etc.

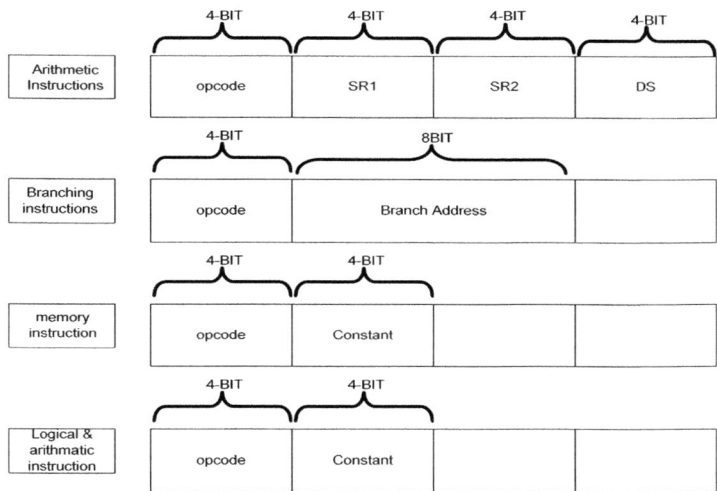

Fig. 3. Instruction set format

The logical instructions are SHIFTL, SHIFTR, DEC and COMPARE. SHIFTR and SHFTL are used to performing shifting by one bit right or left respectively for a register. A DEC instruction decrements the content of a register by one. A COMPARE instruction compares the content of a register with a constant value. Three additional memory instructions (COPY, INPUT, OUTPUT) are employed to control the data memory. A COPY instruction copies the content of a memory to another. This instruction is needed during the execution of the conversion from the projective to the affine coordinates. An INPUT instruction is used to input the curve parameters and private key to the storage elements, in addition to reading the operating mode (decoded field information).

Table 1 shows the specialized instruction set to control the ECC point.

Table 1. Instruction Set

Mnemonic	Op-code	Source address-1	Source address-2	Destination address	Operation
NOP	0000	0000	0000	0000	No operation
MUL	0001	SR_x	SR_y	DS	Multiply $(SR_x) \times (SR_x) \rightarrow$ result in (DS)
SQR	0010	SR_x	SR_x	DS	Square $(SR_x) \rightarrow$ result in (DS)
ADD	0011	SR_x	SR_y	DS	ADD $(SR_x + SR_y) \rightarrow$ result in (DS)
COPY	0100	SR_x	SR_x	DS	COPY $(SR_x) \rightarrow$ into (DS)
MASK	0101	K	0000	0000	Hide the (sw −1 to m) bits in K
COMPARE	0110	R_x	K_y	0000	Compare register R_x with constant K_y
SHIFTR	0111	R_x	0000	0000	Shift right register R_x
SHIFTL	1000	R_x	0000	0000	Shift left register R_x
SET	1001	R_x	0000	0000	Set register R_x with its intial value
DEC	1010	R_x	0000	0000	Decrement register R_x by one
JMPNZ	1011	$Addr_x$	$Addr_x$	0000	Jump to $addr_x$ if the tested value is not equal to zero
JMPZ	1100	$Addr_x$	$Addr_x$	0000	Jump to $addr_x$ if the tested value is equal zero
JMP	1101	$Addr_x$	$Addr_x$	0000	absolute Jump to $addr_x$
OUTPUT	1110	SR_x	0000	0000	Output the data content in memory SR_x
INPUT	1111	SR_x	0000	0000	Input data into memroy SR_x
END	1111	1111	1111	1111	End program

An OUTPUT instruction is used to output the x and y coordinates of the public key. A COPY instruction has 16-bit length as finite field arithmetic instruction. An END instruction terminates the program execution. A SET instruction loads a value to a register. A MASK instruction is adopted at the beginning of the microprogram to zero out the (sw-1 to m) bits in the most significant word of K and keep the remaining bits intact for testing the '0' and '1' conditions, where s, m, w denote the number of words, field order and data-path width respectively.

Point Multiplication Microprogram

We have adopted the Montgomery's point multiplication method using Lopez-Dahab projective coordinates to compute the ECC point multiplication [10]. This technique

SET	k_addr	
MASK	k	
Check_new_k_bit:		
SHIFTL	k	; START NEW ITERATION
COMPARE	bit_mask,00	
JMPNZ	Test_k_bit	
DEC	k_addr	
COMPARE	k_addr,7f	
JMPZ To_affine		; CONVERT TO AFFINE
SET k		
Test_k_bit:		
COMPARE	K , bit_mask	; TEST K BIT-ODD-EVEN
JMPNZ	k_bit_ONE	; JUMP TO K=1 – ROUTINE
MUL	X2,Z1,T1	
MUL	X1,Z2,X2	
MUL	T1,X2,T2	
SQR	T1,T1,Z2	
MUL	Z2,PX,X2	
ADD	T1,X2,X2	;END OF POINT ADDK='0'
;----------------------		
SQR	Z1,Z1,T1	
SQR	X1,X1,T2	
SQR	T2,T2,X1	
MUL	T2,T1,Z1	
SQR	T1,T1,T2	
MUL	T2,b,T1	
ADD	T1,X1,X1	;END OF POINT DOUB
JMP	Check new k bit	;START NEW ITERATION

Fig. 4. A Microcode snippet

exhibits a high degree of symmetry, where both addition and doubling of a point are performed in the same iteration irrespective of the tested binary value of the private key k, yielding a simplification in the control operation and reduction in its circuit's complexity. In terms of microcode design, this is reflected in decreasing the required number of microinstructions to emulate Montgomery's point multiplication algorithm.

A snippet of the proposed microcode instructions to realize the ECC point multiplication is shown in figure 4. Our microprogram is divided into three parts. The first part is dedicated to input the curve parameters and Field information. The second part consists of routines to load, test the binary value of the key, test the end of key, and implement addition and doubling of the point. The third part performs the conversion of the final projective result to the affine coordinate and outputs the public key. The whole ECC point multiplication is microcoded in 105 instructions.

4.2 Microcode Control Unit

The microcode unit is the senior control, which is responsible of manipulating the point operations and the coordinate's conversion. The architecture of this unit is similar to the traditional micro-programmed control unit in [1] and sketched in figure 1.

At the start, The INPUT instruction is used to load the required field information from the mode pins. Then, the field information is passed automatically to the ALUC to adjust its parameters according to the working field. Next, the INPUT instruction is used again to read the input port to load the base point P coordinates and the curve parameter. The SET instruction preloads the internal counters of the

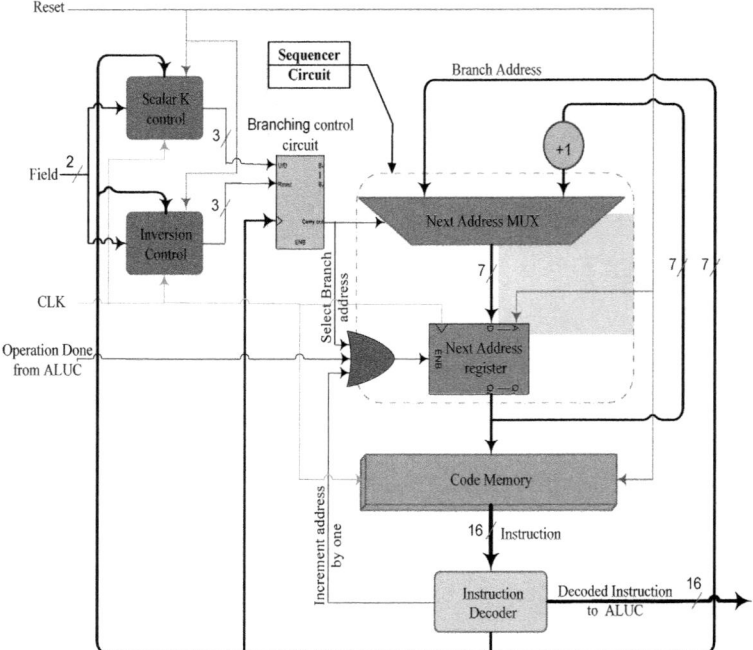

Fig. 5. Microcode control unit

private key control and the finite field inversion control unit with the value of the uppermost address $s = \lceil m / w \rceil$ of the private key K, and the required squaring count $= \lfloor m / 2 \rfloor$ respectively.

Figure 5 shows the key elements of the microcode unit. To execute a microinstruction, the sequencer logic circuit issues the address of the required instruction to the code memory; next, the microinstruction whose address is specified is generated and then decoded in the instruction decoder unit. The contents of the decoded instruction are used to generate commands, which are relative to the category of the instruction. The feedback from the ALUC, the other hardware units and the content of the current instruction is used to specify the next address value and so on. The next address value will be always an incremented value of the current address or an absolute address value. The microcode unit includes a sequencer circuit, branching circuit, private key test unit, finite field inversion control, code memory, and instruction decoder.

Sequencer Circuit

The main function of the sequencer circuit is to produce the required position address of the next instruction to the code memory. The generated next address can be an incremental address, unconditional branching as required from the address field in the previous microinstruction, conditional branching from logic or registers (nonzero/zero

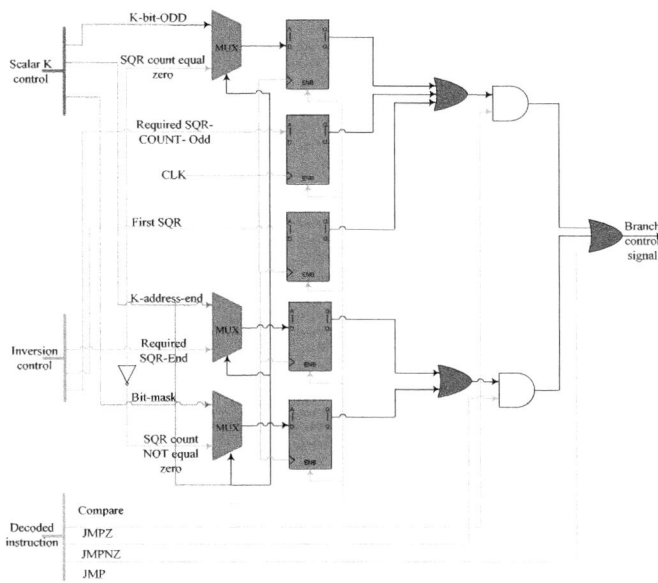

Fig. 6. Branch control circuit

condition, etc), to return to an arbitrary address from subroutine calls. The microcode sequencer in this design is very simple and has the smallest area overhead among all the design units. It consists of a multiplexer and address register. The inputs to this unit are the output from the address register incremented by one and the next address from the current microinstruction (second and third field). The decision to select either of the two inputs is based on the select address signal from the branching circuit.

Branch Control Circuit

The branching circuit provides the decision making capabilities to control the flow of the microprogram according to feedback signals from the hardware units (ALU control unit), the control circuits (private key control, finite field inversion control), or instruction decoder circuit. Three branching instructions are used to support the functionality of this unit, namely, JMP, JMPZ and JMPNZ. The inputs to this unit are the status signals from the private key control (e.g. k-bit is odd-even, k address-end) and the finite field inversion circuits etc. The output signal of this unit is a select next address signal, which is either logic '0' to increment the current address by one or '1' to jump to a specified address in the microinstruction if any of the three branching instructions is fulfilled. The JMPZ and JMPNZ instructions are always preceded by the COMPARE instruction. The status bits from the private key control or the finite field inversion circuits are tested by the COMPARE instructions.

Accordingly, a zero-flag is checked after every COMPARE instruction by JMPZ or JMPNZ instructions to perform a branch, otherwise; the code memory's address is incremented by one. The JMP instruction executes an unconditional branching by loading the next address of the microinstruction from the current one. Figure 6 depicts the branching unit structure.

Code Memory

The code memory is organized as 128×16-bits ROM and has been implemented as DBRAM. The microcode unit controls the execution of the point multiplication through generating the instructions to the sublevel control unit, which in turn manipulates the computations of underlying finite field arithmetic.

Instruction Decoder

The instruction decoder unit maps a microinstruction into a group of control signals to test a value and return a result or to command other hardware units to perform certain actions. In the microcode approach, two types of instruction are used. They are horizontal and vertical instructions [1, 15]. Both terms are related to the way that the microinstruction is used to generate the control signals. In the horizontal instruction, each bit in the microinstruction pattern executes a certain micro-operation i.e. each bit in the microinstruction is attached to a control signal. This technique maximizes the usage of the hardware resources. Hence, this comes at the expense of increasing the complexity of the microcode circuit due to increasing the length of the control word.

On the other hand, the vertical instruction encodes a field of t control bits in a microinstruction to provide 2^t micro-operations. Each field of t bits has a corresponding decoding circuit is placed vertically to produce the controlling circuit. The instruction decoder unit decodes (translates) each op-codes, which upon decoding, activates one or more control signal. The latter technique is opted to implement the instruction decoder circuit, whereas, it has the advantage of reducing the microinstruction width in comparing with former approach at the expense of additional amount of logic and time delays.

4.3 Arithmetic and Logic Unit Control (ALUC)

The ALUC is employed to carry out the multi-precision multiplication algorithm in [9], squaring, and the reduction procedure in [11], which is interleaved with both multiplication and squaring. Furthermore, it controls either the data transactions between the block memories and the ALU and block memories with each other. The ALUC supports the execution of the arithmetic instructions MUL, SQR, ADD, in addition to, performing the data memory instructions (COPY, INPUT, OUTPUT). Upon receiving the decoded instruction from the instruction decoder unit, the ALUC starts to execute the required arithmetic or memory operations e.g. input or output data, copy from memory to another. After the execution of any operation is accomplished, the ALUC sends a DONE signal to the microcode unit to fetch and execute the next instruction and waits for the next command from microcode unit. The ALUC is realized as a finite state machine in 107 states.

4.4 Data Path

As shown in figure 7, the data path includes the arithmetic and logic units (ALU), storage memories and the data multiplexing. All the units on the data-path are supervised by the arithmetic and logic unit control (ALUC) to carry out the all the arithmetic operation over $GF(2^m)$. The data-path between the data memories and the ALU is implemented as a bus structure to remove the need to multiplex the operands from different memory elements at the ALU input [4]. In addition, all the other multiplexers in this design are implemented as tri-state buffers-based multiplexers to reduce their area overheads. The ALU consists of multi-precision multiplier-squarer and adder unit over GF(2), (MSAU) and scalable reduction circuit (SRU).

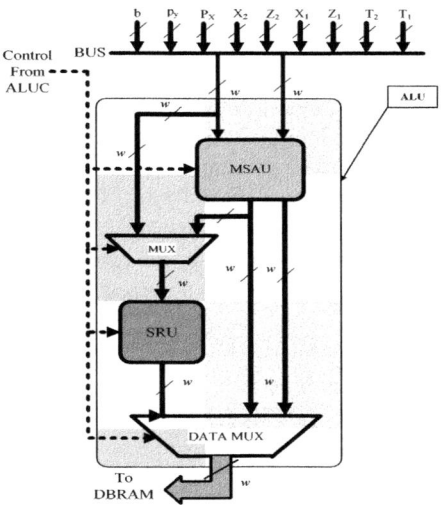

Fig. 7. Scalable ALU for area optimized ECC processor over $GF(2^m)$

Arithmetic and Logic Unit (ALU)

The ALU unit depicted in figure 7, consists of two novel arithmatic circuits namely, a multi-precision multiplier-squarer and adder unit over GF(2) (MSAU) augmented with a scalable reduction unit (SRU) and two data multiplexers. Figure 8 illustrates the proposed MSAU circuit.

The MSAU incorporates three identical instances namely, mulgf2 units, and accumulator circuits. The individual mulgf2 unit, consists of 2×w-bit registers to load the multiplicand and the multiplier operands and w×w multiplier unit.

The individual mulgf2 unit performs a w×w-bit single precision multiplication over GF(2). The output of each mulgf2 unit is 2w-1 bit. The accumulating circuit has a 2w-XOR gate and 2w register. Its function is to accumulate the sum of the partial products of the mulgf2 units with the accumulated 2w-carry from the preceding single-precision multiplications. In addition, it produces the w-bit partial product at the end of each inner loop iteration as a part of the whole multiplication result. The

MSAU performs the partial product in each iteration of Comba's algorithm in the following style. The ALUC unit generates three successive load signals to both multiplicand and the multiplier in the three-mulg2 units.

According to the flow of Comba's algorithm, if the required number of load operations is less than three, the remaining 2w registers are zeroed out. Afterwards the three-mulgf2 units are completely loaded, the ALUC generates a multiply signal to perform three partial products simultaneously. The outputs of the mulgf2 units is XORed with the 2w-bit carry. Thus, the (accumulation) XOR operation by design is performed simultaneously with the multiply operation, yielding a further reduction of the latency of the multiplication operation by $((\# \text{ mulgf2}) -2)$ clock cycles, where # mulgf2 represents the required total number of multiply operations to perform a full precision multiplication. The ALUC issues a store signal at the end of each iteration to store back the partial product result in a DBRAM or to the SRU. The MSAU completes one multiplication over $GF(2^m)$ in $(2s^2 + \#\text{mulgf2} + 2s)$ clock cycles.

Due to the restrictions on the resources in our design and to exploit the contribution of the three-mulgf2 units, the squarer functionality is realized through the same circuitry by bypassing the accumulating circuit which is not needed in the squaring mode owing to the linear nature of the squaring operation over $GF(2^m)$. As shown in figure 3, two additional multiplexers are placed in the multiplier data path to realize the squaring operation. In this case, the squaring is accomplished in $(2s + \lceil s/3 \rceil + 2s)$ clock cycles. The MSAU is used to carry out the modular addition by simply loading two operands into the first mulgf2 unit and a corresponding store signal to store back the result into the DBRAMs.

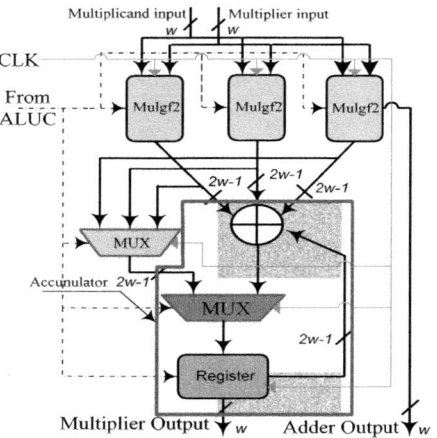

Fig. 8. The novel MSAU architecture

The reduction operation in this work is hardwired and interleaved with the operation of the MSAU circuit. The SRU portrayed in figure 9 adopts the right to left reduction algorithm in [11]. The proposed circuit is designed such that it has the capability to work with four different field extensions, namely, 131, 163, 283, 571

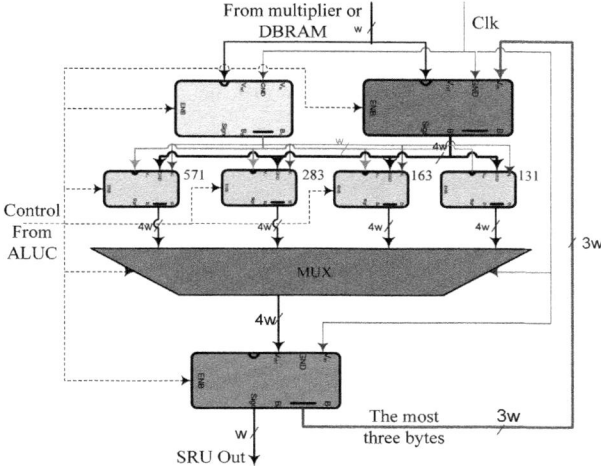

Fig. 9. Scalable reduction unit (SRU)

bits without the need to modify the unit. The operation of the SRU starts after the completion of the computation of the partial product of index s in both multiplication and squaring operations, until the word 2s-2. The ALUC switches the output word from the MSAU into the input of the SRU to reduce it as illustrated in figure 7. The whole reduction process is executed in (21 + 3s) clock cycles.

The storage elements in this design are realized through the available dual block RAM (DBRAM) and distributed memories. The field parameters, the point coordinates, and the private key are implemented using the distributed memories. The projective coordinates (X_1, Z_1, X_2, Z_2) and the two temporary variables (T_1, T_2) consumed three DBRAMs. The code memory is realized on additional DBRAM. Noteworthy, employing the DBRAM reduces considerably the design budget via saving the FPGA resources and leads to a faster and compact circuit architecture.

5   Implementation and Results Analysis

The main design goals are to implement a scalable ECC processor that achieves a small area footprint. The microcode unit, ALUC, arithmetic circuits and other units are modeled in VHDL language. We implemented the proposed microcode ECC processor on the smallest chips from XILINX FPGA, namely, XC3S50 and XC3S200. The design was parameterized for three different data-path widths, namely, 8, 16, 32-bit to quantify the performance gain in each case. The microcode program for the four different curves namely, 131, 163, 283, 571 recommended by [7-8] is developed in assembly language. Xilinx ISE 9.2.04i tool is used to synthesize and place & route the designs for different data paths. ModelSim 6.2 performed all the simulations and verification across all the different levels of the design flow.

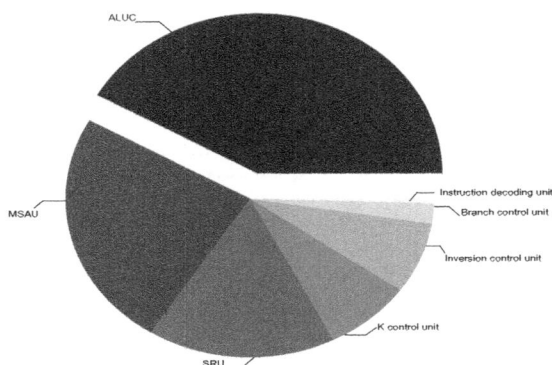

Fig. 10 Area utilization of the individual unit in the microcoded ECC processor for w = 8-bit

The pie chart in figure 10 depicts the area utilization for individual unit in the novel microcode ECC processor for w=8-bit. Clearly, the ALUC has the most contribution (23% of the total design's slice count) in the total area consumption amongst the other units.

To support our claims in this work, a baseline design for ECC processor has been provided. The baseline design implements the senior level control as a hardwired control and keeps the other units. We implemented the baseline design for data-path w=8-bit only, as the senior control circuit in our proposed processor is independent on the data path width and the main purpose is to compare the consumed resources to determine the efficiency in terms of area resources in each case. Table 2, shows the actual cost for the microcode-control- based ECC processor for different data widths and the hardwired controlled-based design for 8-bit data path after place and route process.

Table 2. The whole utilization of the scalable ecc processor on Xilinx FPGA Spartan3 XC3S50-S200 for microcode and hardwired control-based designs

Control	Chip	Data-Path	D-BRAM	LUT	F.F	Slices	Period (nsec)
Microcode	XC3S50	8		847	417	543/70%	13.083
		16		1056	543	619/80%	13.402
			4				
	XC3S200	32		2220	839	1180/61%	14.731
Hardware	XC3S50	8		1087	597	617/80%	12.787

Noticeably, the microcode-based design achieves 10% lower area consumption than the fully hardwired controlled-based design for the same w = 8-bit. The expected performance for microcode-design may be less than the hardwired control; this is attributed to including the code memory that is realized as DBRAM, which contributes in slight increase in the overall design time delay. Moreover, the microcode-based design for w = 16-bit is roughly has the same area overheads as the hardwired control-based design for w=8-bits. The improvements in the area are owing to the microcode technique, which decomposed the senior level control into higher level of circuit abstraction, yielding simplification of the implementation of the logic circuits. Table 3 shows the proposed ECC processor timing for different data widths.

Table 3. The ECC processor timing for different data widths in (msec)

w \ m	131	163	283	571
8	9.578	17.157	76.04	587.6
16	3.51	5.36	21.7	162.16
32	1.72	2.7	8.3	51.38

Table 4 compares the existing works that implemented the ECC point multiplication using both microcode-control and hardware-control methods. In the literature of the ECC, not much works addressed the adoption of the microcode technique to realize the ECC point multiplication. For the sake of fairer comparison, the comparison with the state of the art will be restricted to the designs that implemented microcode control and those that have relatively low FPGA resources utilization. Noteworthy, the equivalent LUT metric used in table 4 is computed based on the fact that any 4-input LUT can be configured also as a 16-bit RAM (Random Access Memory) [26, 27].

Noticeably, our implementation is the distinctive in incorporating the field with 571-bit length. Moreover, it has the smallest code memory utilization. The works in [4, 12] implemented an ECC processor for both standardized and arbitrary curves over $GF(2^m)$. The maximum reported field lengths in [4] and [12] are 256 and 283 bits respectively. In [4], the ECC processor consumed much larger area resources due to the adoption of digit serial multiplier and divider circuit to accelerate the computation of the ECC point multiplication. Compared to [12], our processor for the same word length of w=32 bits takes less resources whilst supporting more and larger fields. In [5], an ECC processor is reported that works over optimal normal controlled by a microcoded unit with relatively low FPGA resources utilization. Our processor outperforms [5] in two aspects; firstly, it achieves 89.6%, 88.2%, 77.6% lower slice count usage for w= 8, 16, 32-bits respectively while supporting higher security levels (571 bits compared to 473 bits); secondly, incorporates scalability across many standard curves for different path widths.

The design in [6] employed two dedicated block memories to hold the instructions to control the point multiplication algorithm and the execution of the finite field arithmetic. Our solution has a smaller area overheads by 62.2%, 55.2%, 16.2% for w=8, 16, 32-bit respectively in terms of equivalent LUT count, with fewer number of

Table 4. Comparison with the State of the Art

	[12] VIRTEX-II XCV2000E	[4] VIRTEX-II XCV2000E	[5] Virtex-I-XCV1000	[13] VIRTEX II	[6] VIRTEX XCV400E	This work Spartan III XC3S50		This work Spartan III XC3S200
Data path	32	256	473	192	167	8	16	32
D-BRAM	8	10	8	2	10	3		
Code Memory	256×16	512×16	512×16	NA	512×16 512×24	128×16		
Memory Eq. LUTs	380	656	719	24	1364	284		
LUTs	2556	20068	8422	4729	1627	847	1056	2220
Total eq. LUTs	2936	20724	9141	4753	2991	1131	1340	2504
Scalability	Yes	Yes	No	No	No	Yes		
Max. (m)	283	255	473	192	167	571		
Max Freq MHZ	150	66.4	18	50	85.7	76	75	70.5
Through put[kbps]/ (m)	83.6/ (163) 43.6/ (283)	1164.2/ (163) 1015.7/ (193) 1013/ (233)	NA	32/ (192)	303.6/ (167)	13.8 (131) 9.5 (163) 3.72 (283) 0.972 (571) 10	37.32 30.41 13.04 3.52 20	76.16 60.37 34.1 11.1 12
Normalized Efficiency = Through put[kbps]/ LUT^2	9.6/ (163) 5/ (283)	2.7/ (163) 2.3/ (193) 2.3/ (233)		1.4/ (192)	34/ (167)	7 (131) 3 (163) 0.8 (283) (571)	17 7 2	9 6 1.7
		Fastest				Smallest		Smallest-32-bit

DBRAMs and support for higher fields. An example for designs that adopts the hardwired control to implement a compact ECC processor over GF(p) on FPGA was presented in [13]. Their compact processor with 192-bit data-path is larger than our scalable ECC processor by 76.2%, 71.8%, 47.3% for w= 8, 16, 32-bit respectively.

In terms of the efficiency, which corresponds to the capability of a design in terms of time per square of the equivalent LUT unit [14], the design in [6] has the best efficiency in the entries of table 4. This is can be accounted for the effect of extending the storage capacity of our design to adapt to the maximum field length (571-bit), which contributed to increasing the total equivalent LUTs count for different data paths. Furthermore, the processor architecture in [6] employed a full-length digit serial multiplier. Consequently, if we presume that our design would scale to m=163 only, which is a comparable to the field order in [6], additionally, the performance of our processor could be further enhanced by augmenting more mulgf2 units in the MSAU to accelerate the modular arithmetic computation, which would have a corresponding trivial increase in the processor' area complexity. In this case, it is expected that our scalable design would exhibit better efficiency than the designs in [6].

6   Conclusions

We investigated the potential of the microcode technique to implement a low resource scalable ECC processor over $GF(2^m)$ on FPGA without detrimental effect on performance. In this context, a scalable ECC processor was presented that supports the named curves in [7] and [8], 131, 163, 283, 571-bit, in which the point operations are emulated using the microcoded instructions. The microcode instructions are stored on a dedicated block memory to save the design resources. The finite field arithmetic control was implemented as a hardwired control. A multi-precision multiplier circuit mapped as a parallel structure was used to speed up the computation of the finite field multiplication operation. The proposed processor architecture was implemented on the smallest and lowest cost FPGA chips from XILINX namely, XC3S50 and XC3S200, the designs for w = 8, 16-bit occupied 70% and 80% of the available resources from XC3S50 while the designs for w = 32-bit consumed 60% of slices of the XC3S200. It is believed that the proposed ECC processor architecture in this work achieves the smallest area consumption amongst the state of the art ECC implementation on FPGA whist still supports scalability across a set of standards curves with practical performance. This design has its importance in numerous low resource applications such as smart cards, mobile handsets, and wireless sensor nodes to implement key exchange protocols.

References

1. Wilkes, M.V.: The Best Way to Design an Automatic Calculating Machine. In: Proc. Manchester Univ. Computer Inaugural Conf., pp. 16–18. Ferranti Ltd. (1951)
2. Hankerson, D., Menezes, A., Vanstone, S.: Guide to Elliptic Curve Cryptography. Springer, Heidelberg (2004)
3. Eslami, Y., Sheikholeslami, A., Gulak, P.G., Masui, S., Mukaida, K.: An area-efficient universal cryptography processor for smart cards. IEEE Transactions on Very Large Scale Integration Systems 14(1), 43–56 (2006)
4. Eberle, H., Gura, N., Chang-Shantz, S.: A cryptographic processor for arbitrary elliptic curves over $GF(2^m)$. In: Proceedings of IEEE International Conference on Application-Specific Systems, Architectures, and Processors, ASAP 2003, June 2003, pp. 444–454 (2003)

5. Leong, P.H.W., Leung, I.K.H.: A microcoded elliptic curve processor using FPGA technology. IEEE Transactions on Very Large Scale Integration (VLSI) Systems 10(5), 550–559 (2002)
6. Orlando, G., Paar, C.: A High-Performance Reconfigurable Elliptic Curve Processor for $GF(2^m)$. In: Paar, C., Koç, Ç.K. (eds.) CHES 2000. LNCS, vol. 1965, pp. 41–56. Springer, Heidelberg (2000)
7. NIST.: Recommended elliptic curves for federal government use, `http://csrc.nist.gov/encryption/.2000`
8. SEC 2. Standards for Efficient Cryptography Group: Recommended Elliptic Curve Domain Parameters. Version 1.0 (2000)
9. Comba, P.: Exponentiation cryptosystems on the IBM PC. IBM Systems Journal 29(4), 525–538 (1990)
10. Lopez, J., Dahab, R.: Fast multiplication on elliptic curves over $GF(2^m)$ without precomputation. In: Koç, Ç.K., Paar, C. (eds.) CHES 1999. LNCS, vol. 1717, pp. 316–327. Springer, Heidelberg (1999)
11. Hassan, M.N., Benaissa, M.: Embedded Software Design of Scalable Low-Area Elliptic-Curve Cryptography. IEEE Embedded Systems Letters 1(2), 42–45 (2009)
12. Benaissa, M., Lim, W.M.: Design of flexible $GF(2^m)$ elliptic curve cryptography processors. IEEE Transactions on Very Large Scale Integration (VLSI) Systems 14(6), 659–662 (2006)
13. Shuhua, W., Yuefei, Z.: A Timing-and-Area Tradeoff GF(P) Elliptic Curve Processor Architecture for FPGA. In: IEEE International Conference on Communications, Circuits and Systems, ICCCAS 2005, pp. 1308–1312 (2005)
14. Rodriguez, F., Saqib, N.A., Diaz-Perez, A., Koç, Ç.K.: Cryptographic Algorithms on Reconfigurable Hardware. Springer, Heidelberg (2006)
15. Vassiliadis, S., Wong, S., Cotofana, S.: Microcode processing positioning and directions. IEEE Micro 23(4), 21–30 (2003)

Author Index